# Sensory and Consumer Preference of Food: Innovative Technologies in the Food Industry

# Sensory and Consumer Preference of Food: Innovative Technologies in the Food Industry

Editor: Logan Lee

Callisto Reference
www.callistoreference.com

**Callisto Reference,**
118-35 Queens Blvd., Suite 400,
Forest Hills, NY 11375, USA

Visit us on the World Wide Web at:
www.callistoreference.com



ISBN: 978-1-64116-839-7 (Hardback)



**Cataloging-in-Publication Data**

Sensory and consumer preference of food: innovative technologies in the food industry / edited by Logan Lee.
p. cm.
Includes bibliographical references and index.
ISBN 978-1-64116-839-7
1. Food industry and trade--Technological innovations. 2. Food--Sensory evaluation.
3. Consumers' preferences. I. Lee, Logan.

TP372.8 .S46 2023
664--dc23

# Table of Contents

# Preface

The purpose of the book is to provide a glimpse into the dynamics and to present opinions and studies of some of the scientists engaged in the development of new ideas in the field from very different standpoints. This book will prove useful to students and researchers owing to its high content quality.

Consumer preference refers to the subjective tastes of consumers and it is determined by how satisfied they are with the food products after making a purchase. Factors such as packaging, price and convenience determine consumer acceptance of food products. In the food industry, one important factor that requires careful consideration is the sensory experience of the consumers with the food. Sensory testing evaluates food products using the senses of touch, sight, taste, and smell. It is frequently utilized to ascertain whether a food product would be accepted by consumers or not. The factors which influence food acceptability may be related to the environment, the food or the person consuming it. This book unravels the recent studies on consumer preferences in the food industry. It presents researches and studies performed by experts across the globe. This book is a vital tool for all researching and studying this field.

At the end, I would like to appreciate all the efforts made by the authors in completing their chapters professionally. I express my deepest gratitude to all of them for contributing to this book by sharing their valuable works. A special thanks to my family and friends for their constant support in this journey.

**Editor**

1

# Investigating Italian Consumer Preferences for Different Characteristics of Provolone Valpadana using the Conjoint Analysis Approach

**Niculina Iudita Sampalean [1], Tiziana de-Magistris [2,3,*] and Daniele Rama [1]**

[1] Department of Agri-Food Economics, Faculty of Agriculture, Food and Environmental Sciences, Campus Piacenza-Cremona, Catholic University of Sacred Heart, via Emilia Parmense, 84, 29122 Piacenza (PC), Italy; niculinaiudita.sampalean@unicatt.it (N.I.S.); daniele.rama@unicatt.it (D.R.)

[2] Unidad de Economía Agroalimentaria y de los Recursos Naturales, Centro de Investigación y Tecnología, Agroalimentaria de Aragón, 50059 Zaragoza, Spain

[3] Instituto Agroalimentario de Aragón (IA2), CITA-Universidad de Zaragoza, 50013 Zaragoza, Spain

* Correspondence: tmagistris@aragon.es

**Abstract:** The objective of this paper was twofold. First, we estimated consumer preferences for an Italian cheese (Provolone Valpadana) with respect to several attributes and levels, such as price, origin certification, production system, 'free from' labelling, and brand. Second, we identified consumer clusters with similar preferences for various cheese characteristics. Preferences were estimated using the conjoint analysis method. Then, a cluster analysis was used to classify consumers into different (three) clusters followed by a market simulation. In all three clusters, the attribute most preferred by Italian consumers was the brand of the cheese: consumers preferred to purchase Provolone cheese having the lowest price, produced by Auricchio, bearing a European Union (EU) quality certification, produced organically, and non-lactose-free. The results of our study provide helpful information to food companies for better segmenting their market and targeting their consumers, as well as effectively promoting their products using brands, certifications as organic and lactose-free. This study contributes to the literature on consumer preference for the EU labelling scheme (voluntary and mandatory). To our knowledge, this is the first study to investigate this combination of multiple labels displayed on the front of Italian cheese packaging.

**Keywords:** Provolone Valpadana; cheese; conjoint analysis; cluster analysis; market shares

## 1. Introduction

Consumers have become highly conscious of food product characteristics when shopping. They search for products of higher quality made with environmentally friendly production processes that have connections with the territory and that prevent health issues [1–4]. To keep up with this change in consumer preferences, a greater range of products have emerged on the market. As a result, this abundant variety of products has made food shopping a more difficult task for consumers [5].

Some food-labelling schemes enable manufacturers to disclose the qualities of food products to consumers [6]. Thus, the use of labels allows food companies to signal quality or the presence of specific desirable but not obvious attributes [7]. In this context, during the past two decades, the European Union (EU) has launched different food labelling regulations that food companies can adopt on a voluntary basis. The aim of these food-labelling schemes is to establish the requirements for the use of particular labels and the procedures for controlling the quality of labelled products to protect consumers from being misled. Two examples of these regulations are EEC Regulation No. 834/2007, concerning organic production and the labelling of organic products, and Regulation No. 2081/92 adopted by the European Commission and further amended with several texts which culminated in EEC Regulation No. 1151/2012, which introduced the quality labels Protected Designation of Origin (PDO) and Protected Geographical Indication (PGI) labels. The PDO and PGI labels were designed as quality cues to reduce asymmetric information and to reduce consumer uncertainty associated with food purchases regarding desirable product characteristics. In particular, the PDO label may be used in the case of food products that are produced, processed, and prepared in a given geographical area using local know-how. (EEC Regulation No. 1151/2012).

Many studies on consumer preferences for food labelling have been conducted since the 2000s, and they have especially focused on consumer willingness to pay (WTP) for multiple front-of-package (FOP) food labelling schemes [1,3,8–10] and some have focused on cheese products [11–13].

Although cheese is one of the most important agricultural food products in Italy, accounting for 87.4% of the total value of Italian dairy products sold abroad [14], studies on consumer preferences for food labelling on cheese products are scarce [12,15–17]. In general, these studies have reported that PDO labels are preferred over organic labelling; nutritional claims are preferred over organic labelling; nutritional claims are preferred over the nutritional fact panel. Yet, most studies have also demonstrated that preferences for food labelling schemes are heterogeneous across consumers, with gender, age, educational level, and that the degree of environmental concern influenced consumer WTP for different cheeses [11,18].

At the European level, 53 of 247 types of cheese designated as PDO or PGI are Italian produced. For Italy, these cheeses represent 57% of the national production value and 50% of exports. Likewise, in 2018, sales of Geographical Indication (GI) cheeses grew by 2% in quantity and 1.5% in value compared with non-GI cheeses; those cheeses decreased by 1.6% in volume and increased by 0.6% in value [19]. In this context, Grana Padano PDO and Parmigiano Reggiano PDO together account for 59.3% of the volume and 66.3% of the value of total sales of GI cheeses in large-scale distribution [19].

Four regions of northern Italy (Lombardy included) account for 65% of the PGI production value; 76 PDO/PGI certified products come from Lombardy, and 9896 PDO/PGI enterprises are located in the region, generating a production value of €1958 million [19].

Provolone Valpadana is one of the most consumed GI cheeses in Italy, with a consumption value of €77 million and representing 1.1%, by weight, of the total Italian GI cheese consumption, more than the well-known Pecorino Toscano PDO or Montasio PDO [19].

Istat's 'Report on Family Consumption Expense' shows that the average monthly household expenditure in Italy in 2017 was €2563.94. Of this total average expenditure, €457.12 was dedicated to food and non-alcoholic drinks. Spending on cheese, milk, and eggs was €58.26 (these are measured as an aggregate), which represents 2.27% of the total average monthly household expenditure and 12.75% of the total average monthly expenditure for food and non-alcoholic drinks. Provolone is a spun-paste semihard cheese made from whole cows' milk with natural acidity from fermentation. Milk is collected in the area of origin within 60 h of milking and then undergoes mild heat treatment until the cheese is pasteurised. The maturation period can vary as follows: for up to 6 kg, the minimum maturation period is 10 days; for over 6 kg, the minimum maturation period is 30 days; for over 15 kg of only the piquant variety, the minimum maturation period is 90 days; for over 30 kg of the product labelled PVS (Provolone Valpadana Strong) the maturation period is over 8 months (for the piquant variety). The cheese may also be smoked, formed into different shapes (one of its most important characteristics) and be formed into blocks of various weight [20].

Provolone Valpadana was recognized as a PDO product in 1996. Since then, all its producers must strictly comply with the production regulations. Today, Provolone Valpadana PDO is produced by authorised cheesemakers and can also be sold by packaging companies. In all cases, every single shape must be marked with the Consortium's trademark and the PDO trademark, which are unequivocal guarantees of quality [21].

Provolone Valpadana is a typical premium cheese with unmistakable qualities; it is easy to compare with other generic products (e.g., spun-paste cheese), but consumers must recognise its unique qualities.

In light of the foregoing context, the objective of this study was twofold. The first objective was to assess Italian consumer preferences for Provolone Valpadana that bears multiple food labels, such as those making claims related to their origin (PDO), method of production (organic), and healthiness (free from lactose). Because of the increasing complexity of consumer preference, the second aim of this study was to examine heterogeneity in preferences for labelled cheeses on the basis of consumers' sociodemographic and personal characteristics.

To achieve these objectives, we conducted an online survey of 246 Italian cheese consumers and used both a conjoint and cluster analysis to analyse our data.

Our study contributes to the consumer cheese preference literature because this is the first study to investigate consumer preferences for these multiple food labels on Italian cheese using an approach combining conjoint analysis and cluster analysis. Furthermore, our findings have marketing implications because they disclose useful information to Italian stakeholders along with the cheese supply chain that can help producers and retailers, among others, design the best strategies to differentiate cheeses by adopting various labelling schemes for different segments of consumers.

*Literature Review*

Consumer preferences towards different food labels and different combinations of food labels have been widely investigated during the last decade for several food products. For instance, Bernabeu et al. [22]. analysed the price, origin, type and production system for wine, while Erraach et al. [9] examined the relative importance of European origin labels (PDO) associated with extrinsic (price and packaging) and intrinsic (color) cues of Spanish olive oil on consumer preferences. Aprile et al. [8] provided evidence about extra-virgin olive oil consumers' WTP. Likewise, de Magistris and Gracia [23] investigated consumers' preferences and WTP for almonds carrying different sustainability labels (e.g., food miles and organic certifications). Similarly, Mesias et al. [24] used price, origin, production system and labelling as attributes to study preferences for beef in Extremadura. Finally, Cilla et al. [25] and Garavaglia and Mariani [26] investigated consumers' preferences for cured ham in Spain.

These studies in general showed that the PDO and organic farming labels are most important. In particular, Aprile et al. [8] showed that the highest premium price is associated with PDO labels, followed by organic farming labels, and then PGI labels. Erraach et al. [9] indicated that price and origin labelling (PDO label) were the attributes that most affect consumers' preferences. De Magistris and Gracia [23] suggested that consumers were willing to pay a positive price premium for 'locally grown' and 'organically produced' attributes. Finally, Mesias et al. [24] showed that the origin of the product is the most important attribute in the choice of beef, followed by quality labelling, production system and price. In the literature, some studies have focused on consumer preferences for several food labels appearing on cheeses sold in Europe. For example, Garavaglia and Marcoz [12] investigated the Fontina cheese bearing the PDO certification. The authors reported differences in consumer preferences depending on the respondent's place of residence. Likewise, Skubic et al. [13] conducted a study in Slovenia and found that price was the most important attribute for consumers. De Magistris and Gracia [27] found that Spanish consumers were willing to pay more for PDO cheese than for organic or light cheeses. Moreover, gender, age, educational level, and the degree of the environmental concern influenced consumer WTP for different cheeses.

Several studies have investigated consumer preferences for cheese products based on production, processing, trading or branding. For example, Tendero and Bernabeu [28] demonstrated that consumers most highly value the PDO label as a guarantee of quality and food safety and that they prefer cheeses that are appropriately priced, aged, and if possible, certified. Similarly, Bernabeu et al. [29] investigated the perceived quality of cheese from Castilla-La-Mancha compared with that of cheeses from the rest of Spain and foreign cheeses. The authors found that consumers attached the most importance to the origin of the cheese, followed by the cheese type, price, and production system (organic). Moreover, Bernabeu et al. [29] revealed that organic production systems had the lowest relative importance as well as a negative influence on consumer preferences for cheese.

Napolitano et al. [17] assessed the effect of information concerning organic production of Pecorino cheese on consumer WTP. They showed that consumers were willing to pay a higher price than local retail price for organic cheese than regular cheese (€3.00/100 g). Moreover, information regarding organic farming was found to be the main determinant of cheese preference and consumer WTP.

Some studies have focused on consumer WTP for cheeses with FOP nutritional claims. For instance, Gracia and de Magistris [3] determined that the most preferred food label was the PDO indication, closely followed by the nutritional fact panel and the EU organic logo. Likewise, de Magistris and Lopez-Galan [11]. Investigated consumer WTP for cheeses bearing 'reduced fat' and 'low salt' claims in Spain. The authors reported that Spanish cheese consumers were willing to pay a premium for packages of cheese with reduced-fat and low-salt claims.

Another important claim related to the health of consumers that may be present on the FOP of cheese is the 'lactose-free' claim. This claim contains important information for consumers with food intolerances (e.g., lactose intolerance). Hartmann et al. [30] investigated how 'free-from labelling' shapes consumer perceptions of food products and whether the absence of an ingredient is considered an indicator of improved nutritional value of the product. The authors concluded that a 'lactose-free' label was highly correlated with the perception of healthiness as well as the intention to pay a premium price, imagining that consumers attribute inappropriate health benefits to products free of lactose.

Lactose-free cheese occurs when cheese is aged more than about one month, as bacteria added from the beginning to the milk 'digest' lactose converting it into lactic acid; the latter contributes to cheese taste and aroma, and first of all it induces protein coagulation, therefore the cheese production [31].

Finally, studies on consumer preference for private brands of cheese are scarce. To our knowledge, only Arfini [15] conducted such a study, comparing consumer WTP for cheeses bearing Consortium labels and EU PDO labels. The findings showed that the presence of the manufacturer's brand was comparatively important when shopping for grated cheese (18.91%) and ordinary cheese (18.24%). The results further indicated that little attention was paid to the private label of the firm producing or marketing the typical products at issue; however, the Consortium label played an important role in reassuring consumers about the quality of the purchased product. For Parmigiano Reggiano cheese, as many as 75% of the interviewed consumers looked for the Consortium label when shopping [15].

## 2. Materials and Methods

### 2.1. Conjoint Analysis and Choice Task Procedure

Conjoint analysis is one of the most widely used marketing research methods for analysing consumer trade-offs and is used to examine survey responses concerning preferences and intentions to buy; it is a method for simulating how consumers might react to changes in current products or new products introduced into an existent competitive array [32]. The model assumes that the substitute products notion can be defined as a series of specific levels of a common set of attributes. It also assumes that the total utility the consumer derives from a product is determined by the utility or part-worth contributed by each attribute level. The conjoint analysis starts with the consumer's overall or global judgements about a set of complex attributes. It then performs a decomposition of the original evaluations of the consumer into separate and compatible utility scales, according to which the original overall judgements can be reconstituted [33].

The most commonly used composition rule adopted in this study is additive because it is the one considering most (80% or 90%) of the variation in preference in almost all cases. Following Mesias et al., 2005 [24] we used the additive composition rule:

$$Y = \sum_{i=1}^{n} X_{ij} \sum_{j=1} v_{ij}$$

where Y is the total utility, Vij is the utility associated with level j (j = 1, 2 . . . , m) of attribute i (i = 1, 2 . . . , n), and Xij is a dummy variable that takes the value of 1 (or 0) in the case of presence (or absence) of the jth level of the ith attribute.

For the qualitative attributes, the part-worth model was used because of its flexibility [31]. For the price attribute, a Linear Less relationship was used because in general, higher prices correspond to lower utility or preferences.

Here, the conjoint analysis (CA) procedure consisted of a choice task to establish the trade-offs that Italian consumers make between price, quality certification (PDO), system production (organic), 'free from' labelling, and brand of Provolone Valpadana Dolce.

Italian consumers were presented 16 varieties of Provolone Valpadana Dolce with different combinations of attributes levels. The attributes (and their levels) used in the study are listed in Table 1 and were identified (as most important when buying cheese: price (€2.20/200 g, €2.86/200 g, and €3.20/200 g), quality certification (with or without PDO), production system (organic or conventional), brand (Auricchio, Latteria Soresina (both national brands), Coop (private label), or no brand) and lactose content (lactose-free or non-lactose-free). To select the price levels, information on the Provolone Valpadana sold in different Italian supermarkets was used, and three price levels were set up (€2.20/200 g, €2.86/200 g, and €3.20/200 g). The most frequent lowest price of a package of 200 g was around €2.20. Therefore, this price was chosen as the lowest price. The second level was 30% higher than the most frequent lowest price, and the third level was 45% higher than the most frequent lowest price. Moreover, the Coop private label was chosen since it could represent any Italian private label of Provolone Valpadana. As for the national brands, two brands were selected: Auricchio, the leader brand [34], and Latteria Soresina, another national brand of Provolone Valpadana cheese. In other words, any other private label and national brand could have been used instead of COOP and Latteria Soresina.

**Table 1.** Attributes and their corresponding levels.

| Attributes | Level |
|---|---|
| Price | 2.20 €<br>2.86 €<br>3.20 € |
| European Quality Certification (Protected Designation of Origin) | Yes<br>No |
| Production System | Organic<br>Conventional |
| Lactose content | Contains Lactose<br>Lactose-Free |
| Brand | Latteria Soresina<br>Auricchio<br>Coop<br>No Brand |

With these five attributes and their 13 levels, 96 potential profiles were obtained. Because this is a particularly high number to show to consumers, an orthogonal design (SPSS) was applied to the profiles to reduce the number to 16, a number more likely to encourage participation of consumers.

An example of one of the 16 obtained profiles is shown in Figure 1.

**Figure 1.** Example of evaluated cheese profile.

The respondents evaluated a picture that contained the different attribute levels in this study. The packaging used was the same as that on the market, but some attribute-level combinations shown are not present in the marketplace. Once the profile's cards were formed, they were shown to the respondents who evaluated them, giving a score from 1 to 10 according to their stated preference. Participants had the option to repeat the same number on different cards. The number 1 and 10 corresponded to the lowest and highest levels of preference, respectively (thus employing the complete profile method). The main argument in support of the full-profile approach is that it gives a more realistic description of stimuli by defining the levels of each of the attributes and considering the potential environmental correlation between factors in real stimuli [35].

Using SPSS software, the utility estimates generated by the conjoint analysis were used in a k-means cluster analysis to classify the consumers into homogeneous preferences groups and classify them into clusters.

### *2.2. Data Gathering*

A self-administered structured electronic survey was completed by 245 people in Italy. The survey was distributed online (Facebook, Messenger, WhatsApp, LinkedIn) between September 2019 and March 2020. All subjects gave their informed consent for inclusion before they participated in the study. The study was conducted in accordance with the Declaration of Helsinki, and the protocol was approved by the Ethics Committee of Catholic's University of the Sacred Heart. A pilot study with Valpadana cheese consumers was conducted to test the validity and the comprehension of the survey. Table 2 shows the characteristics of the final sample of respondents.

Approximately 81% of the respondents were women and only 19% were men. Compared with the Italian population, women were overrepresented, which might be explained by the fact that, in general, women oversee food shopping for households. Roughly 24% of the participants were less than 35 years old, 53% were between 36 and 55 years old, and 23% were older than 55 years. The sample may have skewed toward a younger demographic because the questionnaire was distributed online, and the older population tends to not have access to the Internet or computer skills. Roughly 43% of the participants had completed high school degrees, and 57% had a university degree. More than half were married (53%), which is similar to the Italian population overall (47%). In more than 40% of our sample, the family was composed of three or four members, which is also comparable with the Italian population overall (35%). More than half the respondents (66%) were from southern or island regions, with the central region being the least represented territory (7%) of participants compared with 20%

of the Italian population. For participant occupations, most (40%) were office workers, followed by housewives (14%), freelancers (13%), students or unemployed persons (present in the same proportion of 5%), teachers (2%), and entrepreneurs (2%). Concerning annual income, 26% of the respondents earned less than €10,000 per year, 31% earned between €10,000 and €20,000 per year, and 11% earned more than €40,000 per year.

**Table 2.** Sociodemographic distribution of the collected sample in comparison with the Italian population (%).

| Variables | Levels | Sample | Population |
|---|---|---|---|
| Age (in years) | 18–35 | 24 | |
| | 36–55 | 53 | |
| | >55 | 23 | |
| Gender | Female | 81 | 51 * |
| | Male | 19 | 49 * |
| Education | Elementary and middle school | 2 | |
| | High school Diploma | 41 | |
| | University degree | 57 | |
| Marital status | Bachelor/Maiden | 33 | 42 * |
| | Married | 53 | 47 * |
| | Divorced | 9 | 3 * |
| | In a relationship | 3 | |
| | Widow | 1 | |
| Family members | 1 | 12 | 33 * |
| | 2 | 36 | 27 * |
| | 3–4 | 40 | 35 * |
| | >4 | 12 | 5 * |
| Geographical Distribution | North | 27 | 46 * |
| | South and Islands | 66 | 34 * |
| | Centre | 7 | 20 * |
| Occupation | Office worker | 40 | |
| | Freelance | 13 | |
| | Student | 5 | |
| | Housewife | 14 | |
| | Teacher | 2 | |
| | Entrepreneur | 2 | |
| | Unemployed | 5 | |
| | Other | 19 | |
| Average annual income (€) | <10.000 | 26 | |
| | 10,000–20,000 | 31 | |
| | 20,000–40,000 | 32 | |
| | 40,000–50,000 | 4 | |
| | >50,000 | 7 | |

* ISTAT (National Statistics Institute) data extracted in April 2020. http://dati.istat.it/Index.aspx?QueryId=18460.

## 3. Results

The results section is divided into two parts. The first part describes the results of consumer preferences for different Provolone Valpadana cheese profiles and the clusters formed by these preferences. The second part, using the simulation process, identifies market shares for the most preferred cheeses as well as for some existent cheeses on the market.

### *3.1. Cheese Consumer Preference Structure*

Table 3 presents consumer preferences for various FOP food labelling. We found that brand is the most highly valued attribute (58.24%), followed by price (13.89%), production system (10.65%), lactose content (8.94%), and finally, the quality certification (8.29%). Additionally, based on the utility estimates for each level of attributes, consumers preferred to purchase Provolone cheese having the lowest price, bearing the Auricchio brand, bearing the EU quality certification, having been organically produced, and non-lactose-free.

**Table 3.** Mean part-worth and relative importance for all respondents.

| Attributes | Levels | Utility Estimate | Importance Values (%) |
|---|---|---|---|
| Quality Certification | NO PDO | −0.077 | 8.292 |
| | PDO | 0.077 | |
| Production System | Organic | 0.144 | 10.652 |
| | Conventional | −0.144 | |
| Brand | No Brand | −0.287 | 58.239 |
| | Aurrichio | 0.966 | |
| | Coop | −0.378 | |
| | Latteria Soresina | −0.301 | |
| Lactose Content | Contains Lactose | 0.039 | 8.941 |
| | Lactose-Free | −0.039 | |
| Price | 2.2 | −0.068 | 13.877 |
| | 2.86 | −0.136 | |
| | 3.2 | −0.204 | |
| (Constant) | | 6.723 | |

Notes: Pearson's R = 0.992; Kendall's tau = 1.000.

Using SPSS software, the utility estimates generated by the conjoint analysis were used in a k-means cluster analysis to classify the consumers into homogeneous preferences groups and classify them into clusters. Three clusters of consumers were identified with different preference structures.

The analysis of variance (ANOVA) showed that all clusters differed significantly from each other with respect to price ($p < 0.05$), lactose content ($p < 0.1$) as well as for the other variables were the significance was the highest ($p < 0.01$).

The utility estimates and their relative importance were then calculated for each of the levels of the attribute. The results, together with the size of each cluster (expressed as %), are listed in Table 4 (A and B).

In all three clusters, the attribute most preferred by the Italian consumers was the brand of the cheese, and this attribute's highest relative importance was in cluster 2 (77.16%) followed by cluster 1 (62.83%).

**Table 4.** A. Mean part-worth for the clusters. B. Relative importance for the clusters.

| A | | | | |
|---|---|---|---|---|
| **Attributes** | **Levels** | **Cluster 1 (44%)** | **Cluster 2 (22%)** | **Cluster 3 (34%)** |
| Price | 2.20 € | −0.14947 | −0.01865 | −0.08053 |
| | 2.86 € | −0.1943 | −0.02424 | −0.10469 |
| European Quality Certification (Protected Designation of Origin) | Yes | −0.0089 | 0.0658 | 0.2014 |
| | No | 0.0089 | −0.0658 | −0.2014 |
| Production System | Organic | 0.0656 | 0.0668 | 0.2971 |
| | Conventional | −0.0656 | −0.0668 | −0.2971 |
| Lactose content | Contains Lactose | 0.0171 | 0.0708 | 0.0519 |
| | Lactose-Free | −0.0171 | −0.0708 | −0.0519 |
| Brand | Latteria Soresina | −0.3137 | −0.8606 | 0.0878 |
| | Auricchio | 0.9577 | 2.7149 | −0.1778 |
| | Coop | −0.5533 | −1.0091 | 0.2683 |
| | No brand | −0.0900 | −0.8440 | −0.1800 |

| B | | | |
|---|---|---|---|
| **Attributes** | **Relative Importance** | | |
| | **Cluster 1** | **Cluster 2** | **Cluster 3** |
| Price | 13.57 | 6.49 | 19.13 |
| European Quality Certification (Protected Designation of Origin) | 6.43 | 4.53 | 13.40 |
| Production System | 8.79 | 5.30 | 16.67 |
| Lactose Content | 8.38 | 6.52 | 11.23 |
| Brand | 62.83 | 77.16 | 39.57 |

Regarding the largest and first cluster (44%), the most preferred attribute was the brand (62.83%), followed by price (13.57%). However, these consumers equally preferred the production system and lactose content (8.79% and 8.38%), whereas the quality certification was the least preferred attribute (6.43%).

The second cluster consisted of 22% of the sample who most preferred the brand of cheese (77%); lactose content and price exhibited similar relative importance of preference (6.52% and 6.49%, respectively). As in cluster 1, EU quality certification was the least valued attribute (4.53%). Comparable to the first and the second cluster, in the third cluster (representing 34% of all consumers), the most preferred attribute was brand, followed by price (19.13%); however, in contrast to the other two clusters, the lactose content (11.23%) was the least preferred attribute.

Finally, as expected, the level of price in all clusters exhibited a negative value, especially for the highest price (€3.20/200 g). Moreover, the utility assigned to the 'free from lactose' level exhibited a negative value in all clusters, suggesting that consumers did not prefer lactose-free cheese.

Consumers in cluster 1 showed a clear preference for cheese without a quality certification, in contrast to clusters 2 and 3. Likewise, regarding the brand attribute, cheese with the Auricchio brand presented the highest utility in the first and second clusters but negative utility in the third cluster, in contrast with Latteria Soresina and Coop, both of which exhibited positive utility. Thus, the results suggest that consumers in cluster 1 most preferred the brand Auricchio while in the cluster 3, consumers most preferred the cheese brand Coop.

Regarding lactose content, in cluster 2, consumers ranked the lactose content in the cheese as the second most important attribute in contrast to consumers in clusters 1 and 3. As for the production system, except for cluster 1, all consumers in clusters 2 and 3 ascribed a positive value to the EU quality certification.

### 3.2. Cheese Consumer Profiles

To better define consumers from each cluster, consumer profiles were interpreted by looking at their socioeconomic characteristics (Table 5).

**Table 5.** Sociodemographic characteristics of clusters.

| Variables | Levels | Cluster 1 | Cluster 2 | Cluster 3 |
|---|---|---|---|---|
| Age (in years) | 18–35 | 20% | 11% | 39% |
| | 36–55 | 61% | 61% | 36% |
| | >55 | 19% | 28% | 25% |
| | Sig = 0.001 < 0.01 (1%) | | | |
| Education | Elementary and middle school | 1% | 8% | 6% |
| | High school Diploma | 49% | 43% | 24% |
| | University degree | 51% | 49% | 70% |
| | Sig = 0.003 < 0.01 (1%) | | | |
| Marital status | Bachelor/Maiden | 27% | 23% | 48% |
| | Married | 58% | 4% | 42% |
| | Divorced | 10% | 11% | 7.5% |
| | In a relationship | 5% | 58% | 2.5% |
| | Widow | 0% | 4% | 0% |
| | Sig = 0.020 < 0.05 (5%) | | | |
| Family members | 1 | 14% | 7% | 15% |
| | 2 | 18% | 21% | 32% |
| | 3–4 | 56% | 53% | 6% |
| | >4 | 12% | 19% | 47% |
| | Sig = 0.026 < 0.05 (5%) | | | |
| Geographical Distribution | North | 23% | 13% | 36% |
| | South and Islands | 70% | 83% | 53% |
| | Centre | 7% | 4% | 11% |
| | Sig = 0.010 < 0.05 (5%) | | | |
| Gender | Female | 85% | 91% | 72% |
| | Male | 15% | 9% | 28% |
| | Sig = 0.010 < 0.05 (5%) | | | |

Table 5 shows the Pearson chi-square results of the three clusters and the consumer characteristics that were found to be statistically significantly different across clusters at a 5% significance level (at least). Results indicated that all clusters included a higher proportion of women with a university degree and living in the south or island regions. Although in all three clusters the most predominant gender was female, in cluster 3, we found the highest percentage of men (28%). Cluster 1 consisted of the largest percentage of people aged between 35 and 55 years, and in contrast to clusters 2 and 3, the lowest percentage of older people (>55 years old). Further, 85% were women with high school degrees (49%) or a university degree (51%). Finally, 70% of the cluster consisted of people who lived in the southern or island regions of Italy and living in a family with three or four family members.

Cluster 2 consisted of the largest percentage of women between 35 and 55 years of age and, in contrast to clusters 1 and 3, of the lowest percentage of young people (<35 years old). As in the first cluster, 49% of sample had university degrees, and 83% lived in the southern or island regions in families of three or four members. In cluster 3, 35% of the individuals were between the ages of 18 and 35, 36% between the ages of 36 and 55, with a university degree (70%) and living in a family composed of more than four family members. In contrast with clusters 1 and 2, cluster 3 had the highest percentage of men (28%) and people living in the southern and island regions (53%). If we compare our results with the ones found in the literature for different food groups, we see that the price was one of the most important attributes, which is similar to our results. In contrast to our results, the origin was also one of the most valued attributes. For example, Bernabeu et al. [22] identified three consumer segments: one that shop for wine mainly by price, one by origin, and another one by a combination of price and type of wine. Erraach et al. [9] identified four consumer segments, two of which involve liking the origin label. De Magistris and Gracia [23] suggested that consumers' preferences for locally grown and organically produced almonds were heterogeneous, with three consumer segments identified: 'conventional consumers', 'short distance consumers', and 'sustainable consumers'. Through cluster analysis, Mesias et al. [24] identified three groups of consumers. One that gave least importance to a product's origin and thus obviously valued less whether beef was produced in Extremadura or in other areas. The other clusters gave more importance to the origin, and with one of these clusters, 62% of the weight of the purchasing decision was based on the origin of the product.

### *3.3. Simulation of Market Shares*

The results of the conjoint analysis were utilised to simulate choices among the ideal cheese products for each cluster and three other existent Provolone cheese profiles. This analysis was performed using the part-worth utility function for each respondent. The simulation summary gives the probabilities of choosing a particular simulation profile as the most preferred profile under three different probability-of-choice models. The maximum utility model represents the probability of choosing a profile as the most preferred. The BTL (Bradley–Terry–Luce) model computes the probability of choosing a profile as the most preferred by dividing the profile's utility by the sum of the total utility of the simulation. The logit model is like the BTL model but uses the natural log of the utilities instead of the utilities themselves [28].

Results from the simulation of market shares are shown in Table 6.

**Table 6.** Simulation of market share.

| Card | Maximum Utility | Bradley Terry-Luce | Logit | Quality Certification | Production System | Price | Quality Certification | Production System |
|---|---|---|---|---|---|---|---|---|
| 1 | 29.8% | 19.0% | 27.1% | PDO | Organic | 2.20 | Aurrichio | Contains Lactose |
| 2 | 19.9% | 18.5% | 23.4% | NO PDO | Organic | 2.20 | Aurrichio | Contains Lactose |
| 3 | 12.6% | 15.8% | 13.2% | PDO | Organic | 2.20 | Coop | Contains Lactose |
| 4 | 5.4% | 15.1% | 9.1% | PDO | Conventional | 2.20 | Coop | Contains Lactose |
| 5 | 26.1% | 17.2% | 19.3% | NO PDO | Conventional | 3.20 | Aurrichio | Lactose-Free |
| 6 | 6.2% | 14.4% | 7.8% | NO PDO | Conventional | 3.20 | Latteria Soresina | Contains Lactose |

The largest market share corresponds to the ideal product (cheese number 1) for both the study's total sample and cluster 2, and this share fluctuates between 29.8% and 19.01%. Cheese number 5, an existent cheese, is next, with a market share that fluctuates between 26.1% and 17.2%, and cheese number 2 follows, with a variable market share between 18.5 % and 23.4%, depending on the method used for estimation. Between the first and the second cheese, there was a decrease in market share between 0.5% and 9.9%, depending on the simulation method employed; the decrease was presume it was because the EU quality certification was missing. Thus, adding the quality certification to an existent cheese profile could increase its market share by up to 9.9%. If we look at cheese profiles number 3 and number 4, we can see that the use of the organic production system could increase the market share from 0.7% to 7.2%, building on the distinctive estimating methods.

## 4. Discussion

The first objective of this study was to estimate consumer preference for Italian cheeses with various levels of different attributes such as price, origin certification (PDO), production system (conventional or organic), 'free from' labels (lactose-free), and brand (national brands/private labels) on Italian consumer preferences for cheese.

Based on the analysis of these attributes, the results showed that cheese is a highly differentiated food product. Cheese preferences were most positively affected by being produced by a national brand followed by private label brand production, being produced organically, and non-lactose-free.

Examination of the relative importance of the various attributes indicated that brand was the most important attribute followed by price in two of the three clusters. This result is in contrast to the results found by Skubic et al. [13] who showed that price was the most important attribute for cheese. There were different price levels used in the two studies, and Skubic's study had four levels, one more than the other. This might have an influence on the final results, but a bigger influence on the price effect could be the set of attributes that were used, considering that the brand (which has the highest importance in our study) is completely missing from Skubic's study. Our findings also suggest a positive preference for the organic label. This result is in agreement with Napolitano et al. [17] who found that 'organic' can be a potential tool for cheese differentiation, particularly for small scale and traditional farms. However, our results are in contrast with those of Bernabeu et al. [29], who suggested that the organic attribute is an inadequate differentiation strategy in the cheese market.

Similar to Arfini [15] our results indicated that national brand was the most preferred cheese attribute when analysing different market clusters.

In contrast to Hartmann et al. [30], we found that consumers did not prefer to buy lactose-free cheese, we found that, in all three clusters, Italian consumers preferred to buy non-lactose-free Provolone cheese.

Additionally, the EU quality certification had a positive impact on cheese preference, except for cluster 1, where consumers preferred noncertified to certified cheese. As for the price, for two out of the three clusters, it was the second most important attribute. These results accord with Tendero et al. [23] who discovered that regular cheese consumers value the cheese having the EU quality guarantee but also ascribe much importance to price. Similarly, Garavaglia and Marcoz [12], studying the value of PDO certification of Fontina cheese and its impact on consumer preferences and WTP, found that the PDO certification has substantial importance on consumer purchasing decisions, persuading them to pay a premium price for products with PDO.

Our findings indicated that consumers value the organic production system more than the PDO certification. This result contrasts with other studies [1,3,8,36,37] that reported that consumers value PDO more than organic certification.

Finally, with regards to consumer heterogeneity, our results coincide with de Magistris and Gracia [27], Krystallis and Ness [38] and Fotopoulos and Krystallis [39], which all found that consumer age, education, and income level are important factors when it comes to selecting food products. In particular, our findings demonstrated significant differences between respondents in terms of geographical distribution, area of origin, sex, age, and education. Similar results were found by de Magistris and Gracia [27], which revealed that female respondents, older respondents, and those with a university education were willing to pay more for quality-labelled cheeses.

## 5. Conclusions

We analysed consumer preference for Italian cheese products with respect to FOP origin certification (PDO), production system (conventional or organic), 'free from' labels (lactose-free), price, and brand (national brands/private labels brands), and we explored heterogeneity of consumers. In general, the results indicate that in all three clusters, the attribute most preferred by Italian consumers is the brand of cheese. Italian consumers prefer to purchase Provolone cheese at the lowest price, that carries the Auricchio brand marking, that is EU quality certified, that is organically produced, and which is non-lactose-free. It is expected that most consumers that dislike the 'lactose-free' claim are conscious that Provolone is naturally lactose-free (as they are Provolone Valpadana cheese consumers) and consider any manipulation connected to the 'lactose-free' claim as a possible modification of the typical taste of the cheese.

The identification of certain segments of the Italian Provolone cheese market could help cheese producers better understand consumer needs and their most preferred attributes; consequently, producers could adapt or diversify their marketing mix, especially in the areas of new products development and communication strategy, in order to better match consumer needs. For example, the best way to attract consumers in the first cluster to buy Provolone Valpadana cheese is to use an organically labelled national brand at a fair price. Moreover, because this cluster consists of those people who are married and with children, the communication strategy should accentuate the benefits of organic products on health, especially for children and older individuals.

In the second cluster, people most preferred the national brand and cheese with lactose, and thus, producers should highlight the uniqueness of the flavour, appearance, and texture of their Provolone cheese to indicate how lactose content affects these characteristics. Also, they should emphasize the enjoyment of Provolone cheese in favour of the presumed health benefits that lactose-free cheese could offer.

For the last cluster, comprised of young people holding a university degree, the national brand of the cheese, its organic production, and its PDO certification should be publicised. The quality guaranteed by these labels should be highlighted as well as the unique character and taste of the cheese due to the local environments; moreover, how production of such cheeses supports the local economies could be emphasised.

Given that the market share simulation showed that being a national brand helped to acquire the biggest market share, national producers, as well as modern distribution companies, should create effective strategies to take advantage of this preference.

Organic and EU quality certifications earned significant shares of the market; therefore, different messages that promote a traditional and environmentally friendly production method and the origin of cheeses should be used in promotion and advertising of these attributes.

Although lactose-free cheese was not preferred by consumers, companies that nevertheless decide to use this label should concentrate their advertising strategies on the segment of lactose-intolerant people, communicating to them the possibility of enjoying the taste of dairy without the uncomfortable gastrointestinal symptoms that result from the ingestion of lactose.

Our study has some limitations. Investigating Provolone Valpadana cheese as the focal product prevents generalisation of the results to the entire category of quality-certified cheese, and thus, further research should be conducted on other cheeses. A more widely known cheese that could represent the entire category should be considered because evaluation of it would allow for inferences to all types of cheese. Moreover, a future pan-EU study could be conducted to reveal national market differences, which would allow for different strategies to be adopted in specific national and international contexts.

As for the methodology, we suggest the use of a questionnaire that would investigate not only consumer preference and knowledge of labels but also the relationships between these aspects; this could be followed by a sensory analysis (trained sensory panel) and integrated by an experimental auction where consumers would be asked to bid on the different cheese types and the top bidders on the binding product in the binding round would have to acquire the product.

Considering that exhortations to 'buy national' occur in times of economic turmoil in an attempt to protect domestic industry and jobs, in the future it would be interesting to study the role of protecting domestic industry and jobs through buying local products [40].

Finally, as consumers base their consumption decisions not only on economic reasons but also identities, culture, values, worldviews, and group memberships [41], the long-run benefits of purchasing domestic products and their support of the local economy [42] would be an area recommended for future study.

Finally, further studies should consider voluntary nutritional labels on FOP, named "Nutriscore" or "NutrInform battery "and other ones, on cheese products in combination with quality labelling in order to investigate the trade-off between these types of food labelling.

**Author Contributions:** Conceptualisation, N.I.S. and D.R.; methodology, N.I.S. and D.R.; formal analysis, N.I.S.; data collection, N.I.S., D.R. and T.d.-M.; original draft preparation, N.I.S.; writing—review and editing, N.I.S., D.R. and T.d.-M.; supervision, T.d.-M. and D.R. All authors have read and agreed to the published version of the manuscript.

**Acknowledgments:** This study was supported by the Doctoral School in conjunction with the Agro-Food System (Agrisystem) of the Università Cattolica del Sacro Cuore (Italy). The authors wish to thank all the participants of

## References

1. de Magistris, T.; Gracia, A. Do consumers care about organic and distance labels? An empirical analysis in Spain. *Int. J. Consum. Stud.* **2014**, *38*, 660–669. [CrossRef]
2. de Magistris, T.; Pascucci, S.; Mitsopoulos, D. Paying to see a bug on my food. *Br. Food J.* **2015**, *117*, 1777–1792. [CrossRef]
3. Gracia, A.; de-Magistris, T. Consumer preferences for food labelling: What ranks first. *Food Control* **2016**, *61*, 39–46. [CrossRef]
4. Grunert, K.G. How changes in consumer behaviour and retailing effect competence requirements for food producers and processors. *Econ. Agrar. Recur. Nat.* **2006**, *6*, 3–22.
5. Schwartz, B. *The Paradox of Choice: Why More is Less*; CCC, HarperCollins Publishers: New York, NY, USA, 2004.
6. Caswell, J.A.; Noelke, C.M.; Mojduszka, E.M. Unifying two frameworks for analyzing quality and quality assurance for food products. In *Global Food Trade and Consumer Demand for Quality*; Krissoft, B., Bohman, M., Caswell, J.A., Eds.; Kluwer Academic/Plenum Publishers: New York, NY, USA, 2002; pp. 43–61.
7. McCluskey, J.; Loureiro, M.L. Consumer preferences and Willingness to Pay for Food Labelling: A Discussion of Empirical Studies. *J. Food Distrib. Res.* **2003**, *34*, 95–102.
8. Aprile, M.C.; Caputo, V.; Nayga, R., Jr. Consumers evaluation of food quality labels: The case of the European geographic indicator and organic farming labels. *Int. J. Consum. Stud.* **2012**, *36*, 158–165. [CrossRef]
9. Erraach, Y.; Sayadi, S.; Gómez, A.C.; Parra-López, C. Consumer stated-preferences towards Protected Designation of Origin (PDO) labels in a traditional olive-oil-producing country: The case of Spain. *New Medit* **2014**, *13*, 11–19.
10. Schröck, R. Valuing country of origin and organic claim. *Br. Food J.* **2014**, *116*, 1070–1091. [CrossRef]
11. de Magistris, T.; Lopez-Galan, B. Consumers' willingness to pay for nutritional claims fighting the obesity epidemic: The case of reduced-fat and low salt cheese in Spain. *Public Health* **2016**, *135*, 83–90. [CrossRef]
12. Garavaglia, C.; Marcoz, E.M. Preferenze dei Consumatori e Certificazioni di Origine: Un'indagine del caso della Fontina tramite la Conjoint Analysis. *Mercati Compet.* **2012**, *2*, 65–88.
13. Skubic, K.; Klopčič, M.; Erjavec, K. Consumer preferences regarding national and EU quality labels for cheese, ham and honey: The case of Slovenia. *Br. Food J.* **2018**, *120*, 650–664. [CrossRef]

14. Rama, D. *Il Mercato del Latte Rapporto 2018*; L' Osservatorio sul Mercato dei Prodotti Zootecnici SMEA: Cremona, Italy, 2019.
15. Arfini, F. The value of typical products: The case of prosciutto di Parma and Parmigiano Reggiano cheese. In Proceedings of the 67th EAAE Seminar, Le Mans, France, 28–30 October 1999; Actes et Communications, n.17. INRA: Paris, France, 1999; pp. 77–98.
16. Arfini, F.; Pazzona, M. The coexistence of PDO and brand labels: The case of the ready-sliced Parma ham. In Proceedings of the 8th International European Forum on System Dynamics and Innovation in Food Networks, Igls, Austria, 17–21 February 2014.
17. Napolitano, F.; Naspetti, S.; Zanoli, R.; Favotto, S.; Braghieri, A.; Piasentier, E. Cheese liking and consumer willingness to pay as affected by information about organic production. *J. Dairy Res.* **2010**, *77*, 280–286. [CrossRef] [PubMed]
18. de Magistris, T.; Gracia, A.; Barreiro-Hurle, J. Do consumers care about European food labels? An empirical evaluation using best-worst method. *Br. Food J.* **2017**, *119*, 2698–2711. [CrossRef]
19. Ismea-Qualivita. *Rapporto Ismea-Qualivita Sulle Produzioni Agroalimentari e Vitivinicole Italiane DOP*; IGP e STG; Fondazione Qualivita: Rome, Italy, 2019.
20. EC No C 64/2012, ''On the protection of geographical indications and designations of origin for agricultural products and foodstuffs" Production Regulation. *Off. J. Eur. Union C* **2012**, *64*, 19–24.
21. Provolone Valpadana Consortium Website. Available online: https://www.provolonevalpadana.it/en/characteristics-of-provolone-valpadana/shapes-of-provolone-valpadana (accessed on 10 October 2019).
22. Bernabéu, R.; Brugarolas, M.; Martínez-Carrasco, L.; Díaz, M. Wine origin and organic elaboration, differentiating strategies in traditional producing countries. *Br. Food J.* **2008**, *110*, 174–188. [CrossRef]
23. de Magistris, T.; Gracia, A. Consumers' willingness-to-pay for sustainable food products: The case of organically and locally grown almonds in Spain. *J. Clean. Prod.* **2016**, *118*, 97–104. [CrossRef]
24. Mesias, F.J.; Escribano, M.; de Ledesma, A.R.; Pulido, F. Consumers' preferences for beef in the Spanish region of Extremadura: A study using conjoint analysis. *J. Sci. Food Agric.* **2005**, *85*, 2487–2494. [CrossRef]
25. Cilla, I.; Martínez, L.; Beltrán, J.A.; Roncalés, P. Dry-cured ham quality and acceptability as affected by the preservation system used for retail sale. *Meat Sci.* **2006**, *73*, 581–589. [CrossRef]
26. Garavaglia, C.; Mariani, P. How much do consumers value protected designation of origin certifications? Estimates of willingness to pay for PDO dry-cured ham in Italy. *Agribusiness* **2017**, *33*, 403–423. [CrossRef]
27. de Magistris, T.; Gracia, A. Consumers' willingness to pay for light, organic and PDO cheese: An experimental auction approach. *Br. Food J.* **2016**, *118*, 560–571. [CrossRef]
28. Tendero, A.; Bernabéu, R. Preference structure for cheese consumers: A Spanish case study. *Br. Food J.* **2005**, *107*, 60–73. [CrossRef]
29. Bernabéu, R.; Díaz, M.; Olmeda, M. Origin vs organic in Manchego cheese: Which is more important? *Br. Food J.* **2010**, *112*, 887–901. [CrossRef]
30. Hartmann, C.; Hieke, S.; Taper, C.; Siegrist, M. European consumer healthiness evaluation of 'free-from' labelled food products. *Food Qual. Prefer.* **2018**, *68*, 377–388. [CrossRef]
31. Gatti, M.; Bottari, B.; Lazzi, C.; Neviani, E.; Mucchetti, G. Invited review: Microbial evolution in raw-milk, long-ripened cheeses produced using undefined natural whey starters. *J. Dairy Sci.* **2014**, *97*, 573–591. [CrossRef]
32. Green, P.E.; Kriger, A.M.; Wind, Y. Thirty Years of Conjoint Analysis: Reflections and Prospects. *Interfaces* **2001**, *31*, 56–73. [CrossRef]
33. Green, P.E.; Wind, Y.; Claycamp, J.R. Brand Features Congruence Mapping. *J. Mark. Res.* **1975**, *12*, 306–313. [CrossRef]
34. Sckokai, P.; Soregaroli, C. Impact of private label development across retail formats: Evidences from the Italian dairy market. *Rev. Agric. Environ. Stud.* **2008**, *87*, 27–47.
35. Green, P.E.; Srinivasan, V. Conjoint analysis in consumer research: Issues and outlook. *J. Consum. Res.* **1978**, *5*, 103–123. [CrossRef]
36. Gracia, A.; Barreiro-Hurle, J.; López-Galán, B. Are local and organic claims complements or substitutes? A consumer preferences study for eggs. *J. Agric. Econ.* **2014**, *65*, 49–67. [CrossRef]
37. Scarpa, R.; del Giudice, T.D. Market segmentation via mixed logit: Extra-virgin olive oil in Urban Italy. *J. Agric. Food Ind. Org.* **2004**, *2*, 1–20. [CrossRef]

38. Krystallis, A.; Ness, M. Consumer preferences for quality foods from a South European perspective: A conjoint implementation on Greek olive oil. *Int. Food Agribus. Manag. Rev.* **2005**, *8*, 62–91.
39. Fotopoulos, C.; Krystallis, A.; Ness, M. Wine produced by organic grapes in Greece: Using means-end chains analysis to reveal organic buyers' purchasing motives in comparison to the non-buyers. *Food Qual. Prefer.* **2003**, *14*, 549–566. [CrossRef]
40. Barnes, V. Brands, Geographical Origin, and the Global Economy: A History from the Nineteenth Century to the Present. *J. Leg. Hist.* **2019**, *40*, 326–328. [CrossRef]
41. Bourdieu, P. *Distinction: A Social Critique of the Judgement of Taste*; Harvard University Press: Cambridge, MA, USA, 1984; ISBN 0674212770.
42. Topçu, U.C.; Kaplan, M. Willingness to Buy Foreign Products in Relation to Ethnocentric Tendencies and Worldminded Attitudes of Consumers. *Procedia Soc. Behav. Sci.* **2015**, *207*, 157–164. [CrossRef]

2

# Eliciting Egg Consumer Preferences for Organic Labels and Omega 3 Claims in Italy and Hungary

**Ching-Hua Yeh [1], Davide Menozzi [2] and Áron Török [3,*]**

[1] Institute for Food and Resource Economics, University of Bonn, 53115 Bonn, Germany; chinghua.yeh@ilr.uni-bonn.de

[2] Department of Food and Drug, University of Parma, 43124 Parma, Italy; davide.menozzi@unipr.it

[3] Department of Agricultural Economics and Rural Development, Corvinus University of Budapest, 1093 Budapest, Hungary

* Correspondence: aron.torok@uni-corvinus.hu

**Abstract:** This paper investigates consumers' preferences for egg purchase in two European countries, Hungary and Italy. We utilize random parameter logit models to interpret the results of discrete choice experiments (DCE) for the elicitation of preference of the egg consumers. A sample of 403 in the Hungarian survey and 404 in the Italian survey were recruited in summer 2018. The DCE questionnaire includes the following product and process characteristics: organic labels, nutrition and health claims, and price. Our results show that for Hungarian and Italian consumers, the price is the most important attribute, followed by the nutrition and health claim and the organic production labelling. Three egg consumer segments can be identified via latent class models for each country. In both countries, we found similar consumer groups, the Price Sensitive and Quality Optimizing Opportunist Consumers and Health Conscious Buyers, respectively. Particularly, compared to the other segments the Health Conscious Buyers (46% in Hungary and 49% in Italy) exhibited stronger preference for and are willing to pay a higher price premium for eggs with organic label and nutrition claims. In Italy, we identified a third segment with consumers preferring simpler labelling approach, whilst in Hungary we found a consumer segment distrusting the EU organic logo.

**Keywords:** egg; production method; nutrition and health claims; choice experiment; random parameter logit model; latent class analysis

---

## 1. Introduction

### *1.1. Organic Labelled Foods in the European Union*

In the European Union (EU), organic food must comply with the principles of organic production, certification and labelling of Regulation (EU) 2018/848 of the European Parliament and of the Council. Since the first European organic regulation (Council Regulation EEC 2092/91 of 24 June 1991), organic food production plays an essential role in both the EU's food quality systems and rural development. On the community level, since July 2010 the organic logo (green leaf), introduced by Commission Regulation (EU) No 271/2010, is compulsory for all the pre-packed organic foods and voluntary for non-pre-packed foods. Besides the EU logo, in several European countries other voluntary organic certification labels are managed by different organizations, e.g., the French "AB—agriculture biologique", or the national "Bio"—label in Germany. Though the EU organic label has been in use for almost a decade, its recognition among European consumers is not too high and significantly differs among member states. According to recent surveys by the European Commission [1–4], generally Northern and Western European citizens know the label the most (34–60%), while among Southern and Eastern European the level of recognition was quite low (9–17%).

Not only the label recognition, but also the market share of organic foods differs significantly in the EU. Based on the yearly per capita spending on organic food, Denmark, Sweden, Luxemburg and Austria are the countries where the demand significantly exceeds the EU average (EUR 50 per capita per year). Contrarily, in some Eastern and southern EU member states (e.g., Estonia, Greece, Malta and Portugal) demand for organic labelled food almost does not exist [5].

### 1.2. *Egg Market and Consumption*

Hen egg is consumed worldwide as a valuable and cheap source of protein. Egg producers are in possession of practices required for producing eggs with specific attributes; therefore, they are easy to be classified into distinct product categories [6]. After a setback in the early '90s, per capita egg consumption in Europe had achieved around 13.3 kg [7] and expected to increase to 14.3 kg per capita per year by 2030 [8].

However, in the second half of the 20th century, household egg consumption faced a significant fall in several developed countries (e.g., UK, USA) for several reasons, including changing diets and habits, animal welfare concerns, and health-related issues [9–11]. Additionally, an increase in egg consumption was identified among vegetarians and consumers who avoid red meat [11], and as a consequence of the income decline, given the inelastic demand [10]. Brand, in this sector, is less important than the size, the number of eggs in the pack and variety [11]. However, several studies have analyzed the market segments; based on a consumer survey, in the UK there was a polarization of egg consumers: animal-friendly consumers requiring free-range eggs at one corner and those who care more about functional properties (e.g., size and shell) and the value for money [11]. UK consumers are willing to pay a substantially higher price for eggs with animal-friendly production method than the technology change (e.g., cage-free) that actually would increase the production costs [9].

According to a survey conducted in France, buying organic eggs is not affected by socio-economic factors (e.g., income, age, family size) except for the education level. Purchasing other organic products (e.g., milk) increases the probability of buying organic eggs [12]. Spanish consumers are willing to pay a positive price premium for eggs produced organically as well as for the specification of the production origin of eggs [13]. In Norway, there is a segment of consumers willing to pay a substantial premium for organic eggs, but this segment is limited in size. Other consumers do not seem to appreciate the value added of organic eggs, presenting a decreasing marginal utility for added attributes. In other words, organic labelled eggs are competing with, e.g., free-range eggs, where less costly labelled eggs are able to gain a competitive advantage [14].

In Hungary, the highest egg consumption was realized right before the political change at the end of the 80s, and since then domestic consumption is characterized by a declining trend, due to several reasons, including the lowering purchasing power [15]. Price-sensitive Hungarian consumers seek for the fresh and sound eggs, and consider egg consumption as part of a preferable and healthy diet [16].

In Italy, the average egg consumption is in line with the EU statistics (13.4 kg per capita) and has been relatively stable over the past 20 years [17]. It is estimated that approximately 35% of this quantity is used in the food industry in the form of egg products. Italy is self-sufficient for the consumption of eggs, i.e., produces enough eggs to cover the entire national needs [18].

In general, the EU's egg supply is enough for domestic consumption, the self-sufficiency was around 103% in recent years [8], and the level of national production has covered the level of domestic consumption both in Hungary and Italy [8]. In addition, the supply of organic egg production is highly concentrated, where the top five biggest egg producers in the EU have a market share of 76%. In recent years, Italy was the seventh biggest organic egg producer (6.9%), while the Hungarian production was almost negligible [19].

### 1.3. *Consumers' Preference and Perception of Organic Labelled Food*

In the past few decades, food consumption patterns have been changing rapidly due to the increasing concerns about food quality and safety, health benefits, environmental impacts and animal welfare concerns. This leads to consumer interest in organic food which is free from pesticides and chemical residues and boosts consumer demand for differentiated agricultural products in the food markets.

Kiss et al. [20] and Gergely et al. [21] found that, in general, organic food is considered more expensive compared to conventional ones among Hungarian consumers, and this is the most significant barrier why it is not consumed widely. The premium price level was also confirmed by previous field research that the organic foods cost more and had a 56–85% price premium compared to their conventional counterparts [22]. However, families with children consume more organic foods [21] and vegetarians are more likely to become organic consumers because their primary motivation is to avoid bad quality products, though natural-like products are usually confusing whether they are organic [23]. The typical Hungarian organic food consumers are female, live in a larger city, belong to the upper-income level segment and are committed to a healthy lifestyle and environmental protection [22,24]. Hungarian's organic consumption is highly influenced by food prices and individuals' income levels, and their main motivations are to follow a healthy diet and to avoid diseases and other health-related reasons [25]. Based on the case of the organic milk market in Hungary, Szente et al. [26] suggested that low price category and local/regional product strategy together with prestige product strategy should be implemented in order to increase the level of organic consumption. Among organic consumers, the average willingness to pay (WTP) is 30–50%; however, a low level of recognition for such products is common [27]. Regarding the Hungarian consumers' attitude towards organic labelled products, it has been found that they are sensitive for sustainability issues [22]. However, organic food consumption has remained relatively low due to the lack of trust in organic products, and consumers cannot distinguish them from regular ones [28]. A recent consumer survey highlighted that freshness, flavor, positive impacts on health, ingredients and to be free from additives are the most important characteristics that influence consumers when buying organic foods [24].

Italy is the fifth organic market worldwide, following the US, Germany, France and China, with more than EUR 3.1 billion retail sales and 3.2% of market share in 2017. Italy is also a net exporter of organic products, with a value of total exports higher than EUR 2 billion in 2017 [5]. In general, Italian women and young people show a higher intensity of organic food consumption and WTP for organic products [29,30]. The presence of children in the family results in a higher WTP for organic products too [30]. Familiarity is another factor influencing organic consumption and WTP, since consumers who regularly purchase organic foods are also willing to spend more for these products [30]. The organic label is more familiar among young adults, in terms of both visibility and understanding. It was suggested that this higher visibility and familiarity can be justified by the high market penetration of organic products at the national level [31].

A cross national study has shown that in Italy the EU logo was widely used, also because neither a governmental nor a dominant private organic logo existed. This factor could also explain the positive attitude and trust of Italian consumers towards the EU logo [32]. The trust in organic labels also has a positive impact on attitude towards and WTP for organic food. For this reason, it was studied that providing additional production information about organic farming practices did not affect the WTP for organic products [30]. In other words, the EU organic logo was for these consumers self-explanatory. Among the different supply chain actors, it was demonstrated that trust in farmers is a significant predictor of the intention to purchase organic products [33]. Italian organic consumers exhibit a higher level of sustainability concern in their general food choices, adhere to more sustainable consumption principles and have a more sustainable lifestyle [29,34]. Those consumers attaching a higher importance to health aspects and naturalness of food, and are concerned over food safety, exhibit a higher organic consumption intensity [29], and are willing to pay more for organic products [30].

In general, consumers usually perceive organic eggs as being healthier than conventional products, and they are willing to pay more for them [13,35–37]. Andersen [36] found a high and positive correlation between Danish consumers' revealed WTP a premium for organic and free-range eggs, indicating a similar perception of organic eggs with eggs carrying labels indicating improved animal welfare. Consumers' WTP was higher for organic, especially among wealthier, urban and more educated consumers [36]. In Denmark, consumers are ready to pay a significant price premium (58%) for organic eggs [38]. In Spain, 80% of consumers were willing to pay a price premium for organic eggs (EUR 0.85 for a six package) [39] and organic and local claims were complements for many of the consumers [13]. For Italian occasional organic consumers, locally produced foods (including eggs) are considered to have a higher quality level than organic, as the latter was found to be more distant [40]. In the survey of Gerini, Alfnes and Schjøll [14], the Norwegian consumers were divided into three segments according to their organic purchase frequency (always/occasionally/never). They found that only regular organic buyers willing to pay more for organic eggs, while those are never buying organic foods are avoiding organic eggs even when they cost the same as the other eggs. Similar results were obtained by Heng and Peterson [41] with a US sample.

In the USA due to a Salmonella outbreak, half a billion eggs were recalled in 2010. Li et al. [42] compared results of WTP using auction experiments before and after this food safety scare. The results show that the scandal had no significant effect on the preferences for organic and conventional eggs. However, respondents that were supplied with negative information of the recall were willing to pay significantly more for organic eggs. Among Polish egg consumers, price and farming system had the most important effect on preferences, and free-range eggs were preferred more than organic ones [43]. In Germany, the organic attribute of eggs was ranked after "free of antibiotics" both in raw and processed (pasta containing egg) format: WTP estimate for the organic egg was EUR 0.23 while for pasta only EUR 0.02 [44]. Swiss consumers would pay almost 32% more for organic eggs [45]. Teixeira et al. [46] investigated emerging countries in terms of consumers' views on egg farming and purchasing habits. In Brazil and Chile, they found that egg consumers care the most for animal welfare, and more than half of them are willing to pay at least 5% more for eggs produced with animal-friendly methods.

### *1.4. Importance of Nutrition and Health Claims for Eggs*

To avoid a situation where product claims could mask the overall nutritional status of a food product and mislead consumers when purchasing food, Regulation (EC) No 1924/2006 introduced the concept of nutrient profiles. This regulation seeks to protect consumers from misleading or false claims and make it easier for producers and processors to identify nutrition and health claims that can be used on specific food products [47].

Eggs usually have many desirable attributes that might attract consumers' attention (among others, cage-free, organic, locally produced, nutrient enhanced, additive-free, etc.). Heng and Peterson [41] tried to investigate how these attributes interact and affect the overall price premium in the US market. They identified four consumer groups and found that the consumer segment of the "attribute seekers" are in favor of all labelled attributes except the health-related (omega 3) claim. A previous cross-country study investigated that almost 40% of the EU citizens stated that they buy eggs from animal welfare production (free-range or outdoor) and less than 20% admitted not to pay attention to the type of production system [48]. A US survey found that American egg consumers think that animal-friendly production methods contribute to the higher egg quality level, and a majority of them would pay more for such products. In another US study, Heng et al. [35] reported that most respondents in their study willingly paid a price premium for eggs from organic farms, outdoor access, and cage-free housing. Additionally, consumers paid more attention to animal friendly conditions than to organic production in the choice of eggs. However, consumers' attitude towards nutrient improved eggs (e.g., omega 3 enhanced) are not clear in terms of WTP [41]. Only health-conscious Canadian consumers are willing to pay a remarkable price premium for omega 3 enhanced eggs, and their stated WTP is lower than the actual market price difference compared to generic eggs [49].

Given the above background, this paper first aims to assess consumers' attitude towards, interest to, and willingness to pay for differentiated eggs with respect to production methods (i.e., organic) and nutrition and health claims, in two European countries, Hungary and Italy. This work uses the stated preferences approach through hypothetical choice experiments to estimate consumer demand and can help advance the understanding of the synergies or trade-offs between organic label with other nutrition and health claims. Secondly, the current study also aims to segment the country-specific consumers' egg purchase decisions, based on the two attributes: organic label, and nutrition and health claims. We explicitly considered consumers' heterogeneous preferences, in the light of individuals' choices and characteristics such as socio-demographics, food purchase habits, and consumers' food value regarding motives of food purchase, etc., identifying different consumer segments in each country. Thirdly, the final aim of this study is to estimate consumers' WTP for differentiated eggs at the segmentation level in order to evaluate the product's position in the market. Although the results of our study do not claim to cover the whole complex eggs market, they are expected to help practitioners and actors in the egg industry to better understand consumer motivations when buying eggs. This might further help them to target their customer segments, to develop and implement strategies in differentiated egg markets, as well as to evaluate the potential opportunities. In sum, this study extends the understanding of the trade-offs or synergies between organic labels with other nutrition and health claims, identifying consumer segments with similar behavioural and preference patterns.

In the following sections, we describe the experimental design and producers (Section 2), analytic results (Section 3), and a discussion and conclusion (Section 4).

## 2. Materials and Methods

### 2.1. Theoretical Framework

In the present study, an egg-purchasing scenario was simulated using discrete choice experiments (DCE). DCE methodology has been frequently conducted in the last decades in the marketing field e.g., [50–52] as a popular method to elicit consumer preferences for a product or a service and the attributes that comprise such a product or service. A DCE simulates consumers' purchase decision and has proven to be predictive for consumers' behavior on the market [53]. Participants are usually confronted with different sets of the choice situation and requested to indicate which of the several egg alternatives they would like to choose or to purchase. When making their choice or purchase decision, participants are forced to make a trade-off between different attributes and levels, thereby revealing their preferences.

The DCE is analyzed within a random utility framework [54,55], which, in our case, assumes that preference for eggs is a function of the utility or value of that egg's individual attributes plus an error term and allows the analysis of the stated choice under utility maximization. For each individual $i$, the utility function of an alternative $j$ is a function of the product and process characteristics $x_{ijt}$ in the choice set $t$. Each individual $i$ maximises his/her utility when choosing between $J$ alternatives can be specified as follows:

$$U_{ijt} = \beta_i x_{ijt} + \varepsilon_{ijt} \tag{1}$$

where $U_{ijt}$ is the utility that individual $i$ obtains from alternative $j$ at the situation $t$; $\beta_i$ is a vector of parameters of variables for individual $i$ representing his/her preferences; $x_{ijt}$ is a vector of observed attributes, and $\varepsilon_{ijt}$ is the i.i.d. extreme value type 1 stochastic error term. Principally, the standard analytic practice is to pool DCE choice data from individuals and estimate an aggregated model applying multinomial logistic regression. Because participants are assumed to have heterogeneous preferences and differ in error variances, it is important to consider these individual preferences and heterogeneity in the modelling process. To date, there has been number of modelling attempts to allow for eliciting heterogeneous preferences in DCE analysis. The state-of-the-art practice is the random parameters logit modelling (RPL) approach [56], which extends the traditional multinomial logit models by allowing parameters to randomly vary across individuals. This is computed by involving a respondent-specific stochastic component that specifies the individual specific deviation from the overall utility mean [56,57]. Following Train [56] and Hensher and Johnson [58], the probability of individual $i$ choosing alternative $j$ is an integral of standard logit probabilities over the parameter densities, which can be denoted as:

$$P_{itj}\,(\theta) = \int \frac{\exp\left(\beta_i x_{ijt}\right)}{\sum \exp\left(\beta_i x_{ijt}\right)} f(\beta_i|\theta)d\beta_i \tag{2}$$

where $\theta$ is the parameter vector that specifies the distribution of $\beta$ across sampling participants. In the present study, we applied the application of the RPL models using hierarchical Bayesian estimation on the effect-coded choice data. This hierarchical Bayesian choice modelling approach was motivated by the fact that compared to the classical maximum likelihood RPL models, the hierarchical Bayesian approach is known for its higher accuracy when heterogeneity in preference of the investigated population increases [59]. The hierarchical Bayesian RPL model involves two hierarchical steps [56,60,61]: in the first hierarchical step, the individual-level parameters are computed via a multivariate normal distribution characterized by a vector of mean values and a matrix of covariances. In the second hierarchical step, given an individual-level parameter, participant's likelihood of selecting specific products/alternatives can be further estimated by an aggregated logit model.

In addition, another alternative modelling strategy to discover the heterogeneous preference is Latent Class Analysis (LCA) [56]. In the present study, we follow the LCA procedure implemented by Boxall and Adamowicz [61], which is also called as finite-mixture model, where the finite mixture accommodates interpersonal heterogeneity in preference for the observed attributes describing choice alternatives LCA assumes that the overall preference distribution is constituted by a combination of unobservable latent segments that heterogeneous in their utility between the segments but are homogeneous within the segment [56,62,63]. For the LCA, the individuals are assumed to belong to a class s with a certain probability $C_{is}$ for $s = 1, \ldots, S$ (where $C_{is} > 0$ and $\sum C_{is} = 1$; $S$ denotes the total number of classes). Thus, the probability of an individual's membership of segment s will take the following form:

$$C_{is} = \frac{\exp(\alpha\lambda_S)}{\sum_{s=1}^{S}\exp(\alpha\lambda_S)} \tag{3}$$

where $\lambda_S$ denotes a vector of the segment specific parameters and $\alpha$ is the scale factor that is assumed to be equal to one, so each participant has a probability of belonging to a particular segment [61]. For conducting LCA, individual $i$'s choice probability for alternative $j$ in choice situation $t$ can then be given as [53,56,58]:

$$P_{ijt} = \sum_{s=1}^{S} C_{is}\frac{\exp\left(\beta_s x_{ijt}\right)}{\sum_{j=1}^{J}\beta_s x_{ijt}} \tag{4}$$

The present paper conducted classical LCA using the maximum likelihood approach. Besides estimating preferences for different consumer classes, the LCA models also provide the probability of each class membership for each individual. This is modelled as a function that participant $i$ is classified into class $s$ under the assumption that the stochastic error terms in membership likelihood function are i.i.d. across individuals and classes. We need to note that the LCA does not assign each individual to a particular class, but assigns to each individual a probability of membership in every class [56]. Thus, due to the properties of the RPL and LCA methods, the RPL and LCA were selected for the analysis of the country-specific DCE data in this paper to simultaneously approximate part-worth utility parameters and class membership from the DCE choices.

Moreover, we further compute country-specific segment-specific WTP estimates for each attribute level for the identified consumer segments by dividing the respective attribute level coefficient by the price coefficient. Because of involving the effect coding process in the discrete choice model, we calculated the mean WTP according to the following specification reported in Bech and Gyrd-Hansen [64]:

$$WTP = \left( \frac{-2\ \beta_{attribute\ level}}{\beta_{price}} \right) \tag{5}$$

Finally, we utilized the Kruskal–Wallis non-parametric test [65] to further examine whether different (country-specific) consumer segments significantly differs with respect to the demographic information, purchase behaviour and food value. Kruskal–Wallis test is widely used to compare three or more independent groups that are the same or different on respective variables of interest. When the statistics of the Kruskal–Wallis statistic is calculated as statistically significant, it indicates that at least one of the compared groups is significantly different from the others.

### 2.2. Survey Instrument and Design

The selection of eggs as the studying object in this study was based on the fact that eggs are usually used as a food ingredient, represent an important part of the human diet, and are highly differentiated by the product and process attribute, e.g., nutrition and health claims. The nutrition claims allowed for regular eggs are regulated in the EU by the Regulation (EC) n. 1924/2006 of the European Parliament and of the Council. Given the role that consumers play in the decision-making process of egg purchase, it is important to strengthen our understanding of their preference for, and respond to, different information to the labelling of egg products. The online cross-country questionnaire with respect to egg purchase was devised together with academic researchers specializing in the field of agricultural and nutrition science, and consisted of three parts.

The questionnaire started with a welcome address and statement of research motivation. The first part of the questionnaire included screening questions whether the participant was (partly) responsible for their household food purchase, currently lives in the respective country, and consumed egg in the last three months. The DCE questions constituted the second part of the survey. For the choice experiment, the appropriate eggs' attributes and their respective levels as well as an adequate egg-purchasing scenario were designed and defined. By conducting an internal discussion among academic researchers and market experts, we defined egg's attributes that were repeatedly discussed as influencing the purchase decision of the egg on the market, were proved influential in previous studies (e.g., Carlsson et al. [66], Heng et al. [67]), and relatively independent of one another and could be modified to impact consumers' decision making. Three attributes were included in the present cross-country study (namely, organic labels, nutrition and health claims, and price), which vary by defined levels in order to describe the choice situations. Table 1 presents the attributes and the respective levels used in the cross-country DCE setting.

**Table 1.** Attributes and respective levels used in the choice experiment.

| Country | Attributes and Respective Levels |
|---|---|
| **Hungary** | 1. **Organic labels:**<br>• No-label<br>• EU organic label<br>• EU organic label and Hungarian organic label<br>2. **Nutrition and health claims:**<br>• No claim<br>• High of Omega 3 fatty acids (nutrition claim)<br>• High of Omega 3 fatty acids, contributes to maintenance of normal function of the heart and normal blood pressure (nutrition and health claim)<br>3. **Price (HUF/6 eggs box):**<br>• 189 HUF (c.a. EUR 0.62)<br>• 279 HUF (c.a. EUR 0.91)<br>• 369 HUF (c.a. EUR 1.19)<br>• 459 HUF (c.a. EUR 1.48) |
| **Italy** | 1. **Organic labels:**<br>• No-label<br>• EU organic label<br>• EU organic label and Hungarian organic label<br>2. **Nutrition and health claims:**<br>• No claim<br>• High of Omega 3 fatty acids (nutrition claim)<br>• High of Omega 3 fatty acids, contributes to maintenance of normal function of the heart and normal blood pressure (nutrition and health claim)<br>3. **Price (HUF/6 eggs box):**<br>• EUR 1.50<br>• EUR 1.90<br>• EUR 2.30<br>• EUR 2.70 |

In the introduction to the DCE question, participants were explained briefly about the upcoming purchase task using the "cheap talk script" [68,69]. For the operation of DCE questions, participants were asked to make a purchase decision among the three presented egg products and an opt-out option. The inclusion of an opt-out option in the DCE has been considered to improve the realism of the DCE. Based on an artificially created image of egg products, the combination of attributes and levels of the DCEs were graphically represented. Figure 1a,b provide examples of the DCE choice situation used in the Hungarian and Italian online surveys, respectively. The DCE applied in the present study has an unlabelled design, with three alternatives and an opt-out egg alternative. Due to a large number of total possible choice sets ($(3 \times 3 \times 4)^3 = 46,656$), a reduced D-optimal design based on three egg product alternatives was employed using Ngene software (Version 1.2, ChoiceMetrics, Sydney, Australia) [70] to construct the DCE design plan. The D-optimal design seeks to maximize the logarithm of the determinant of the information matrix and consequently minimizes the determinant of the covariance matrix of the parameter estimates. In total, 20 blocking versions of the DCE questionnaire were generated in order to limit the context effect and response fatigue. Each version of the DCE survey was randomly assigned to each participant and each participant was requested to answer six DCE purchasing questions.

**Figure 1.** (**a**) An example of Hungarian discrete choice experiment (DCE) question. (**b**). An example of Italian DCE question.

The third part of the questionnaire contained Likert scaling questions about participants' perception and purchase habits with respect to egg shopping, as well as the food value questions regarding the importance of motives underlying the food choice. In the fourth part of the questionnaire, we also include items measuring demographic information and socioeconomic status. After completion of the online questionnaire, the participants were thanked for their participation.

The cross-country questionnaire was firstly developed in English in close collaboration between Hungarian and Italian researchers, and subsequently translated to the local language. Back-translations were carefully examined and led to a minor modification of the questionnaire. Finally, the online questionnaire was then pretested in order to ensure that no further changes appeared necessary in the survey.

### 2.3. Data Collection

A random and nationwide sample was drawn from population defined as adult shoppers above 18 years of age, living in the studied country (Hungary and Italy), at least partly responsible for their household food shopping, and have bought eggs in the last three months. Data were collected via cross-country online surveys in summer 2018 via the marketing research institute, LiGHTSPEED (http://www.lightspeedresearch.com/). The average completion time was about 10 min.

## 3. Results

In total, 807 usable questionnaires were obtained and deemed valid for further analysis. Table 2 shows the demographics statistics of the cross-country sample. In general, participants in both countries are half female, mostly well-educated, and living in the urban area. According to our descriptive outputs, the mean age of our participants was higher than the national statistics. Moreover, we notice that the share of female Hungarian respondents was slightly lower than the national average. Overall, the differences of the sample structure between Hungarian and Italian data are not substantial except that, in both countries, participants consisted of more post graduated respondents living in urban areas and with less children in their households. These sample characteristics should be noted when interpreting the estimated results of the DCE data, considering all the well-known limitations of online surveys highlighted in the literature (e.g., [71]).

**Table 2.** Sample structure across the three countries: Hungary and Italy.

| | **Hungary** | | **Italy** | | **All** |
|---|---|---|---|---|---|
| | **Sample** | **Nat. Avg.** [a] | **Sample** | **Nat. Avg.** | |
| Valid *N* | 403 | | 404 | | 807 |
| Gender | | | | | |
| Female (%) | 49.6 | 52.5 | 52.2 | 51.3 | 50.9 |
| Male (%) | 50.4 | 47.5 | 47.8 | 48.7 | 49.1 |
| Average age | 42.2 | 41.4 | 43.2 | 44.9 | 42.7 |
| Living area | | | | | |
| Rural area (%) | 15.4 | 30.5 | 13.9 | 24.3 | 14.6 |
| Urban medium town (%) | 37.5 | 34.4 | 45.3 | 42.4 | 41.4 |
| City (%) | 47.1 | 35.1 | 40.8 | 33.3 | 44.0 |
| Education | | | | | |
| Lower secondary/primary education or below (%) | 1.7 | 4.9 | 6.9 | 39.9 | 4.3 |
| Upper secondary education (%) | 31.3 | 48.1 | 39.9 | 41.5 | 35.6 |
| University or college entrance qualification (%) | 24.6 | 30.1 | 17.3 | 0.9 | 20.9 |
| Bachelor's degree or equivalent level (%) | 26.1 | 10.1 | 13.4 | 3.6 | 19.7 |
| Master, Postgraduate or doctoral degree (%) | 16.4 | 6.8 | 22.5 | 14.1 | 19.5 |
| Household monthly net income | | 222,000 HUF | | EUR 2616 | |
| (HU) < 150,000 HUF (c.a. EUR 486)/(IT) < EUR 900 (%) | 10.2 | | 5.9 | | 8.1 |
| (HU) 150,001–205,000 HUF (c.a. EUR 487–EUR 664) /(IT) EUR 900–EUR 1,500 (%) | 14.9 | | 20.8 | | 17.8 |
| (HU) 205,001–235,000 HUF (c.a. EUR 665–EUR 761) /(IT) EUR 1501–EUR 2500 (%) | 9.4 | | 28.7 | | 19.1 |
| (HU) 235,001–380,000 HUF (c.a. EUR 762–EUR 1231) /(IT) EUR 2501–EUR 3500 (%) | 31.0 | | 18.3 | | 24.7 |
| (HU) 380,001–835,000 HUF (EUR 1232–EUR 2705) /(IT) EUR 3501–EUR 4500 (%) | 25.3 | | 9.2 | | 17.2 |
| (HU) ≥ 835,001 HUF (c.a. EUR 2706)/(IT) ≥ EUR 4501 (%) | 2.7 | | 3.2 | | 3.0 |
| Prefer not to answer (%) | 6.5 | | 13.9 | | 10.2 |
| Household size | 2.96 | 2.86 | 2.94 | 2.40 | 2.95 |
| Number of children (<18 year) in a household | 0.59 | 1.06 | 0.54 | 0.50 | 0.56 |

[a] Nat. Avg. = National average in Hungary and in Italy, respectively.

### 3.1. RPL Model Estimates

In the first step of the DCE analysis, a country-specific random parameter logit model (RPL) was estimated for the Hungarian and Italian DCE data, assuming that there is consumer preference heterogeneity. Table 3 presents the results for the attribute importance measures of the cross-country

DCE choice data. Our results reveal that in both countries price is the most important attribute, followed by the information of nutrition and health claims with respect to omega 3, while the organic labels are seen as the least important.

**Table 3.** Attribute importance scores for Hungary and Italy.

| Country | Hungary | Italy |
|---|---|---|
| *N* | 403<br>Avg. Imprt. (S.D.) | 404<br>Avg. Imprt. (S.D.) |
| Organic labels | 17.34 (12.23) | 17.17 (12.63) |
| Nutrition and health claims | 23.87 (15.70) | 29.82 (17.00) |
| Price | 58.79 (21.61) | 53.01 (22.78) |

Avg. Imprt. = Average Importance, S.D. = Standard Deviation.

Table 4 shows the results for the country-specific RPL models. For each random parameter, the estimated average utility mean value and standard deviation are reported using zero-centred measures. They represent the relative attractiveness of the levels within each attribute in the numerical format expressing that the higher the number the more preferable it is to participants and conversely, the lower (more negative) the number, the less preferable it is to participants. The average utility of the opt-out alternative is computed as the mean value of the individual specific constants, and as expected, the constants in the bottom row of Table 4 are all negative and significant. This implies that consumers in both countries generally prefer selecting one of the egg alternatives in the DCE tasks. Our RPL results in both countries also show that participants are in favor of organic eggs in relation to the conventional ones, and prefer the eggs promoted by information of nutrition and health claims compared to the ones that have no nutrition and health claims. Particularly, in both countries we also found a stronger preference of eggs associated with a detailed (full) information of nutrition and health claims compared to eggs that carry only a simplified (short) nutrition claim. More specifically, this preference is considerably stronger for Italian consumers compared to Hungarian consumers. In addition, as expected our RPL outputs also reveal that participants' preference utility decreases with an increase in price which is in line with a negative price elasticity of demand. Finally, the estimated standard deviation of the random coefficients, i.e., "No label" in Hungarian data, shown in Table 4 is 32.54 indicating heterogeneity across participants in the population with respect to the effect of "No label" information. Additionally, our results also find that the standard deviation of the random parameters are statistically significant across countries, which implies that there is a substantial heterogeneity in preference regarding the respective characteristics when purchasing eggs across consumers, which can actually be considered an informative indication because it may lead to a beneficial segmentation strategy.

**Table 4.** Empirical results from random parameter logit models of the cross-country DCE data.

| Country | Hungary | Italy |
|---|---|---|
| *N* | 403 | 404 |
| Null Log-likelihood | −3352.05 | −3360.37 |
| Restricted Log-likelihood | −2538.72 | −2447.81 |
| AIC | 5091.45 | 4909.63 |
| BIC | 5119.44 | 4937.62 |
| McFadden Pseudo $R^2$ | 0.24 | 0.27 |
| | **Avg. Utilities (S.D.)** | **Avg. Utilities (S.D.)** |
| Organic labels | | |
| No label | −14.48 (32.54 ***) | −21.54 (23.58 ***) |
| EU organic | 0.34 (19.47 ***) | 2.53 (14.49 **) |
| EU organic + national organic | 14.13 (20.38 ***) | 19.01 (24.49 ***) |
| Nutrition and health claims | | |
| No-claim | −24.04 (42.32 ***) | −32.36 (44.58 ***) |
| Nutrition claim | 10.05 (21.78 ***) | 5.30 (23.01 ***) |
| Nutrition and health claim | 13.99 (30.52 ***) | 27.06 (37.24 ***) |
| Price | | |
| | −54.12 (31.56 ***) | −47.20 (33.21 ***) |
| Constant | −190.80 (194.06 ***) | −160.75 (181.33 ***) |

**,***; $p$-value of the random parameter's standard deviation < 0.01, 0.001; S.D. = Standard Deviation. Note 1: The reported average utilities are zero-centred; Note 2: The WTP measure derived from the RPL models can be found in Appendix A, Table A1.

### 3.2. *Latent Class Analysis (LCA) Estimates and Profiles*

We further analyzed the DCE choice data applying latent class analysis (LCA) to identify country-specific consumer segments with different product and process characteristics change preferences, and to compute parameter estimates quantifying each segment's preference for the levels of each attribute. In contrast to the RPL models, LCA assumes unobserved preference heterogeneity after a discrete distribution, identifying for a distinct number of underlying classes of participants with similar preferences. In the present study, the Akaike information criterion (AIC), Bayesian information criterion (BIC), Chi-square, and log-likelihood measures were used to select the number of classes in this LCA (Table 5).

**Table 5.** Summary of measures for selecting the optimal number of classes.

| | Number of Groups | Log-Likelihood | AIC | BIC | Chi-Square |
|---|---|---|---|---|---|
| | 2 | −2203.38 | 4432.75 | 4508.03 | 2297.37 |
| Hungary | 3 | −2037.72 | 4115.44 | 4231.26 | 2628.67 |
| (*N* = 403) | 4 | −1954.10 | 3962.20 | 4118.54 | 2795.92 |
| | 5 | −1930.24 | 3928.49 | 4125.37 | 2843.63 |
| | 2 | −2280.17 | 4586.34 | 4661.65 | 2160.41 |
| Italy | 3 | −2097.40 | 4234.80 | 4350.66 | 2525.96 |
| (*N* = 440) | 4 | −2026.22 | 4106.43 | 4262.85 | 2668.32 |
| | 5 | −1989.34 | 4046.68 | 4243.65 | 2742.07 |

However, according to Boxall and Adamowicz [62] and Hole [72], there are no established absolute statistical solutions to select an optimal number of latent classes. Based on the interpretability and comparability of the DCE outputs between Hungary and Italy, the three latent classes solution in the present cross-country study was identified. The results of the three-class LCA models in both countries are reported in Table 6, based on the three attributes: Organic labels, Nutrition and health claims, and Price. Moreover, Table A2 in the Appendix A presents the results of the Kruskal–Wallis tests for country-specific consumer groups.

**Table 6.** Latent class model.

| | Hungary | | | | | | Italy | | | | | |
|---|---|---|---|---|---|---|---|---|---|---|---|---|
| *N* | 403 | | | | | | 404 | | | | | |
| Null log-likelihood | −3352.05 | | | | | | −3360.37 | | | | | |
| Restricted log-likelihood | −2037.72 | | | | | | −2097.39 | | | | | |
| AIC | 4115.44 | | | | | | 4234.80 | | | | | |
| BIC | 4231.26 | | | | | | 4350.66 | | | | | |
| Chi-Square | 2628.67 | | | | | | 2525.95 | | | | | |
| | Group 1 Health Conscious Buyers | | Group 2 Price Sensitive and Quality Optimizing Opportunist Consumers | | Group 3 Unlabelled Eggs Seekers | | Group 1 Health Conscious Buyers | | Group 2 Price Sensitive Consumers | | Group 3 Clean Labelling Consumers | |
| Group size | 46.4% | | 45.6% | | 8.0% | | 49.0% | | 39.5% | | 11.6% | |
| | Relat. Imprt. (%) | Part-worth utilities (t-stat.) | Relat. Imprt. (%) | Part-worth utilities (t-stat.) | Relat. Imprt. (%) | Part-worth utilities (t-stat.) | Relat. Imprt. (%) | Part-worth utilities (t-stat.) | Relat. Imprt. (%) | Part-worth utilities (t-stat.) | Relat. Imprt. (%) | Part-worth utilities (t-stat.) |
| Organic labels | 20.30 | | 9.70 | | 21.85 | | 23.86 | | 4.59 | | 19.82 | |
| No label | | −25.48 *** | | −18.82 *** | | 35.49 *** | | −33.51 *** | | −8.78 *** | | −33.26 *** |
| | | (−3.08) | | (−7.02) | | (2.99) | | (−6.90) | | (−3.11) | | (−2.60) |
| EU organic | | −9.95 | | 10.28 *** | | −30.07 * | | −4.58 | | 5.00 | | 26.20 * |
| | | (−1.09) | | (3.49) | | (−2.03) | | (0.88) | | (1.55) | | (2.18) |
| EU organic and national organic | | 35.43 *** | | 8.54 *** | | −5.42 | | 38.09 *** | | 3.79 | | 7.06 |
| | | (4.50) | | (3.32) | | (−0.40) | | (8.95) | | (1.32) | | (0.56) |
| Nutrition and health claims | 31.40 | | 14.69 | | 13.88 | | 45.91 | | 16.08 | | 56.32 | |
| No-claim | | −54.78 *** | | −24.85 *** | | 15.24 | | −76.23 *** | | −27.99 *** | | 110.70 *** |
| | | (−6.30) | | (7.90) | | (1.23) | | (−13.32) | | (−8.09) | | (9.72) |
| Nutrition claim | | 15.37 | | 5.64 * | | 11.14 | | 14.74 *** | | 7.72 ** | | −58.26 *** |
| | | (1.78) | | (1.98) | | (0.87) | | (2.98) | | (2.43) | | (−3.75) |
| Nutrition and health claim | | 39.41 *** | | 19.21 *** | | −26.39 | | 61.49 *** | | 20.27 *** | | −52.44 *** |
| | | (5.11) | | (7.01) | | (−1.89) | | (14.75) | | (6.69) | | (−3.39) |
| Price | 48.30 | −48.30 *** | 75.61 | −75.61 *** | 64.27 | −64.27 *** | 30.23 | −30.23 *** | 79.32 | −79.32 *** | 23.86 | −23.86 *** |
| | | (-9.60) | | (18.41) | | (5.89) | | (−10.53) | | (−18.53) | | (−3.16) |
| Constant | | −475.76 *** | | −39.90 *** | | 161.25 *** | | −264.96 *** | | −11.32 | | 90.35 *** |
| | | (−12.63) | | (4.87) | | (0.97) | | (−11.46) | | (−1.82) | | (6.61) |

Relat. Imprt. = Relative importance; *, **, ***; $p < 0.05, 0.01, 0.001$.

For all classes of the Hungarian and Italian sample, as expected, the coefficients for the price attribute were negative and statistically significant, suggesting that higher egg prices generate disutility. Participants in both countries prefer to purchase eggs carrying labels with the combination of the EU organic label and the national ones, with full nutrition and health claim are more likely to be in Class 1, namely Healthy Conscious Buyers. Participants from the segment of Healthy Conscious Buyers represent 46.4% of the Hungarian sample and 49.0% of the Italian sample. The coefficients of the constant in this consumer segment are strongly negative and significant in both countries, suggesting dislike for the opt-out alternatives shown in the DCE.

Participants in Class 2 in both countries, namely Price Sensitive and Quality Optimizing Opportunist Consumers, care mostly for the price and concern little for the information of nutrition and health claims, and rarely pay attention to organic labelling. Particularly, participants from Price Sensitive Consumers in Italy even exhibit insignificant part-worth utility for organic labelling. The relative importance of the nutrition and health claim is also quite low, although statistically significant, indicating that for these consumers, price represents really the main factor influencing choices.

The consumer segment of Class 3 in Hungary groups individuals who prefer not having the organic logo. Indeed, they show a statistically significant positive preference for eggs that carry no organic certification and a statistically significant negative preference for eggs labelled by the EU organic logo. Therefore, we named this segment Unlabelled Eggs Seekers. Regardless the presence of the nutrition and health claims, these consumers simply consider (low) prices for their choices, particularly if not associated with the EU organic logo label. Differently to Hungarian Class 3, participants from the Class 3 in Italy are in favor of buying EU organic certified eggs with no information regarding nutrition and health benefits. We named this segment Clean Labelling Consumers, since these individuals significantly dislike the labels with more (nutrition and health) information, preferring eventually the product with a simple EU organic logo. More importantly, the coefficients of the constant in Class 3 in both countries are positive and statistically significant, indicating that consumers prefer to select the "opt-out" alternative in the DCE.

According to the results of Kruskal–Wallis test presented in Table A2 in the Appendix A, we are able to further explore whether the three consumer segmentations significantly differ with respect to socio-demographics, behavior, and food value among the groups in the country-specific data. It is found that in Hungary, Health Conscious Buyers consist of participants who are mostly living in the urban area and in general purchase eggs infrequently compared to the other groups. For participants belong to the group of Unlabelled Eggs Seekers the share of people mainly dwelling in small size town, and rarely purchasing eggs with high in omega 3 fatty acids is significantly higher. Particularly, participants from this small segment significantly depreciate environmental-friendly and animal-friendly eggs (e.g., not paying attention to organic labels), and those high in omega 3 fatty acids. This can be explained that the Unlabelled Eggs Seekers might fulfil their needs for eggs by purchasing eggs unpacked, e.g., like those unlabelled purchased directly in rural areas from small scale producers (with up to maximum 50 laying hens in the farm). Moreover, the Price Sensitive and Quality Optimizing Opportunist Consumers involve participants who mainly live in large cities, buy organic eggs frequently, often purchase eggs with high in omega 3 fatty acids, and especially prefer to consume foods produced in an environmental friendly and animal-friendly manner.

The results of Kruskal–Wallis test in Table A2 in the Appendix A also show that in Italy Healthy Conscious Buyers purchase more often eggs with high in omega 3 fatty acids than the two other groups. Moreover, for these consumers it is important that the food eaten on a typical day is familiar, animal friendly, and a way of managing their mood (e.g., a good feeling or coping with stress) than the other classes. Price Sensitive Consumers buy less often organic eggs, and eggs at a farmers market or farm shop. For them, it is more important that the food eaten on a typical day is affordable, whilst it is less important that this food is environmentally friendly and produced and traded in a fair manner. Finally, Clean Labelling Consumers, buy more often eggs at a farmers market or farm shop, and find less relevant that the food eaten is a way of managing their mood.

### 3.3. Willingness to Pay Estimates

The results of country-specific segment-specific Willingness-to-Pay (WTP) estimates are presented in Table 7. In Hungary, Health Conscious Buyers are willing to pay a price premium of EUR 1.47 per 6-pack of eggs for the combination of the EU and national organic labelled eggs, and a WTP of EUR 1.63 for eggs with both nutrition and health claims, respectively. In addition, compared to the other groups the Price Sensitive and Quality Optimizing Opportunist Consumers demonstrate positive and relatively minor WTP for eggs carry either single EU organic label (EUR 0.27 per 6-pack of eggs) or a combination of the EU and national organic label (EUR 0.23 per 6-pack of eggs), as well as for eggs with either simplified nutrition claim (EUR 0.15 per 6-pack of eggs) or the full nutrition and health claim (EUR 0.51 per 6-pack of eggs). However, they would only buy eggs sold with no organic label (EUR −0.50) and no nutrition and health claims (EUR −0.66) with discounts. More interestingly, we also found that the Unlabelled Eggs Seekers are willing to pay a price premium (EUR 1.10 per 6-pack of eggs) for eggs with no label of production, which can be considered a significant premium as 80% of them pays less than EUR 2.0, while for these consumers to have the EU organic label on the package is disincentive as they would pay less for such eggs.

**Table 7.** Mean WTP estimates for cross-country latent consumer segments.

| Country | Hungary | | | Italy | | |
|---|---|---|---|---|---|---|
| *N* | 403 | | | 404 | | |
| Segment label | Group 1 Health Conscious Buyers | Group 2 Price Sensitive and Quality Optimizing Opportunist Consumers | Group 3 Unlabelled Eggs Seekers | Group 1 Health Conscious Buyers | Group 2 Price Sensitive Consumers | Group 3 Clean Labelling Consumers |
| Segment size | 46.40% | 45.60% | 8.00% | 49.00% | 39.50% | 11.60% |
| | WTP | WTP | WTP | WTP | WTP | WTP |
| Organic labels | | | | | | |
| No label | −1.06 *** | −0.50 *** | 1.10 *** | −2.22 *** | −0.22 *** | −2.79 *** |
| EU organic | −0.41 | 0.27 *** | −0.94 * | −0.30 | 0.13 | 2.20 * |
| EU organic and national organic | 1.47 *** | 0.23 *** | −0.17 | 2.52 *** | 0.10 | 0.59 |
| Nutrition and health claims | | | | | | |
| No-claim | −2.27 *** | −0.66 *** | 0.47 | −5.04 *** | −0.71 *** | 9.28 *** |
| Nutrition claim | 0.64 | 0.15 * | 0.35 | 0.98 *** | 0.19 ** | −4.88 *** |
| Nutrition and health claim | 1.63 *** | 0.51 *** | −0.82 | 4.07 *** | 0.51 *** | −4.40 *** |

Relat. Imprt. = Relative importance; *, **, ***; $p < 0.05, 0.01, 0.001$.

In Italy, the three classes presented a negative WTP for the product with no organic label, whilst only the Clean Labelling Consumers express a positive WTP (EUR 2.20 per 6-pack of eggs) for the EU organic label alone. This premium is higher than the price usually paid by half of the respondents in this group, i.e., EUR 1.70 per 6-pack of eggs. The Healthy Conscious Buyers report a positive and significant WTP of EUR 2.52 per 6-pack of eggs for the combination of EU organic and national (private certification body) label. This premium is relatively high considering that 55% of the respondents in this group usually pay an average price lower than EUR 2.00 per 6-pack of eggs. Considering the nutrition and health claim, the Healthy Conscious Buyers and the Price Sensitive Consumers report a negative WTP for the product with no label, indicating a preference for having at least the nutrition claim displayed on the package. In particular, Healthy Conscious Buyers expressed a high premium for the full nutrition and health claim (EUR 4.07 per 6-pack of eggs), especially if compared to the price normally paid by these consumers. The Price Sensitive Consumers report a moderately positive and significant WTP for the product with the nutrition and health claim (EUR 0.51). On the other hand, the Clean Labelling Consumers indicate a positive and significant WTP only for the product without any nutrition and health claim, showing a disutility for providing this information on the package.

## 4. Discussion

The present study utilized DCEs to assess Hungarian and Italian consumers' interest in and WTP for organic label and nutrition and health claims considering a standard 6-pack of eggs as a specific food object. This paper contributed significantly to the understanding of the synergies between organic food labels with nutrition and health claims, and of the heterogeneous consumers' preferences. Using an LCA modelling approach, we have identified three distinct segments among respondents in each country, revealing heterogeneous preferences and WTP for organic labels and nutrition and health claims, as well as different socio-demographic and behavioral characteristics.

The results in both countries show a general preference of consumers for the EU organic logo combined with the national one, as compared to the EU organic logo alone or to the product with no label. Similarly, in both countries, consumers prefer the full nutrition and health claim (i.e., High in Omega 3 fatty acids, contributes to the maintenance of normal function of the heart and normal blood pressure), compared to the nutrition information alone (i.e., High of Omega 3 fatty acids) or to the product with no additional information. However, we should consider that more educated and urban consumers, like those included in the samples, might understand and use nutrition and health claims, as well as organic labels, more than less educated and rural ones [73,74].

However, a more detailed LCA indicates significant differences across segments of respondents. In particular, in Hungary and Italy, respectively, 46.4% and 49.0% of the sample has been grouped in the Healthy Conscious Buyers with high and positive WTP estimates for the EU organic and national organic labels as well as the nutrition and health claims. In Italy, in particular, this segment presents a stronger familiarity with the organic labelled products and with foods carry nutrition claims. Other studies have documented that familiarity affects both organic consumption and WTP for organic food [30]. In other words, those regularly purchasing organic products and foods with nutrition claims are also willing to spend more on these products. The EU organic logo is generally more familiar in the Italian market, where no other governmental nor a dominant private organic logo existed. It was suggested that this factor could also explain the positive attitude and trust of Italian consumers towards the EU logo [32], in particular if associated with the certification body one. Possibly the non-significant effect of this attribute for 51% of the sample, represented in the Price Sensitive and Clean Labelling segments, could be due to the absence of dominant organic labels of private companies in the Italian organic market.

Therefore, synergies between organic and nutrition labels have emerged in both countries, at least in the Healthy Conscious Buyers segment. In Hungary also the Price Sensitive and Quality Optimizing Opportunist Consumers (46% of the sample) exhibit positive and significant effects for both labels. In other studies, organic consumers have shown a higher level of sustainability concern in their general food choices, including principles related to a more sustainable lifestyle, as well as health aspects and naturalness of food [29,34]. Similarly, consumers often perceive organic eggs healthier than conventional products, and they are willing to pay more for them [13,35–37]. Synergies between sustainability and health-related labels have also been reported in Italy and other European countries for several fish species, including seabream, seabass and herrings [75].

On the other hand, consumers grouped in the Unlabelled Eggs Seekers segment in Hungary and in the Clean Labelling Consumers segment in Italy reported non-significant or negative effects. This finding is in line with the study reported by [14] that consumers do not appreciate the value added of organic eggs, presenting a decreasing marginal utility for added attributes. As suggested in other cases (see, e.g., Grunert, et al. [76]), low values of WTP for eco-friendly and healthy-related attributes could be due to consumers' perception of the effectiveness of these attributes, and not by the low interest or value per se. In Hungary, the unfavorable attitude towards the EU organic logo and nutrition claims of the Unlabelled Eggs Seekers segment can be also explained by their shopping behavior of rarely purchasing eggs that are organic certified and high in omega 3 fatty acid, and by their need for unlabelled eggs to be available in Hungarian farmers' market and at small scale producers in short food supply chains. Thus, labels (both trademarks and quality marks) only have a very limited impact on consumers' attitude as motivators to purchase organic food [24] in Hungary.

Gender does not represent a factor significantly affecting the probability of belonging to a specific segment in both countries, the same finding holds for both age and educational level. Particularly, in Hungary rather few socio-demographics are able to explain the probability of belonging to one of the three segments. Although the Unlabelled Eggs Seekers show considerably low interest of buying eggs certified organically, with additional nutritional value (e.g., omega 3 fatty acid) and produced environmental-friendly and animal-friendly, there is still unclear evidence whether the groups between Healthy Conscious Buyers and Price Sensitive and Quality Optimizing Opportunist Consumers in Hungary possess distinguishable attitudes towards organic labelled eggs or eggs with nutrition claims. This partially contradicts the results of previous studies indicating that a typical Hungarian organic food consumer is female, belongs to the upper-income level segment, and is committed to a healthy lifestyle and environmental protection [22,24]. In Italy, although few socio-demographics are significantly different across the three segments, a different attitude was reported. Healthy Conscious Buyers, for instance, are more interested that the food eaten on a typical day is familiar, animal friendly, and able of managing their mood than the other classes, whilst Price Sensitive Consumers prefer that the food eaten on a typical day is affordable. This latter segment is less interested that this food is environmentally friendly and produced and traded in a fair manner. These two segments, representing almost 90% of the sample, identify two different approaches to food choices. On the one hand, a segment of consumers more involved in the nutritional and ethical dimension of food purchase decisions, and on the other hand, a segment more interested in the convenience and economic effects of food choices. These behaviors are not uncommon and have already been reported in other studies showing that Italian organic consumers are more concerned with the sustainability dimensions of their general food choices, and have a more sustainable lifestyle [29,34].

## 5. Conclusions

The study presented is insightful in the scope of consumer behavior research as well as in practice for producers, traders and other stakeholders in the food market; there are several implications of these findings. Sustainable food choices must be affordable and able to meet the need of different segments of society. In Italy and Hungary, approx. 40–45% of the sample is more interested in the affordability of food choices than in their environmental and health consequences. The sustainability of dietary choices must necessarily face all the environmental, social and economic aspects [77]. It is not uncommon that organic products are perceived as being more expansive than conventional ones, and this is the most significant barrier why they are not consumed widely [20,21]. Moreover, the fact that the organic certified labels seem to be less important should be considered seriously, as this might indicate that there is a lack of trust in the organic certification system in general and/or a label of knowledge with respect to what organic implies. Subsidies or voucher programs should encourage healthy and sustainable dietary choices, particularly to enable their adoption in low-income and middle-income segments of the population [78,79]. For companies, the synergies found in the Healthy Conscious Buyers segments for eco-friendly and healthy-related choices suggest the possibility of combining these labels to meet the expectations of consumers more involved in the ethical and nutritional dimensions of purchase decisions [75]. As it is found that not all (nutrition and health) claims are similarly appealing to consumers from different countries such as in Hungary and Italy, food producers and processors should consider tailoring food claims as well as food labelling of their products to country-specific demands, refining in that way the effectiveness of claim-based/label-based marketing communications. Finally, the identified segments of the Healthy Conscious Buyers, Price Sensitive and Quality Optimizing Opportunist Consumers, the Clean Labelling Consumers, and the Unlabelled Eggs Seekers represent a structured profile of the European (egg) shoppers, suggesting the proportions of consumers holding similar consumption patterns of preferences around which marketing strategy could be designed that would further facilitate communication development towards sustainable and healthier food choices.

This might suggest that companies should implement strategies combining these elements (e.g., in communication campaigns) for selling their product in premium segments of the market.

Some limitations of the study should be also highlighted. The use of hypothetical choice experiments may have inflated the WTP estimates, and future analysis could benefit from triangulation with revealed preference data obtained, for instance, in real settings (e.g., auction-based experiments). On the other hand, this allowed us to obtain a large amount of data; moreover, the use of cheap talk scripts should have minimized any hypothetical bias. Second, we focused only on three attributes, ignoring other elements usually used by consumers in their food choices, e.g., brands, other production practices (e.g., free-range breeding). Therefore, the results of this research do not claim to cover the whole complex eggs market, whereas they provide practitioners and actors in the egg industry with more insights to understand consumer motivations when buying eggs carrying less investigated features, such as sustainability and health-related attributes. Third, in the present study, we only focus on one product, eggs. Thus, our outputs hold only for this animal product. For a better understanding of the relevance of organic labels and nutrition claims on consumers' purchase decision of across other product categories, additional investigations are needed in the future research. Finally, the authors suggest that for future research, it is worth addressing the attribute non-attendance (ANA) incidence, as the ANA can occur when a respondent simplifies the choice task by ignoring an attribute [80], and this simplification could become an issue, because ignoring an attribute may alter the effect of the DCE. Confidently, our theoretical and experimental approach can be indeed applied to other product categories in the future.

**Author Contributions:** Conceptualization, C.-H.Y., D.M. and Á.T.; methodology, C.-H.Y.; software, C.-H.Y.; validation, C.-H.Y.; formal analysis, C.-H.Y.; investigation, C.-H.Y., D.M. and Á.T.; resources, C.-H.Y., D.M. and Á.T.; data curation, C.-H.Y.; writing—original draft preparation C.-H.Y., D.M. and Á.T.; writing—review and editing, C.-H.Y., D.M. and Á.T.; funding acquisition, C.-H.Y., D.M. and Á.T. All authors have read and agreed to the published version of the manuscript.

**Acknowledgments:** We are grateful to Monika Hartmann's insightful guidance in the working package of the STRENGTH2FOOD project. Additionally, we would like to thank the four anonymous reviewers for their valuable comments and suggestions.

## Appendix A

**Table A1.** WTPs derived based on the RPL models.

| Country | Hungary | Italy |
|---|---|---|
| *N* | 403 | 404 |
| | WTP | WTP |
| Organic labels | | |
| No label | −0.54 | −0.91 |
| EU organic | 0.01 | 0.11 |
| EU organic and national organic | 0.52 | 0.81 |
| Nutrition and health claims | | |
| No-claim | −0.89 | −1.37 |
| Nutrition claim | 0.37 | 0.22 |
| Nutrition and health claim | 0.52 | 1.15 |

**Table A2.** Cross-country consumer segment profiles and the results of the Kruskal–Wallis test.

| N | Hungary | | | | Italy | | | |
|---|---|---|---|---|---|---|---|---|
| | 403 | | | | 404 | | | |
| | Group 1 *Health Conscious Buyers* | Group 2 *Price Sensitive and Quality Optimizing Opportunist Consumers* | Group 3 *Unlabelled Eggs Seekers* | | Group 1 *Health Conscious Buyers* | Group 2 *Price Sensitive Consumers* | Group 3 *Clean Labelling Consumers* | |
| | 46.40% | 45.60% | 8.00% | **K-W test** | 49.00% | 39.50% | 11.60% | **K-W test** |
| Gender | | | | | | | | |
| Female (%) | 46.6 | 49.6 | 50 | | 51.0 | 53.7 | 52.2 | |
| Male (%) | 53.4 | 50.4 | 50 | | 49.0 | 46.3 | 47.8 | |
| Age | 42.2 | 42.2 | 45.6 | | 43.5 | 42.4 | 44.4 | |
| Living location | | | | * | | | | |
| Rural area (%) | 13.6 | 15.4 | 25.0 | | 10.8 | 13.4 | 28.3 | |
| Small size town (%) | 42.4 | 37.5 | 46.9 | | 46.4 | 46.3 | 37.0 | |
| Large city (%) | 44.0 | 47.1 | 28.1 | | 42.8 | 40.2 | 34.8 | |
| Education | | | | | | | | |
| Up to secondary school education (%) | 36.1 | 25.0 | 31.1 | | 52.2 | 48.2 | 44.3 | |
| University degree, vocational training or higher (%) | 63.9 | 75.0 | 68.9 | | 47.8 | 51.8 | 55.7 | |
| How often do you buy organic egg? (1 indicates Never, 7 indicates Every time) | 2.7 | 3.1 | 2.6 | * | 4.6 | 3.8 | 4.5 | *** |
| How often do you buy eggs with high in omega 3 fatty acids? (1 indicates Never, 7 indicates Every time) | 2.6 | 3.0 | 1.8 | *** | 3.4 | 2.6 | 2.3 | *** |
| How often do you buy eggs at a farmers market or farm shop? (1 indicates Never, 7 indicates Every time) | 4.4 | 4.6 | 4.6 | | 3.9 | 3.4 | 4.2 | ** |
| What price do you normally pay for a 6-pack of eggs? | | | | *** | | | | |
| (HU) Below c.a. EUR 0.49/(IT) Below 1.40 EUR (%) | 0.5 | 1.7 | 9.4 | | 3.6 | 23.2 | 13.0 | |
| (HU) Between c.a. EUR 0.50–EUR 0.72/(IT) Between 1.40 EUR and 1.70 EUR (%) | 36.1 | 19.4 | 34.4 | | 21.1 | 42.1 | 37.0 | |
| (HU) Between c.a. EUR 0.72–EUR 0.94/(IT) Between 1.71 EUR and 2.00 EUR (%) | 40.3 | 33.3 | 25.0 | | 29.4 | 22.0 | 17.4 | |
| (HU) Between c.a. EUR 0.95–EUR 1.17/(IT) Between 2.01 EUR and 2.30 EUR (%) | 9.9 | 22.8 | 9.4 | | 19.6 | 3.7 | 4.3 | |
| (HU) Between c.a. EUR 1.18–EUR 1.40 (IT) Between 2.31 EUR and 2.60 EUR (%) | 9.4 | 11.1 | 6.3 | | 10.3 | 3.0 | 15.2 | |
| (HU) Between c.a. EUR 1.41–EUR 1.62/(IT) Between 2.61 EUR and 2.90 EUR (%) | 1.0 | 8.3 | 3.1 | | 5.2 | 0.0 | 0.0 | |
| (HU) More than c.a. EUR 1.63/(IT) More than 2.90 EUR (%) | 1.0 | 2.2 | 3.1 | | 2.1 | 0.0 | 2.2 | |
| Do not remember (%) | 1.6 | 1.1 | 9.4 | | 8.8 | 6.1 | 10.9 | |
| How often do you buy eggs? | | | | ** | | | | |
| About or Less than three times a month (%) | 72.8 | 56.7 | 59.4 | | 63.4 | 61.6 | 65.2 | |
| About or more than once a week (%) | 27.2 | 43.3 | 40.6 | | 36.6 | 38.4 | 34.8 | |

**Table A2.** *Cont.*

| N | Hungary | | | | Italy | | | |
|---|---|---|---|---|---|---|---|---|
| | 403 | | | | 404 | | | |
| | Group 1 *Health Conscious Buyers* | Group 2 *Price Sensitive and Quality Optimizing Opportunist Consumers* | Group 3 *Unlabelled Eggs Seekers* | | Group 1 *Health Conscious Buyers* | Group 2 *Price Sensitive Consumers* | Group 3 *Clean Labelling Consumers* | |
| | 46.40% | 45.60% | 8.00% | **K-W test** | 49.00% | 39.50% | 11.60% | **K-W test** |
| How often do you eat eggs? | | | | | | | | |
| Once a week or les (%) | 45.0 | 45 | 50.0 | | 63.4 | 69.5 | 71.7 | |
| More than once a week (%) | 55.0 | 55.0 | 50.0 | | 36.6 | 30.5 | 28.3 | |
| It is important to me that the food I eat on a typical day . . . (1 indicates strongly disagree and 7 strongly agree) | | | | | | | | |
| . . . is healthy. | 5.65 | 5.73 | 5.06 | | 5.79 | 5.71 | 5.57 | |
| . . . is a way of managing my mood (e.g., a good feeling or coping with stress). | 4.95 | 5.07 | 4.47 | | 5.18 | 4.92 | 4.52 | *** |
| . . . is convenient (in buying and cooking). | 5.34 | 5.28 | 4.84 | | 5.19 | 5.20 | 4.91 | |
| . . . provides me with pleasure (e.g., appearance, texture, smell, taste). | 5.67 | 5.57 | 5.13 | | 5.74 | 5.77 | 5.41 | |
| . . . is natural (no additives, only natural ingredients). | 5.35 | 5.56 | 5.19 | | 5.63 | 5.41 | 5.50 | |
| . . . is affordable. | 6.15 | 5.84 | 5.69 | | 4.54 | 5.15 | 4.76 | *** |
| . . . helps me control my weight. | 4.74 | 4.87 | 4.44 | | 5.14 | 4.95 | 4.83 | |
| . . . is familiar. | 5.16 | 5.16 | 4.75 | | 4.94 | 4.60 | 4.35 | ** |
| . . . is environmentally friendly. | 5.09 | 5.41 | 4.81 | * | 5.36 | 5.00 | 5.24 | ** |
| . . . is animal friendly. | 4.68 | 5.23 | 4.38 | *** | 5.51 | 5.15 | 5.28 | ** |
| . . . is produced and traded in a fair manner. | 5.07 | 5.27 | 5.88 | | 5.18 | 4.74 | 5.20 | *** |
| Monthly net household income | | | | | | | | |
| (HU) < 150,000 HUK (c.a. EUR 486)/(IT) < EUR 900 (%) | 11.5 | 3.1 | 3.1 | | 3.6 | 8.5 | 6.5 | |
| (HU) 150,001–205,000 HUF (c.a. EUR 487–EUR 664)/(IT) EUR 900–EUR 1500 (%) | 14.1 | 18.8 | 18.8 | | 19.6 | 22.6 | 19.6 | |
| (HU) 205,001–235,000 HUF (c.a. EUR 665–EUR 761)/(IT) EUR 1501–EUR 2500 (%) | 11.0 | 6.3 | 6.3 | | 26.3 | 31.1 | 30.4 | |
| (HU) 235,001–380,000 HUF (c.a. EUR 762–EUR 1231)/(IT) EUR 2501–EUR 3500 (%) | 29.8 | 28.1 | 28.1 | | 22.2 | 14.6 | 15.2 | |
| (HU) 380,001–835,000 HUF (EUR 1232–EUR 2705)/(IT) EUR 3501–EUR 4500 (%) | 25.7 | 28.1 | 28.1 | | 12.4 | 5.5 | 8.7 | |
| (HU) ≥ 835,001 HUF (c.a. EUR 2706)/(IT) ≥ EUR 4501 (%) | 2.1 | 6.3 | 6.3 | | 3.6 | 3.7 | 0.0 | |
| Prefer not to answer (%) | 5.8 | 9.4 | 9.4 | | 12.4 | 14.0 | 19.6 | |
| Household size | 2.96 | 2.93 | 3.09 | | 2.93 | 2.95 | 2.89 | |
| Number of kids | 0.61 | 0.56 | 0.63 | | 0.57 | 0.47 | 0.65 | |

***, **, * Demographics are significant at 0.1%, 1%, 5%, level with asymptotic method of non-parametric Kruskal–Wallis test among cross-country three latent segment.

## References


1. European Commission. Europeans' Attitudes towards Food Security, Food Quality and the Countryside 2012. Available online: http://ec.europa.eu/public_opinion/archives/ebs/ebs_389_en.pdf (accessed on 12 February 2020).
2. European Commission. Europeans, Agriculture and the Common Agricultural Policy (CAP) 2014. Available online: http://ec.europa.eu/public_opinion/archives/ebs/ebs_410_en.pdf (accessed on 12 February 2020).
3. European Commission. Europeans, Agriculture and the CAP 2016. Available online: https://www.google.com.hk/url?sa=t&rct=j&q=&esrc=s&source=web&cd=&ved=2ahUKEwjl3eXu4cfrAhUlGaYKHRT0DDEQFjAAegQIAxAB&url=http%3A%2F%2Fec.europa.eu%2FCOMMFrontOffice%2FPublicOpinion%2Findex.cfm%2FResultDoc%2Fdownload%2FDocumentKy%2F69756&usg=AOvVaw2gqUNtyndysnLJ2HiCd96s (accessed on 12 February 2020).
4. European Commission. Europeans, Agriculture and the CAP 2018. Available online: https://ec.europa.eu/commfrontoffice/publicopinion/index.cfm/ResultDoc/download/DocumentKy/82123 (accessed on 24 April 2020).
5. Research Institute of Organic Agriculture (FiBL). Data on Organic Agriculture in Europe. Volume 2019. Available online: https://statistics.fibl.org/europe.html (accessed on 18 February 2020).
6. Banks, Q.D. Grade standards and product identification labels influence the demand for eggs and other farm products. *Am. J. Agric. Econ.* **1963**, *45*, 1365–1369. [CrossRef]
7. Food and Agriculture Organization Corporate Statistical Database (FAOSTAT). New Food Balances. Available online: http://www.fao.org/faostat/en/#data/FBS (accessed on 2 December 2019).
8. European Commission. EU Agricultural Outlook for Markets and Income 2019-2030. Available online: https://ec.europa.eu/info/food-farming-fisheries/farming/facts-and-figures/markets/outlook/medium-term_en (accessed on 15 February 2020).
9. Bennett, R. Farm animal welfare and food policy. *Food Policy* **1997**, *22*, 281–288. [CrossRef]
10. Brown, D.J.; Schrader, L.F. Cholesterol Information and Shell Egg Consumption. *Am. J. Agric. Econ.* **1990**, *72*, 548–555. [CrossRef]
11. Fearne, A.; Lavelle, D. Segmenting the UK egg market: Results of a survey of consumer attitudes and perceptions. *Br. Food J.* **1996**, *98*, 7–12. [CrossRef]
12. Monier, S.; Hassan, D.; Nichèle, V.; Simioni, M. Organic Food Consumption Patterns. *J. Agric. Food Ind. Organ.* **2009**, *7*. [CrossRef]
13. Gracia, A.; Barreiro-Hurle, J.; Galán, B.L. Are Local and Organic Claims Complements or Substitutes? A Consumer Preferences Study for Eggs. *J. Agric. Econ.* **2014**, *65*, 49–67. [CrossRef]
14. Gerini, F.; Alfnes, F.; Schjøll, A. Organic- and Animal Welfare-labelled Eggs: Competing for the Same Consumers? *J. Agric. Econ.* **2016**, *67*, 471–490. [CrossRef]
15. Erdős, A.D.; Szőllősi, L. A tojástermelés és-fogyasztás nemzetközi és magyarországi helyzete, főbb tendenciái. *Anim. Welf. Etol. Tartástechnológia* **2018**, *14*, 8–14.
16. Molnár, S.; Szőllősi, L. Az étkezési tojás fogyasztási szokásainak főbb jellemzői Magyarországon. *Táplálkozásmarketing* **2014**, *1*, 133–138.
17. Food and Agriculture Organization (FAO). Per Capita Egg Consumption. 2017. Available online: https://ourworldindata.org/meat-production#per-capita-egg-consumption (accessed on 3 April 2020).
18. Mercati, I. Analisi e Studio Filiere Agroalimentari. Available online: http://www.ismeamercati.it/analisi-e-studio-filiere-agroalimentari (accessed on 3 April 2020).
19. Eurostat. Organic Production of Animal Products. Available online: https://appsso.eurostat.ec.europa.eu/nui/show.do?dataset=org_aprod&lang=en (accessed on 2 December 2019).
20. Kiss, C.; Vincze, J.; Tenk, A.; Paszthy, G.; Toaso, S. Fogyasztói vélemények a bio-sertéshúsról, illetve a sertéshúsfogyasztásról. *Gazdálkodás* **2005**, *49*, 7.
21. Gergely, É.; Szabó, B.; Balázs, K. Az egészség-és környezettudatosság, valamint az értékrend hatása a bioélelmiszer-fogyasztásra. *Mark. Menedzsment* **2014**, *48*, 27–37.
22. Borsos-Repka, N. A mai magyar bio-élelmiszerpiac néhány jellegzetessége. Árak, vevők és eladók. *Tarsadalomkutatas* **2010**, *28*, 225–241. [CrossRef]
23. Dörnyei, K. Bioélelmiszer-fogyasztási szokások: Szegmentálás és a bizalom fontossága. *Mark. Menedzsment* **2008**, *42*, 34–42.

24. Nagy-Pércsi, K.; Fogarassy, C. Important Influencing and Decision Factors in Organic Food Purchasing in Hungary. *Sustainability* **2019**, *11*, 6075. [CrossRef]
25. Horváth, Á.; Kovács, A.; Gyenge, B. A vásárlók biotermékekkel szembeni attitűdje a hiper-és szupermarketekben. *Mark. Menedzsment* **2003**, *37*, 23–34.
26. Szente, V.; Szigeti, O.; Polereczki, Z.; Varga, Á.; Szakály, Z. Towards a new strategy for organic milk marketing in Hungary. *Acta Aliment.* **2015**, *44*, 32–38. [CrossRef]
27. Hofer, M. Bioélelmiszerek Fogyasztásának Tendenciája Magyarországon. Ph.D. Thesis, Széchenyi István Egyetem, Győr, Hungary, 2009.
28. Szente, V. Consumer motivations in the purchase of organic foods in Hungary. *Acta Fytotech. Zootech.* **2015**, *18*, 145–147. [CrossRef]
29. Annunziata, A.; Massimiliano, A.; Mariani, A. Measuring sustainable food consumption: A case study on organic food. *Sustain. Prod. Consum.* **2019**, *17*, 95–107. [CrossRef]
30. Vecchio, R.; Van Loo, E.J.; Annunziata, A. Consumers' willingness to pay for conventional, organic and functional yogurt: Evidence from experimental auctions. *Int. J. Consum. Stud.* **2016**, *40*, 368–378. [CrossRef]
31. Annunziata, A.; Mariani, A.; Vecchio, R.; Annunziata, A.; Angela, M.; Riccardo, V. Effectiveness of sustainability labels in guiding food choices: Analysis of visibility and understanding among young adults. *Sustain. Prod. Consum.* **2019**, *17*, 108–115. [CrossRef]
32. Janssen, M.; Hamm, U. The mandatory EU logo for organic food: Consumer perceptions. *Br. Food J.* **2012**, *114*, 335–352. [CrossRef]
33. Carfora, V.; Cavallo, C.; Caso, D.; Del Giudice, T.; De Devitiis, B.; Viscecchia, R.; Nardone, G.; Cicia, G. Explaining consumer purchase behavior for organic milk: Including trust and green self-identity within the theory of planned behavior. *Food Qual. Preference* **2019**, *76*, 1–9. [CrossRef]
34. Scalvedi, M.L.; Saba, A. Exploring local and organic food consumption in a holistic sustainability view. **2018**, *120*, 749–762. [CrossRef]
35. Heng, Y.; Peterson, H.H.; Li, X. Consumer attitudes toward farm-animal welfare: The case of laying hens. **2013**, *38*, 418–434.
36. Andersen, L.M. Animal Welfare and Eggs—Cheap Talk or Money on the Counter? *J. Agric. Econ.* **2011**, *62*, 565–584. [CrossRef]
37. Monier-Dilhan, S. Food labels: Consumer's information or consumer's confusion. *OCL* **2018**, *25*, D202. [CrossRef]
38. Baltzer, K. Consumers' willingness to pay for food quality—The case of eggs. *Food Econ. Acta Agric. Scand. Sect. C* **2004**, *1*, 78–90. [CrossRef]
39. Galán, B.L.; Gracia, A.; Barreiro-Hurle, J. What comes first, origin or production method? An investigation into the relative importance of different attributes in the demand for eggs. *Span. J. Agric. Res.* **2013**, *11*, 305. [CrossRef]
40. Naspetti, S.; Bodini, A. Consumer Perception of Local and Organic Products: Substitution or Complementary Goods? *Int. J. Interdiscip. Soc. Sci.* **2008**, *3*, 111–121. [CrossRef]
41. Heng, Y.; Peterson, H.H. Interaction Effects among Labeled Attributes for Eggs in the United States. *J. Int. Food Agribus. Mark.* **2018**, *30*, 236–250. [CrossRef]
42. Li, T.; Bernard, J.C.; Johnston, Z.A.; Messer, K.D.; Kaiser, H.M. Consumer preferences before and after a food safety scare: An experimental analysis of the 2010 egg recall. *Food Policy* **2017**, *66*, 25–34. [CrossRef]
43. Żakowska-Biemans, S.; Tekień, A. Free Range, Organic? Polish Consumers Preferences Regarding Information on Farming System and Nutritional Enhancement of Eggs: A Discrete Choice Based Experiment. *Sustainability* **2017**, *9*, 1999. [CrossRef]
44. Dahlhausen, J.L.; Rungie, C.; Roosen, J. Value of labeling credence attributes-common structures and individual preferences. *Agric. Econ.* **2018**, *49*, 741–751. [CrossRef]
45. Gangnat, I.D.M.; Mueller, S.; Kreuzer, M.; Messikommer, R.E.; Siegrist, M.; Visschers, V.H.M. Swiss consumers' willingness to pay and attitudes regarding dual-purpose poultry and eggs. *Poult. Sci.* **2018**, *97*, 1089–1098. [CrossRef] [PubMed]
46. Teixeira, D.L.; Larraín, R.; Hötzel, M.J. Are views towards egg farming associated with Brazilian and Chilean egg consumers' purchasing habits? *PLoS ONE* **2018**, *13*, e0203867. [CrossRef]

47. European Commisson. Regulation (EC) No 1924/2006 of the European Parliament and of the Council of 20 December 2006 on Nutrition and Health Claims Made on Foods. Available online: https://eur-lex.europa.eu/legal-content/en/ALL/?uri=CELEX%3A32006R1924 (accessed on 12 February 2020).
48. European Commission. Attitudes of Europeans towards Animal Welfare 2016. Available online: https://ec.europa.eu/commfrontoffice/publicopinion/index.cfm/ResultDoc/download/DocumentKy/71348 (accessed on 24 August 2020).
49. Asselin, A.M. Eggcentric Behavior—Consumer Characteristics That Demonstrate Greater Willingness to Pay for Functionality. *Am. J. Agric. Econ.* **2005**, *87*, 1339–1344. [CrossRef]
50. Gracia, A.; De-Magistris, T. Preferences for lamb meat: A choice experiment for Spanish consumers. *Meat Sci.* **2013**, *95*, 396–402. [CrossRef]
51. Loureiro, M.L.; Umberger, W.J. A choice experiment model for beef: What US consumer responses tell us about relative preferences for food safety, country-of-origin labeling and traceability. *Food Policy* **2007**, *32*, 496–514. [CrossRef]
52. Van Loo, E.J.; Caputo, V.; Nayga, R.M.; Meullenet, J.-F.; Ricke, S.C. Consumers' willingness to pay for organic chicken breast: Evidence from choice experiment. *Food Qual. Preference* **2011**, *22*, 603–613. [CrossRef]
53. Louviere, J.J.; Hensher, D.A.; Swait, J.D. *Stated Choice Methods: Analysis and Applications*; Cambridge University Press: Cambridge, UK, 2000.
54. Uncles, M.D.; Ben-Akiva, M.; Lerman, S.R. Discrete Choice Analysis: Theory and Application to Travel Demand. *J. Oper. Res. Soc.* **1987**, *38*, 370. [CrossRef]
55. McFadden, D. Conditional Logit Analysis of Qualitative Choice Behaviour. In *Frontiers in Econometrics*; Zarembka, P., Ed.; Academic Press: New York, NY, USA, 1974.
56. Train, K.E. *Discrete Choice Methods with Simulation*; Cambridge University Press: Cambridge, UK, 2009.
57. Jaeger, S.R.; Rose, J.M. Stated choice experimentation, contextual influences and food choice: A case study. *Food Qual. Preference* **2008**, *19*, 539–564. [CrossRef]
58. Hensher, D.A.; Johnson, L.W. *Applied Discrete-Choice Modelling*; Routledge: Abingdon, UK, 2018.
59. Akinc, D.; Vandebroek, M. Comparing the performances of maximum simulated likelihood and hierarchical Bayesian estimation for mixed logit models. In Proceedings of the International Choice Modelling Conference 2017, Cape Town, South Africa, 3–5 April 2017.
60. Rossi, P.E.; Allenby, G.M. Bayesian Statistics and Marketing. *Mark. Sci.* **2003**, *22*, 304–328. [CrossRef]
61. Rossi, P.E.; Allenby, G.M.; McCulloch, R. *Bayesian Statistics and Marketing*; John Wiley & Sons: Hoboken, NJ, USA, 2012.
62. Boxall, P.C.; Adamowicz, W. (Vic) Understanding Heterogeneous Preferences in Random Utility Models: A Latent Class Approach. *Environ. Resour. Econ.* **2002**, *23*, 421–446. [CrossRef]
63. Swait, J. A structural equation model of latent segmentation and product choice for cross-sectional revealed preference choice data. *J. Retail. Consum. Serv.* **1994**, *1*, 77–89. [CrossRef]
64. Bech, M.; Gyrd-Hansen, R.; Gyrd-Hansen, D. Effects coding in discrete choice experiments. *Health Econ.* **2005**, *14*, 1079–1083. [CrossRef]
65. Kruskal, W.H.; Wallis, W.A. Use of ranks in one-criterion variance analysis. *J. Am. Stat. Assoc.* **1952**, *47*, 583–621. [CrossRef]
66. Carlsson, F.; Frykblom, P.; Lagerkvist, C.J. Consumer Benefits of Labels and Bans on Genetically Modified Food-An Empirical Analysis Using Choice Experiments. Available online: https://ideas.repec.org/p/ags/aaea04/20370.html (accessed on 24 August 2020).
67. Heng, Y.; Peterson, H.H.; Li, X. Consumer responses to multiple and superfluous labels in the case of eggs. *J. Food Distrib. Res.* **2016**, *47*, 62–82.
68. Bosworth, R.; Taylor, L.O. Hypothetical Bias in Choice Experiments: Is Cheap Talk Effective at Eliminating Bias on the Intensive and Extensive Margins of Choice? *BE J. Econ. Anal. Policy* **2012**, *12*, 12. [CrossRef]
69. Carlsson, F.; Frykblom, P.; Lagerkvist, C.-J. Using cheap talk as a test of validity in choice experiments. *Econ. Lett.* **2005**, *89*, 147–152. [CrossRef]
70. ChoiceMetrics, Ngene. 1.2 User Manual & Reference Guide, Australia. Available online: http://www.choice-metrics.com (accessed on 15 June 2020).
71. Bethlehem, J. Selection Bias in Web Surveys. *Int. Stat. Rev.* **2010**, *78*, 161–188. [CrossRef]
72. Hole, A.R. Modelling heterogeneity in patients' preferences for the attributes of a general practitioner appointment. *J. Health Econ.* **2008**, *27*, 1078–1094. [CrossRef]

73. Ditlevsen, K.; Denver, S.; Christensen, T.; Lassen, J. A taste for locally produced food—Values, opinions and sociodemographic differences among 'organic' and 'conventional' consumers. *Appetite* **2020**, *147*, 104544. [CrossRef] [PubMed]
74. Nocella, G.; Kennedy, O. Food health claims—What consumers understand. *Food Policy* **2012**, *37*, 571–580. [CrossRef]
75. Menozzi, D.; Nguyen, T.T.; Sogari, G.; Taskov, D.; Lucas, S.; Santiago Castro-Rial, J.L.; Mora, C. Consumers' willingness to pay for fish products with health and environmental labels: Evidence from five European countries. *Nutrients* **2020**, *12*, 2650. [CrossRef]
76. Grunert, K.G.; Hieke, S.; Wills, J. Sustainability labels on food products: Consumer motivation, understanding and use. *Food Policy* **2014**, *44*, 177–189. [CrossRef]
77. Willett, W.; Rockström, J.; Loken, B.; Springmann, M.; Lang, T.; Vermeulen, S.; Garnett, T.; Tilman, D.; Declerck, F.; Wood, A.; et al. Food in the Anthropocene: The EAT–Lancet Commission on healthy diets from sustainable food systems. *Lancet* **2019**, *393*, 447–492. [CrossRef]
78. Afshin, A.; Peñalvo, J.L.; Del Gobbo, L.; Silva, J.; Michaelson, M.; O'Flaherty, M.; Capewell, S.; Spiegelman, N.; Danaei, G.; Mozaffarian, D. The prospective impact of food pricing on improving dietary consumption: A systematic review and meta-analysis. *PLoS ONE* **2017**, *12*, e0172277. [CrossRef] [PubMed]
79. Niebylski, M.L.; Redburn, K.A.; Duhaney, T.; Campbell, N.R. Healthy food subsidies and unhealthy food taxation: A systematic review of the evidence. *Nutrition* **2015**, *31*, 787–795. [CrossRef] [PubMed]
80. Scarpa, R.; Zanoli, R.; Bruschi, V.; Naspetti, S. Inferred and Stated Attribute Non-attendance in Food Choice Experiments. *Am. J. Agric. Econ.* **2012**, *95*, 165–180. [CrossRef]

# 3

# Dynamic Changes in Post-Ingestive Sensations after Consumption of a Breakfast Meal High in Protein or Carbohydrate

**Mette Duerlund *, Barbara Vad Andersen and Derek Victor Byrne**

Department of Food Science, Faculty of Science and Technology, Aarhus University, Kirstinebjergvej 10, DK-5792 Aarslev, Denmark; barbarav.andersen@food.au.dk (B.V.A.); derekv.byrne@food.au.dk (D.V.B.)
* Correspondence: mette.duerlund@food.au.dk.

**Abstract:** The obesity epidemic urges exploration of several parameters that play an important role in our eating behaviours. Post-ingestive sensations can provide a more comprehensive picture of the eating experience than mere satiety measurements. This study aimed to (1) quantify the dynamics of different post-ingestive sensations after food intake and (2) study the effect of protein and carbohydrate on hedonic and post-ingestive responses. Forty-eight participants (mean age 20.4) were served a breakfast meal high in protein (HighPRO) or high in carbohydrate (HighCHO) on two separate days using a randomised controlled crossover design. Post-ingestive sensations were measured every 30 min, for 3 h post intake using visual analogue scale (VAS). Results showed a significant main effect of time for all post-ingestive sensations. HighCHO induced higher hedonic responses compared to HighPRO, as well as higher ratings for post-ingestive sensations such as Satisfaction, Food joy, Overall wellbeing and Fullness. HighPRO, on the other hand, induced higher ratings for Sweet desire post intake. The development of sensations after a meal might be important for consumers' following food choices and for extra calorie intake. More detailed knowledge in this area could elucidate aspects of overeating and obesity.

**Keywords:** post-ingestive sensation; appetite; satiety; consumer; protein; carbohydrate; breakfast

---

## 1. Introduction

Public health issues, like the obesity epidemic currently unfolding, demand action and require multiple approaches [1]. It urges us as researchers to pursue and explore several parameters that could play an important role in our eating behaviours. In particularly, assessing subjective appetite and its relationship to eating can provide important insights into our overall intake behaviours [2,3]. The concepts of satiation (cessation of a meal) and satiety (fullness over time) has been in great focus in recent decades [4,5]. This, among other things, involves the study of 'The Satiety Cascade' which conceptualizes as various signals (sensory, cognitive, post-ingestive, post-absorptive) that relate to eating and the concepts of satiation and satiety [6]. Several modifications over time of 'The Satiety Cascade' reveal a complex and dynamic interplay of numerous factors involved in the process of eating, including satiety and other post-ingestive effects [7,8].

Significant focus in the appetite research field has been on quantifying satiety as well as measuring the satiating effects of different foods, especially foods differing in macronutrient content. Researchers attempt to determine if specific diet manipulations might obtain benefits related to satiation and satiety [4,9–11]. Particularly research on proteins' effect on satiety has gained much attention [4,9,11–17]. In a research study with healthy women, high-protein-low-carbohydrate Greek yoghurt led to increased fullness and delayed subsequent eating compared to low-protein-high-carbohydrate yoghurt (iso-caloric) [15]. Moreover, Anderson and Moore (2004) conclude that the protein content of a food is a strong determinant of short-term satiety [16], and Morell and Fiszman (2017) state that protein is the most effective food macronutrient providing satiating effects [11].

However, asking consumers only about satiation and satiety in research studies does not provide the full picture on post intake experiences nor about the foods effect on these post intake experiences. Studies have shown that other sensations prove important to our hedonic experiences of food [12,18–21] and can explain post-meal desires [22] potentially leading to additional calorie intake. For instance, Andersen et al. (2017) found that, besides the sensation of fullness, also energy and psychological wellbeing were important positive drivers of food satisfaction, and that nausea negatively influenced food satisfaction [18]. Thus, other sensations also contribute to the consumers' eating experience, and by including additional sensations, we obtain a more comprehensive picture of what lies behind and beyond satiety. It enables us to study which sensations contribute positively or negatively to the food experience as well as enables us to study how certain product factors contribute to these different post-ingestive sensations. Post-ingestive experiences therefore offer new premises to predict consumers' food choice and better understand our eating behaviours [10,23]. The contribution of different sensations to food choice demonstrates importance, and researchers within Sensory and Consumer Science subsequently move from one-sip evaluations into a bigger picture focusing on the full consumer experience, including post-ingestive experiences [10,12,18,23–26]. This extended information empowers a deeper understanding of the complex interplay between appetite sensations and food intake. It enables insights into the importance and implications of different sensations to food preference and eating behaviour.

Exploring different post-ingestive sensations within the spectrum of appetite and eating experiences, allows consumers to connect to their ability to introspect on a conscious level. Interoception defines as our ability to sense internal bodily states such as hunger, satiety and energy, but also other perceptions like pain, temperature, muscular and visceral sensations. Interoception focuses on our subjective evaluations as well as our self-awareness e.g., 'how I feel' [27–29]. Evaluating interoceptive states provides us, as researchers, with consumers' current conscious awareness and gives an in-depth picture of the spectrum of appetite [2]. The perception of bodily signals might therefore constitute crucial factors in understanding and regulating our overall food intake [28,30].

Consequently, assessing appetite should not merely consist of hunger and satiety measurements, but could include several other parameters e.g., 'feeling good', energy, satisfaction, relaxation, heartbeat detection, food pleasure and other subjective evaluations [23,24,26,31–33]. For instance, Meiselman (2016) accentuates that wellbeing and quality of life can be central additions when measuring consumer perceptions of food [26]. Research within wellbeing points towards relevance in Consumer Science and several researchers also introduce the concept of wellbeing in relation to food [23,26,32,34,35]. Andersen and Hyldig (2015) found physical wellbeing to function as an important element in food satisfaction, and that physical wellbeing also includes the sense of an appropriate energy level after intake [24]. Food studies have used variables such as psychological wellbeing, desire for other foods, food satisfaction, nausea, fulfilment of expectation, reflux, physical wellbeing, pleasantness, mental hunger, alertness, energy level when evaluating post-ingestive experiences [12,18–20]. Møller (2015) states that there is a need for methods quantifying pleasure and satisfaction obtained from eating food which should go beyond the actual eating event and rely on different types of memory and interoceptive states after eating [31].

Exploring beyond satiety and focusing on post-ingestive sensations also allows us to contribute further on the knowledge we have on macronutrients' effects. Foods macronutrient content influences the satiating power of a food, but does it also influence other post-ingestive sensations besides satiety? Protein's and carbohydrate's role in more elaborated aspects of post-ingestive experiences remain unclear and gives rise to measuring post-ingestive sensations and their dynamics in relation to macronutrient content. One study by Boelsma et al. (2010), sought to measure postprandial wellness after intake of two protein-carbohydrate meals. They found that a liquid high-protein breakfast induced higher levels of specific parameters of wellness, such as satisfaction and pleasantness, than did a liquid high-carbohydrate breakfast. Postprandial wellness was measured using the variables satisfaction, pleasantness, sleepiness, relaxation, mental alertness and physical energy [12]. Additional research in this area is needed to establish further relationships, for example protein and carbohydrate's effect on post meal desires as well as post meal wellbeing. Also, the relevance for food intake behaviour remains to be established [12,23].

From a scientific point of view, this study clarifies protein and carbohydrate's effect on specific post-ingestive sensations. From an industry perspective, it is relevant to know what increased protein or carbohydrate content entail in terms of post intake experiences, as this can determine whether consumers choose a product and consequently consume it repeatedly. The study contributes with new knowledge about the sensations perceived in the time after intake, namely, how these sensations develop over time, defined by *the dynamics* of the sensations. Post-ingestive sensations occur after food intake (per definition) and for this study, post-ingestive sensations are defined as *'the subjective perceptions of the body after eating'* as described by Duerlund et al. (2019) [23].

The present study seeks to unravel and understand the extended eating experience with its related post-ingestive sensations. This research thus situates itself in the area of understanding a more elaborated consumer experience and contributes to the diverse and complex area of appetite research. The overall aims for this research study were thus to:

(1) Quantify the dynamics of different post-ingestive sensations after food intake
(2) Study the effect of protein and carbohydrate on hedonic and post-ingestive responses.

The study involved development of a questionnaire and conducting of a consumer study serving a breakfast meal differing in macronutrient content. The test meals consisted of yoghurt either high in protein content (HighPRO) or high in carbohydrate content (HighCHO).

## 2. Materials and Methods

### 2.1. Participants

Participants included Danish students recruited from Ollerup Sports Academy located on Fyn, Denmark, via voluntarily sign-up after personal contact to the Academy and subsequent advertisement via their intranet ($n_{total}$ = 48). Inclusion criteria involved being above 18 years of age, being a breakfast eater, being a liker of yoghurt, and not suffer from any food allergies. Table 1 displays the characteristics from the 48 participants including standard deviation and range. Prior to the study, participants gave their written consent. Ethical approval is not required for this type of study according to the National Committee on Health Research Ethics in Denmark (Section 14 (2) in the Committee Act) [36]. For their participation, students received a cinema ticket for an optional movie at a cinema close by.

**Table 1.** Participant characteristics.

| Characteristics | |
|---|---|
| $n_{total}$ | 48 |
| Male/female | 31/17 |
| Age (years) | 20.4 ± 1.09 (18–25) * |
| Weight (kg) | 71.5 ± 11.99 (51–108) * |
| Height (cm) | 175.7 ± 8.49 (162–192) * |
| BMI (kg/m$^2$) | 23.0 ± 2.41 (19–30.5) * |

* Mean ± standard deviation (range). BMI—body mass index.

### 2.2. Study Design and Procedure

The study was conducted as a central-location-test at the Ollerup Sports Academy utilizing a randomised controlled crossover design. Each participant came for two separate breakfast sessions on two separate consecutive days. The test meals were served in random order across the participants. On the test day, participants came fasting since 22:00 the night before and the study started in the morning at 7:30 a.m. for breakfast and ran for 3 h until 10:30 a.m. The breakfast meals were obligatory intake to make sure that all participants consumed the same amount of calories. Subjective measurements were collected pre intake, during intake (after three bites), and in 30 min intervals until three h post intake (T = 0, 15, 30, 60, 90, 120, 150, 180 min). See schematic overview of the study design in Figure 1.

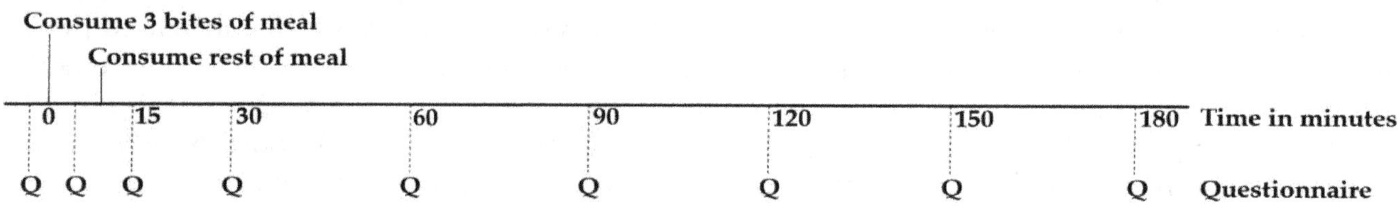


**Figure 1.** Schematic overview of the study with time intervals for answering of questionnaire.

## 2.3. Test Meals

The two test meals consisted of yoghurt with an added topping of plain muesli, almonds, raisins, and fresh blue berries, which is considered a typical breakfast in Denmark. The topping stayed consistent for the two meals, and the purpose was to create an appealing breakfast meal (visual, flavour, texture) rather than just yoghurt. The two meals were iso-caloric (393 Cal), but differed in protein and carbohydrate content, producing a high-protein test meal (HighPRO) and a high-carbohydrate test meal (HighCHO). The yoghurt was commercially bought (Arla® lactose-free yoghurt natural) as the base and then added protein (whey protein isolate, Lacprodan® SP-9225 Instant, Arla Foods, Viby, Denmark) or carbohydrate (glucose syrup, Dansukker®, Copenhagen, Denmark). The test meals were made following standardized procedures to ensure validity and standardization. For specification and content of the two test meals, see Table 2.

**Table 2.** Content of the two test meals: High-protein breakfast and High-carbohydrate breakfast.

| Food Ingredient | High-Protein | | High-Carbohydrate | |
|---|---|---|---|---|
| | Amount (g) | Calories (Cal) | Amount (g) | Calories (Cal) |
| Yoghurt [1] | 300 | 123 | 300 | 123 |
| Lacprodan [2] | 35 | 131.5 | - | - |
| Glucose syrup [3] | - | - | 42.35 | 131.5 |
| Muesli | 30 | 99 | 30 | 99 |
| Almonds | 6 | 30 | 6 | 30 |
| Raisins | 2.5 | 5 | 2.5 | 5 |
| Fresh blueberries | 2.5 | 5 | 2.5 | 5 |
| Total | 376 | 393 | 383.35 | 393 |

[1] Arla® lactose-free yoghurt natural; [2] Lacprodan® SP-9225 Instant (whey protein isolate); [3] Dansukker® glucose syrup.

## 2.4. Questionnaire and Post-Ingestive Response Variables

The questionnaire and chosen post-ingestive response variables were developed based on existing scientific literature as well as results from a previous qualitative focus group study conducted by the authors about consumer reflections on post-ingestive sensations [12,18,20,23]. Measurements were collected using visual analogue scale (VAS) via Compusense®Cloud software (Compusense Inc., Guelph, Ontario, Canada) in randomized order. Data was thus collected on a continuous scale ranging from 0, anchored "not at all" to 10, anchored "extremely". The full questions were developed and phrased in the participants' mother tongue, Danish. Appendix A displays the original Danish phrasing of the questions as well as the translated English phrasings. Additionally, all participants filled out an extra demographics questionnaire including gender, age, weight and height. The included response variables are presented in Table 3. Sensory satisfaction refereed to the hedonic experience of the meal's sensory properties during intake, whereas Satisfaction refereed to a general positive response to the food after intake [19].

**Table 3.** Response variables included in the questionnaire pre-, during-, and post intake.

| Pre Intake (Immediately before) | During Intake (after Three Bites) | Post Intake (0–15–30–60–90–120–150–180 min) |
|---|---|---|
| Energy | | Energy |
| Relaxation | | Relaxation |
| Concentration | | Concentration |
| Sleepiness | | Sleepiness |
| Fullness | | Fullness |
| Hunger | | Hunger |
| Overall wellbeing | Desire to eat the rest | Overall wellbeing |
| Physical wellbeing | Overall liking | Physical wellbeing |
| Mental wellbeing | Sensory satisfaction | Mental wellbeing |
| Desire-to-eat | | Desire-to-eat |
| Sweet desire | | Sweet desire |
| Salty desire | | Salty desire |
| Fatty desire | | Fatty desire |
| In-need-of-food | | In-need-of-food |
| | | Food joy |
| | | Satisfaction |

### 2.5. Data Analysis

Data on self-reported weight and height were used to calculate Body Mass Index (BMI): weight (kg)/(height (m))$^2$. Analysis of variance (ANOVA) was used to test effect of breakfast meal on hedonic response variables. Repeated measures ANOVA was applied to analyse dynamics in the post-ingestive sensations over time for each test meal. We applied the option REstricted Maximum Likelihood (ReML) with an Autoregressive covariance structure, which assumes that correlation is the highest between consecutive measures and declines for measures further apart.

Repeated measures ANOVA is appropriate when repeated scores are made sequentially within products and assumes non-independent measurements over time. Mauchly's Sphericity test was applied to show significant non-independence and thereby justify the use of repeated measures ANOVA. Repeated measures ANOVA, option least squares (LS) was applied on centred data to analyse if the change in post-ingestive sensations over time was different for each test meal. $p$-values $< 0.05$ were considered statistically significant, and Fisher's Least Significance Difference (LSD) test was applied to account for multiple comparisons. Effect sizes were examined using Cohen's d values [37]. All statistical analyses of the data were carried out using XLSTAT by Addinsoft, version 2018.6. (XLSTAT,Long Island, NY, USA)

## 3. Results

### 3.1. Hedonic Differences between Test Meals

After consuming three bites of the test meal, participants evaluated three hedonic questions: Overall liking, Desire-to-eat the rest of the portion, and Sensory satisfaction. A difference was seen between the two test meals with significantly higher ratings for HighCHO compared to HighPRO for all hedonic response variables, Table 4.

**Table 4.** Mean ($n$ = 48) hedonic ratings ± standard deviations of Overall liking, Desire-to-eat the rest and Sensory satisfaction after three bites of the two test meals: High-protein (HighPRO) and High-carbohydrate (HighCHO).

| Response Variable during Intake (after Three Bites) | HighPRO | HighCHO | Difference | | |
|---|---|---|---|---|---|
| | | | F | $p$-Value | d * |
| Overall liking | 3.496 ± 2.1 $^{b}$ | 5.906 ± 1.6 $^{a}$ | 145.5 | <0.0001 | 1.3 |
| Desire-to-eat the rest | 5.004 ± 2.6 $^{b}$ | 6.963 ± 1.6 $^{a}$ | 80.4 | <0.0001 | 0.9 |
| Sensory satisfaction | 4.144 ± 1.8 $^{b}$ | 5.908 ± 1.3 $^{a}$ | 83.8 | <0.0001 | 1.1 |

* Effect size (Cohen's d). Means with different superscript ($^{a,b}$) within a row differ significantly.

### 3.2. *Dynamics of Post-Ingestive Sensations*

A highly significant effect of time ($p \leq 0.0001$) was seen for all post-ingestive sensation variables across test meals: Energized (F = 18.45, $p < 0.0001$), Relaxation (F = 5.38, $p < 0.0001$), Concentration (F = 13.89, $p < 0.0001$), Sleepiness (F = 7.42, $p < 0.0001$), Fullness (F = 88.78, $p < 0.0001$), Hunger (F = 64.22, $p < 0.0001$), Overall wellbeing (F = 12.65, $p < 0.0001$), Physical wellbeing (F = 9.58, $p < 0.0001$), Mental wellbeing (F = 6.95, $p < 0.0001$), Desire-to-eat (F = 53.18, $p < 0.0001$), Sweet desire (F = 4.42, $p < 0.0001$), Salty desire (F = 6.07, $p < 0.0001$), Fatty desire (F = 4.90, $p < 0.0001$), In need of food (F = 48.72, $p < 0.0001$), Food joy (F = 4.92, $p < 0.0001$) and Satisfaction (F = 5.05, $p < 0.0001$).

This shows that all the response variables change significantly over a period of three h after consumption of these breakfast meals. The 3-h (0–180 min) dynamics for post-ingestive sensations for each test meal, HighPRO and HighCHO respectively, are presented in Tables 5 and 6.

Subjective measures for Energized significantly increased with large effect sizes ($p < 0.0001$, HighCHO: d = 0.9, HighPRO: d = 1.1) after consumption for both test meals and peaked with a maximum score (4.55) for HighPRO at T = 15, and a maximum score for HighCHO (4.79) at T = 60, where after the sensation of Energized decreased again.

**Table 5.** Time effects and means values ($n$ = 48) of post-ingestive sensations for **HighPRO (A)** at different time points after consumption.

| A: HighPRO | Pre Intake | 15 min | 30 min | 60 min | 90 min | 120 min | 150 min | 180 min |
|---|---|---|---|---|---|---|---|---|
| Energized | 2.87 $^{a}$ | 4.55 $^{d}$ | 4.43 $^{d}$ | 4.47 $^{d}$ | 3.94 $^{c}$ | 3.66 $^{bc}$ | 3.36 $^{b}$ | 3.56 $^{bc}$ |
| Relaxation | 6.10 $^{c}$ | 5.58 $^{b}$ | 5.34 $^{ab}$ | 5.41 $^{b}$ | 5.64 $^{bc}$ | 5.17 $^{ab}$ | 5.27 $^{ab}$ | 4.90 $^{a}$ |
| Concentration | 3.57 $^{a}$ | 4.65 $^{c}$ | 4.70 $^{c}$ | 4.63 $^{c}$ | 4.41 $^{c}$ | 4.02 $^{b}$ | 3.85 $^{ab}$ | 3.76 $^{ab}$ |
| Sleepiness | 5.93 $^{c}$ | 4.92 $^{ab}$ | 4.76 $^{ab}$ | 4.49 $^{a}$ | 5.07 $^{b}$ | 5.30 $^{b}$ | 5.24 $^{b}$ | 4.81 $^{ab}$ |
| Satiety | 1.97 $^{a}$ | 5.53 $^{f}$ | 4.71 $^{e}$ | 4.38 $^{e}$ | 3.89 $^{d}$ | 3.41 $^{c}$ | 2.63 $^{b}$ | 2.14 $^{a}$ |
| Hunger | 7.24 $^{e}$ | 4.58 $^{a}$ | 4.84 $^{ab}$ | 5.17 $^{b}$ | 5.62 $^{c}$ | 6.25 $^{d}$ | 7.09 $^{e}$ | 7.92 $^{f}$ |
| Satisfaction [1] | - | 3.39 $^{d}$ | 3.48 $^{d}$ | 3.43 $^{d}$ | 3.21 $^{cd}$ | 2.94 $^{bc}$ | 2.66 $^{ab}$ | 2.57 $^{a}$ |
| Overall wellbeing | 4.28 $^{a}$ | 5.02 $^{b}$ | 5.01 $^{b}$ | 5.24 $^{b}$ | 5.00 $^{b}$ | 4.56 $^{a}$ | 4.47 $^{a}$ | 4.26 $^{a}$ |
| Physical wellbeing | 4.61 $^{bcd}$ | 4.92 $^{de}$ | 5.18 $^{e}$ | 5.27 $^{e}$ | 4.70 $^{cd}$ | 4.33 $^{ab}$ | 4.38 $^{abc}$ | 4.09 $^{a}$ |
| Mental wellbeing | 4.80 $^{bc}$ | 5.10 $^{cd}$ | 5.31 $^{de}$ | 5.55 $^{e}$ | 5.06 $^{cd}$ | 4.80 $^{bc}$ | 4.66 $^{ab}$ | 4.33 $^{a}$ |
| Desire to eat | 7.56 $^{ef}$ | 4.78 $^{a}$ | 5.03 $^{a}$ | 5.53 $^{b}$ | 5.99 $^{c}$ | 6.47 $^{d}$ | 7.17 $^{e}$ | 7.93 $^{f}$ |
| Sweet desire | 3.92 $^{a}$ | 4.37 $^{ab}$ | 4.10 $^{a}$ | 3.96 $^{a}$ | 4.03 $^{a}$ | 4.30 $^{ab}$ | 4.67 $^{b}$ | 5.27 $^{c}$ |
| Salty desire | 2.60 $^{a}$ | 2.88 $^{ab}$ | 2.81 $^{ab}$ | 2.61 $^{a}$ | 2.64 $^{ab}$ | 2.96 $^{b}$ | 3.31 $^{c}$ | 3.83 $^{d}$ |
| Fatty desire | 2.81 $^{ab}$ | 2.68 $^{a}$ | 2.70 $^{a}$ | 2.67 $^{a}$ | 2.80 $^{ab}$ | 2.86 $^{ab}$ | 3.10 $^{b}$ | 3.91 $^{c}$ |
| In need of food | 7.35 $^{d}$ | 4.91 $^{a}$ | 4.93 $^{a}$ | 4.99 $^{a}$ | 5.66 $^{b}$ | 6.32 $^{c}$ | 7.18 $^{d}$ | 7.88 $^{e}$ |
| Food joy [1] | - | 3.32 $^{d}$ | 3.29 $^{d}$ | 3.11 $^{cd}$ | 2.93 $^{bc}$ | 2.71 $^{ab}$ | 2.40 $^{a}$ | 2.42 $^{a}$ |

Means with different superscript ($^{a,b,c,d,e,f}$) within a row differ significantly. [1] Satisfaction and Food joy only measured after intake and not pre intake.

**Table 6.** Time effects and means values ($n$ = 48) of post-ingestive sensations for **HighCHO (B)** at different time points after consumption.

| B: HighCHO | Pre Intake | 15 min | 30 min | 60 min | 90 min | 120 min | 150 min | 180 min |
|---|---|---|---|---|---|---|---|---|
| Energized | 2.79 [a] | 4.06 [bc] | 4.42 [cde] | 4.79 [e] | 4.52 [de] | 4.14 [bcd] | 3.79 [b] | 3.88 [b] |
| Relaxation | 5.95 [e] | 5.82 [de] | 5.64 [cde] | 5.78 [cde] | 5.48 [cd] | 5.34 [bc] | 5.00 [b] | 4.41 [a] |
| Concentration | 3.50 [a] | 4.80 [cd] | 4.75 [cd] | 5.02 [d] | 4.72 [cd] | 4.61 [c] | 4.45 [bc] | 4.08 [b] |
| Sleepiness | 6.34 [d] | 5.18 [c] | 5.02 [bc] | 4.64 [ab] | 4.37 [a] | 5.05 [bc] | 5.08 [bc] | 4.91 [bc] |
| Satiety | 1.79 [a] | 5.79 [f] | 5.46 [f] | 4.93 [e] | 4.44 [d] | 3.33 [c] | 2.29 [b] | 1.83 [ab] |
| Hunger | 7.15 [d] | 4.14 [a] | 4.49 [ab] | 4.76 [b] | 5.64 [c] | 6.70 [d] | 7.74 [e] | 7.88 [e] |
| Satisfaction [1] | - | 5.57 [d] | 5.26 [d] | 4.90 [c] | 4.66 [bc] | 4.44 [b] | 4.06 [a] | 3.98 [a] |
| Overall wellbeing | 4.40 [a] | 5.71 [cd] | 5.66 [cd] | 5.76 [d] | 5.37 [c] | 4.97 [b] | 4.68 [ab] | 4.51 [a] |
| Physical wellbeing | 4.45 [a] | 5.56 [cd] | 5.71 [d] | 5.49 [cd] | 5.32 [c] | 4.92 [b] | 4.43 [a] | 4.24 [a] |
| Mental wellbeing | 4.86 [ab] | 5.62 [c] | 5.81 [c] | 5.82 [c] | 5.58 [c] | 5.15 [b] | 4.81 [ab] | 4.55 [a] |
| Desire to eat | 7.28 [d] | 4.67 [a] | 4.93 [a] | 5.06 [a] | 5.80 [b] | 6.72 [c] | 7.72 [d] | 8.37 [e] |
| Sweet desire | 4.18 [b] | 3.68 [a] | 3.66 [a] | 3.63 [a] | 3.73 [a] | 3.75 [a] | 4.47 [b] | 4.96 [c] |
| Salty desire | 2.56 [a] | 3.02 [b] | 2.79 [ab] | 2.98 [b] | 2.97 [b] | 3.11 [b] | 3.54 [c] | 4.02 [d] |
| Fatty desire | 2.69 [b] | 2.33 [a] | 2.60 [ab] | 2.73 [b] | 2.67 [b] | 2.89 [bc] | 3.16 [c] | 3.57 [d] |
| In need of food | 7.14 [d] | 4.31 [a] | 4.69 [a] | 4.77 [a] | 5.39 [b] | 6.62 [c] | 7.64 [de] | 8.15 [e] |
| Food joy [1] | - | 5.05 [d] | 5.01 [d] | 4.73 [cd] | 4.61 [c] | 4.21 [b] | 3.89 [ab] | 3.73 [a] |

Means with different superscript ([a,b,c,d,e,f]) within a row differ significantly. [1] Satisfaction and Food joy only measured after intake and not pre intake.

Subjective measures of Fullness significantly increased ($p < 0.0001$, HighCHO: d = 2.4, HighPRO: d = 2.1) immediately after consumption of both test meals (T = 15), where it reached its highest mean ratings as well (HighPRO: 5.53, HighCHO; 5.79). Thereafter Fullness sensation gradually decreased until T = 180, to the same as the pre intake baseline measurement at T = 0. So comparing T = 0 and T = 180, there was no significant difference in Fullness, suggesting that three h is the time point where Fullness sensation resets itself when consuming these meals in these amounts.

Development over three h for Sleepiness started with the highest mean score at T = 0 for both test meals (HighPRO: 5.93, HighCHO: 6.34), and then significantly decreased ($p < 0.0001$, HighCHO: d = 0.6, HighPRO: d = 0.5) immediately after consumption (T = 15). Hereafter the rating of Sleepiness continuously decreased until T = 60 for the HighPRO meal, and until T = 90 for the HighCHO meal, where it significantly (HighCHO: $p = 0.010$, d = 0.4; HighPRO: $p = 0.042$, d = 0.3) started to increase again and peaked at T = 120 and T = 150, after which rating of sleepiness again started to decline until T = 180.

Subjective measures of Overall wellbeing showed a significantly increase ($p < 0.0001$, HighCHO: d = 0.9, HighPRO: d = 0.5) immediately after consumption (T = 15) for both test meals, and stayed constant for 60 min after consumption, with a significantly decrease 90 min after consumption (HighCHO: $p = 0.021$, d = 0.3; HighPRO: $p = 0.021$, d = 0.2), and returning to similar ratings as pre intake when reaching T = 180.

Subjective measures of Food joy showed a gradual decline in mean ratings for both test meals during the whole period after consumption, starting at their highest mean ratings at T = 15. At 90 min after consumption, measures of Food joy dropped significantly compared to T = 15 (HighCHO: $p = 0.015$, d = 0.2, HighPRO: $p = 0.026$, d = 0.2). Hereafter, ratings of Food joy continued to decrease, and for both test meals there was a significant overall decrease in Food joy from T = 15 until T = 180 ($p < 0.0001$, HighCHO: d = 0.7, HighPRO: d = 0.5). Food joy was not measured pre intake (T = 0), since it was a measure related to the consumed test meal.

For HighCHO, ratings of Sweet desire significantly decreased immediately after consumption ($p = 0.023$, d = 0.2), as opposite to HighPRO. For HighPRO, sweet desire ratings immediately increased after consumption ($p = 0.050$, d = 0.2), where after it decreased again until T = 60. Ninety min after consumption, Sweet desire ratings increased significantly until T = 180 for HighPRO ($p < 0.0001$, d = 0.5), with the biggest increase between T = 150 and T = 180 after consumption ($p = 0.010$, d = 0.2).

From T = 120 until T = 180, Sweet desire ratings significantly increased after consuming for HighCHO ($p < 0.0001$, d = 0.6).

Subjective measures for In-need-of-food significantly decreased ($p < 0.0001$, HighCHO: d = 1.3, HighPRO: d = 1.4) immediately after consumption (T = 15) for both test meals. Hereafter, In-need-of-food showed a steep incline for the rest of the period until T = 180, where it reached its highest mean ratings with significantly higher ratings than pre intake (HighPRO: $p = 0.022$, d = 0.2, HighCHO: $p < 0.0001$, d = 0.8). Between T = 90 and T = 120, the dynamics of In-need-of-food ratings showed a crossover between the two test meals, with HighPRO changing from having higher In-need-of-food ratings to lower In-need-of-food ratings than HighCHO.

Consumer's rating of Satisfaction was the highest immediately after consumption (T = 15), and then decreased throughout the three h. This trend shows for both test meals, and for HighCHO, Satisfaction significantly decreased between T = 30 and T = 60 ($p = 0.024$, d = 0.2), and then again between T = 120 and T = 150 ($p = 0.017$, d = 0.2).

### *3.3. Effect of Test Meal on Post-Ingestive Sensations*

For an overview of the significant differences ($p < 0.005$) between HighCHO and HighPRO see Figure 2, where an asterisk indicates significant differences at that time point. Only selected post-ingestive variables are displayed in the figure. Pre intake (T = 0), no difference for the two test meals were found for any of the post-ingestive variables. For the sensations of Satisfaction and Food joy, results showed significant differences between HighPRO and HighCHO for the whole period of the three-hour study ($p < 0.0001$, d = 0.8), with the HighCHO inducing significantly higher ($p < 0.0001$, d = 0.8) ratings than the HighPRO meal.

Energized sensation showed a significantly higher rating for the HighCHO meal at 90 ($p = 0.014$, d = 0.4) and 120 ($p = 0.049$, d = 0.3) min after consumption, whereas Sleepiness ratings showed significantly higher ratings for HighPRO at T = 90 ($p = 0.023$, d = 0.4). Significantly Fullness differences between the two test meals were found at 30 ($p = 0.013$, d = 0.4) and 60 ($p = 0.031$, d = 0.3) min after consumption, with HighCHO depicting the highest Fullness sensations at these time points.

When reaching T = 180 both test meals showed no differences in Fullness. Both hunger and In-need-of-food variables significantly differed between the two test meals 150 min (Hunger: $p = 0.011$, d = 0.4, In-need-of-food: $p = 0.040$, d = 0.3) after consumption with HighCHO inducing the highest ratings for both variables. Same trend was shown for Desire-to-eat-something, where HighCHO had significantly higher ratings than HighPRO at T = 150 ($p = 0.029$, d = 0.3).

Significant differences between the two test meals were seen at T = 15 for Overall wellbeing ($p = 0.009$, d = 0.5), Physical wellbeing ($p = 0.012$, d = 0.4) and Mental wellbeing ($p = 0.033$, d = 0.3) with HighCHO rating the highest compared to HighPRO. Overall wellbeing showed significant meal differences for the first three time intervals (T = 0, 15, 30), and hereafter no differences were observed between HighCHO and HighPRO. Ratings for Sweet desire differed significantly between the two meals with HighPRO inducing higher Sweet desires at T = 15 ($p = 0.035$, d = 0.3) and again at T = 120 ($p = 0.035$, d = 0.2). For Relaxation, Fatty desire, and Salty desire, no significant test meal differences were observed at any of the time points.

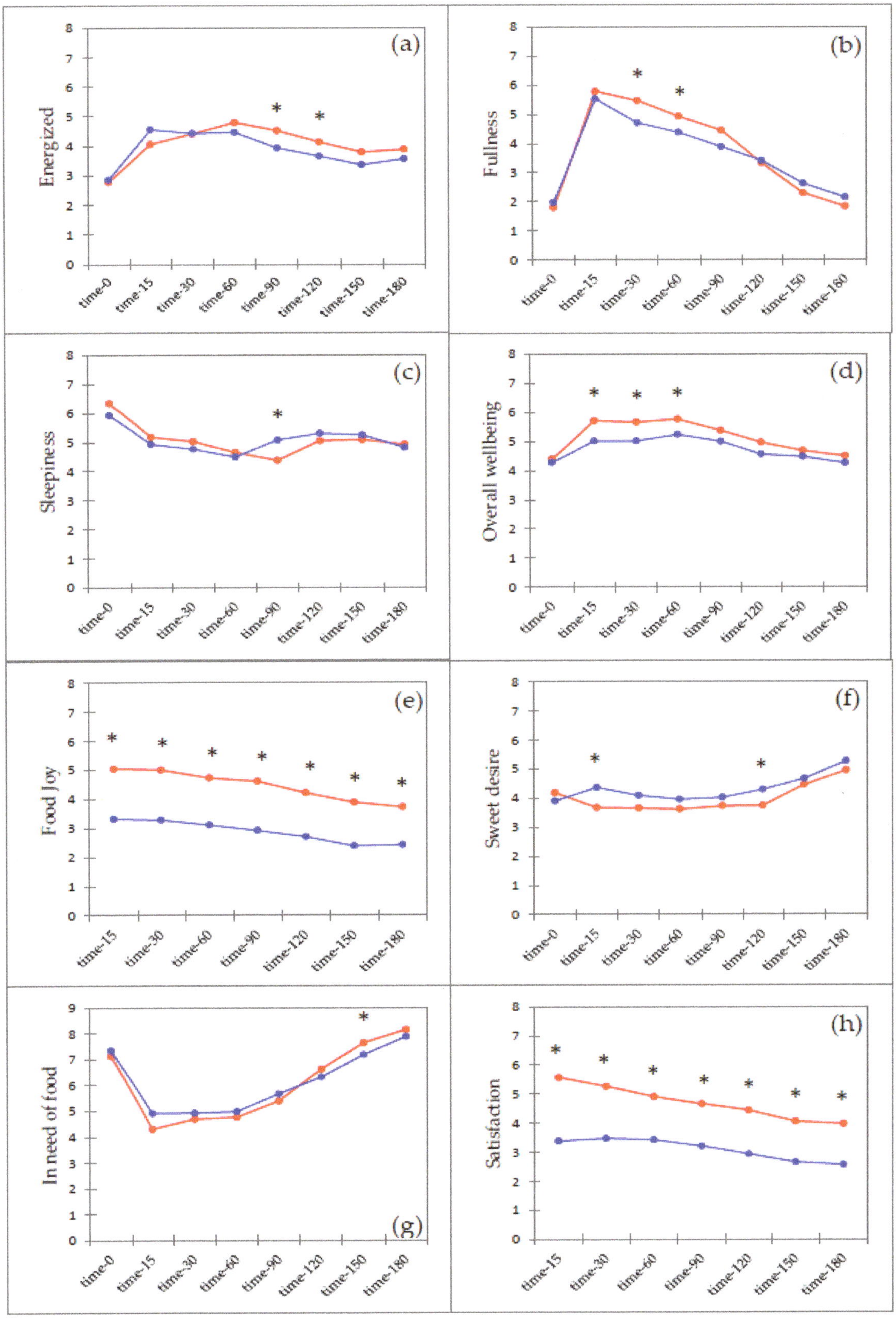


**Figure 2.** Post-ingestive responses at time-0 (pre intake),15,30,60,90,120,150,180 for Energized (**a**), Fullness (**b**), Sleepiness (**c**), Overall wellbeing (**d**), Food joy (**e**), Sweet desire (**f**), In need of food (**g**), Satisfaction (**h**). Red = High-carbohydrate test meal, Blue = High-protein test meal. Asterisk (*) indicates significant difference between the two test meals for the specific time point ($p < 0.05$).

## 4. Discussion

### 4.1. Implications of Hedonic Parameters

Consumers' overall hedonic acceptances of the two test meals revealed a significantly lower acceptance (with large effect sizes [37,38]) of the HighPRO test meal compared to the HighCHO test meal for all three hedonic variables: Overall liking, Desire-to-eat the rest, and Sensory satisfaction. Increasing the protein content of a meal can significantly decrease its palatability [4,39]. A reason in this study might be an off-flavour of whey protein isolate that can carry through to the meal. It is well known that off-flavour of whey protein is a primary negative driver of acceptance [40]. Attributes that could have influenced the consumers' ratings in this study are potential mouth-drying and chalky mouthfeels in the HighPRO meal compared to the sweet and mouth-coating attributes in the HighCHO meal [40,41]. However, these descriptions are not validated from a trained sensory panel. Additionally, sweetness often functions as a strong influencer of pleasantness and palatability with a powerful hedonic appeal [42,43], and thus the HighCHO meal could have induced higher hedonic acceptance due to the sweet glucose syrup added. Martini et al. (2018) found reduced palatability for high-protein pasta formulations using protein sources from egg white and soy isolate indicating that protein is likely to negatively affect sensory properties of pasta except for appearance [13]. Conversely, Boelsma el al, (2010) found no clear difference in enjoyment of two meals differing in protein content being high protein/low carbohydrate or low protein/high carbohydrate meals. Both test meals by Boelsma et al. (2010), however, were added glucose syrup, as opposite to our study, where only the HighCHO test meal was added glucose syrup [12]. Boelsma et al. (2010) succeeded in their intention to keep their two test meals similar in hedonic aspects [12]. For future test meal development involving glucose syrup and/or whey protein isolate, consideration around sensory properties and the effects on consumer's hedonic appreciation should carefully be taken into account. Standardizing test meals for sensory properties and hedonic appeal then enables a clearer picture of product effects on evaluated parameters.

### 4.2. Satisfaction and Food Joy

In general, we saw that the HighCHO test meal induced highest ratings for both Satisfaction and Food joy. For both Satisfaction and Food joy, the questions were phrased so that the consumers should have in mind the test meal they ate e.g., "How satisfied are you with the breakfast meal you ate today?", and "Please rate your sense of joy when thinking of the meal you ate today". The evaluation of these sensations hence related to the memory of the test meal. The results could imply that when consumers rate Satisfaction and Food joy, they generally associate and refer to their hedonic acceptance, here being overall liking and the Sensory satisfaction, which they experienced during consumption. As mentioned, HighCHO rated significantly higher compared to HighPRO for hedonic acceptance in this study. Mattes and Vickers (2018) researched whether food-liking influenced hunger and fullness. They suggest, from their studies, that if people eat a food they greatly enjoy, instead of eating a less-well-liked version, they will experience more pleasure, satisfaction, and satiety [44]. We did see that the well-liked HighCHO test meal, in addition to Satisfaction and Food joy, also induced higher ratings for satiety at time points 30 and 60 min after consumption. Boelsma et al. (2010) found an association between satiety and feelings of satisfaction, indicating that satiety sensations may be important factors influential in the sense of wellness after consumption of food in general [12]. In addition, Andersen and Hyldig (2015) highlight the importance of sensory properties for food satisfaction. They found that sensory satisfaction was highly influential and predictive for food satisfaction in a consumer study with chicken soups [19]. Cardello et al. (2000) moreover suggest that satisfaction is a more appropriate measure of consumers' response to foods than liking [45]. Having said that, it must be emphasized that our results demonstrated the sensations of both Satisfaction and Food joy to last longer after consumption of the HighPRO test meal compared to the HighCHO test meal. For Satisfaction, we saw a significant time effect in decline after 120 min for the HighPRO test meal, whereas this significant decline occurred already after 60 min for the HighCHO test meal. This indicates that more than the hedonic evaluation is part of the consumers' sense of satisfaction. This is also pointed out by research signifying that multiple factors contribute to food satisfaction including hedonic experience as well as other post-ingestive sensations [18,24,46] For Food joy, the significant decline occurred after 90 min with the HighPRO, but already after 60 min with the HighCHO. Nevertheless, the HighCHO test meal rated significantly higher overall for both Satisfaction and Food joy at all time points, so interpretation of these findings should be taken with care. More studies on dynamics and quantification of Food satisfaction, pleasure, and post-ingestive sensations from eating foods are encouraged [18,19,31].

### 4.3. *Measuring Dynamics of Post-Ingestive Sensations*

Consuming the test meals induced various dynamics of post-ingestive sensations over the three-hour period post intake. We saw a significant effect of time for all the evaluated post-ingestive variables, supporting the relevance and use of post-ingestive response variables as measures of post food experiences, as well as showing consumers' ability to sense and report these sensations in general. Andersen et al. (2017) also showed significant time effects for thirst, fullness, and energy level, but not for hunger, physical wellbeing, or psychological wellbeing in a time period of 40 min post intake of four fruit drinks [18]. Measuring satiety and post-prandial wellness, Boelsma et al. (2010) found that differences between their test meals stood particularly evident at about 3–4 h after consumption, this with a high-protein/low-carbohydrate meal rating higher than a low-protein/high-carbohydrate liquid breakfast meal [12]. Our measurements ended after 3 h, but the results display rather the opposite, that especially for Fullness and for Overall wellbeing, the differences were primarily evident earlier, at 30 and 60 min after intake, with highest ratings induced by the HighCHO test meal, not the HighPRO test meal, although only with small to medium effect sizes as classified by Cohen (1988) [37]. We do not know the subsequent development of post-ingestive sensations after the 3-h study period for this study. However, we believe that these studies in general show that consumers can sense and express intensity differences in post-ingestive sensations beyond satiety, and that the intensity discrepancies between studies are caused by the different foods eaten rather than consumers' ability to perceive the sensations.

#### 4.3.1. The Dynamic of Sweet Desire

The dynamics for Sweet desire revealed an immediate incline for the HighPRO test meal and an immediate decline for the HighCHO test meal, showing a significant meal difference instantly after intake. Since the HighPRO test meal probably tasted less sweet than the HighCHO test meal with the added glucose syrup, these results indicate a development of sweet desire after eating something not sweet. Oppositely, after consuming the sweeter HighCHO test meal, the desire for sweet seems to decrease after consumption of sweet foods in this study. An explanation for this difference in Sweet desire dynamics could involve the theory behind sensory specific satiety (SSS) and sensory specific desires (SSD). SSS describes the decline in pleasantness of a food eaten relative to a food not eaten [47–49], but also suggests that it can reflect a decrease in both food liking as well as in food wanting for the eaten food after consumption. The latter was suggested by Haverman et al. (2009) from a study consuming chocolate milk [50]. SSD can be described as an intrinsic motivation to eat and general desire for certain foods [51,52], for instance sweet, salty, fatty desires evaluated in this present study. In a study by Harington et al. (2016), they found that the desire for sweet was maintained for their whole study period of three h after eating two slices of bread, whereas participants' desire for salty, fatty and savoury decreased [53]. Another research study, looking at the effect of adding cayenne pepper to a soup, found an increase in sweet desire after consumption of a spiced soup compared to a non-spiced soup, suggesting alteration in sensory specific desires with the addition of spices to food [22]. In addition, Harington et al. (2016) suggest that our desire for sweet foods is partially disconnected from our appetite, hypothesizing a 'dessert mentality' [53]. The present study showed an immediate increase for both Fullness and for Sweet desire after consumption of the HighPRO test meal, indicating that these two sensations can be present at the same time. Recent work from Duerlund et al.

(2019) found qualitative evidence from focus group interviews, expressing that it is possible to be full and still, at the same time, desire something else [23]. Murray and Vickers (2009) report similar results that one may feel physically full but mentally hungry. They suggest an overlap of hunger and fullness sensations, stating that it is possible to be slightly hungry and slightly full at the same time [54]. In a study by Lowe and Butryn (2007) they distinguish between homeostatic hunger and hedonic hunger driven by either the need for nutrients or a need for pleasure, respectively. They found that satiation did not seem to have an effect on hedonic hunger compared to homeostatic hunger [21]. We believe that mental hunger [54] and hedonic hunger [21] can relate to desires for specific sensory stimuli. Our findings suggest that one can experience an increase in fullness and at the same time feel an increase in sweet desire when consuming a non-sweet meal, this in accordance with the theory of SSS for one food and the development of SSD for something else.

#### 4.3.2. Energized and Sleepiness as Two Opposites?

The post-ingestive sensation variables Energized and Sleepiness each displayed different dynamic curves in the three-hour study period after intake. Exploring consumer views of post-ingestive sensations, Duerlund et al. (2019) proposed that post-ingestive energy includes a positive general energy sensation, and a more negative low energy sensation e.g., feeling sleepy. In the qualitative study, it was mentioned that consuming specific foods could lead to sensations of heaviness, sleepiness, and lack of ability to focus [23]. In the present study, we saw that both Energized and Concentration increased immediately after consumption of food (with large effect sizes), whereas Sleepiness decreased (with medium effect size). Could these results indicate that food-induced post-ingestive sensations like feeling Energized and Sleepy function as two opposites of one pole? This question statement, however, undeniably requires more research for affirmation, validation, and elaboration. In relevance to energy and sleepiness, research suggests a phenomenon called 'post-lunch sleepiness' or a 'post-lunch dip' in energy. This is an innate tendency for sleep during the early afternoon, often more pronounced after consuming a heavy meal and/or a high fat/carbohydrate meal [23,55–58]. According to Monk (2005), this 'post-lunch dip' can occur even without having lunch, or one being unaware of the time of day [57]. In the present study, consumers had breakfast in the early morning and the results demonstrated that for the first 60 min the opposite to what is known as the 'post-lunch-dip' occurred. Here we saw that Sleepiness decreased, and Energized increased. With that said, Sleepiness increased again after 1 h post intake, but with the protein-rich meal inducing the highest Sleepiness ratings compared to the high-carbohydrate meal, suggesting that the 'post-lunch sleepiness' phenomenon did not apply to the present study design and results. Interpreting these results, this could have connection to the time of day and participants' metabolic state. The participants came in fasting early morning, hence energy deficient, suggesting that any energy addition logically should lead to more energy.

### *4.4. Product Factors and Their Effects on Post-Ingestive Sensations*

The largest product effects in this study were seen for variables Food joy, Satisfaction and Overall wellbeing - all holistic variables, where the HighCHO rated significantly higher than the HighPRO, possibly leading back to consumer's hedonic responses to the two test meals as discussed. Proteins' role in this holistic area of post-ingestive sensations is not well-researched. Boelsma et al. (2010) sought to measure postprandial wellness after intake of two protein-carbohydrate meals, and found that a liquid high-protein breakfast induced specific parameters of wellness such as satisfaction and pleasantness more than a liquid high-carbohydrate breakfast did. This is the opposite for our study, where the high-carbohydrate meal induced higher Satisfaction, Mental-, Physical-, and Overall wellbeing than the HighPRO test meal. This difference may be due to a liquid product rather than a more solid product and the appropriateness of this. Often, high protein content causes a more viscous product, which may be more appropriate than a thin product for liquid breakfasts. Together with Boelsma et al. (2010) we state that more research is needed to establish further relationships and that the relevance for food intake behaviour remains to be established [12].

On the contrary, protein's role on appetite and especially fullness/satiety is a well-researched area, and, in summary, high-protein diets seem to provide a tool for appetite control. Ergo, meals higher in protein tend to increase satiety compared to meals lower in protein, at least in the short term [4,11,16]. Interestingly, we found that the HighCHO meal induced higher Fullness ratings for the first two h after consumption, where after we observed a crossover to HighPRO inducing the highest Fullness, however not with a significant effect. For Hunger, In-need-of-food and Desire-to-eat, we saw this crossover just after 90 min and only with a significant product effect at 150 min after consumption. These results indicate a subjective shift in appetite state somewhere between 90 and 120 min after consumption dependant on intake of protein or carbohydrate content in this study. The HighCHO test meal started with a higher Fullness rating but also ended with a higher Desire-to-eat rating compared to the HighPRO test meal. There is still much we do not know about protein-induced satiety. This present study contributes with additional knowledge to unravel the many aspects of protein-induced satiety. It also point towards the need for studying more individual satiety sensations in order to gain insights into quantitative or qualitative effects of different types of protein as well as other macronutrients. For instance, Karalus and Vickers (2016) suggest 41 items to study within satiety questionnaires [20]. Many factors appear to influence satiety including palatability and type of protein. It is still not clear the amount and/or the type of protein that is required to promote satiety, and long term data are still limited and inconclusive within this research area [4,11,14,59].

### 4.5. *Measuring Food-Related Wellbeing—Towards a Holistic Approach*

The present research study goes beyond mere satiety and hunger measurements and holds a more holistic approach to consumer food experiences including the aspect of wellbeing. Wellbeing is a growing area of interest in research, and especially food-related wellbeing has in recent years gained much focus [22,26,34,35]. There exists no one universal definition or way to measure wellbeing as the concept proves complex, holistic and multidimensional [34,60–62] and moreover differs cross-culturally [32,35]. King et al. (2015) have developed a method, The WellSense Profile$^{TM}$, which addresses consumer wellbeing in association with food. The method consists of five dimensions that encompasses wellbeing i.e., emotional, physical, social, intellectual, and spiritual [34]. In this study, we included certain and selected aspects of wellbeing, for instance Mental-, Physical-, and Overall wellbeing as well as Satisfaction and Food joy, all subjective and holistic measurements as suggested by Duerlund et al. (2019) and inspired by existing research on post-ingestive sensations, food satisfaction, and food reward/pleasure [12,18,21,23,63]. We also argue that variables such as Energized and Relaxation also contribute to food-related wellbeing. To support this, Andersen and Hyldig (2015) found that included in physical wellbeing is the sense of an appropriate energy level after intake and that physical wellbeing functions as an important determinant in food satisfaction [24]. The findings from this study adds to the relevance, importance and applicability of evaluating subjective food related wellbeing in consumer research, especially post food intake. Continuing research on food-related wellbeing helps us gain more knowledge into this complex and multifaceted concept as well as to how to best measure food-related wellbeing. Merely asking consumers to rate their mental and physical wellbeing post intake may not provide the most accurate measures, and perhaps more implicit measures of food-related wellbeing rather than just asking are needed. Coming studies from the authors will expand on the findings of post-ingestive sensations in relation to food reward and food joy/pleasure.

### 4.6. *Study Considerations, Future Perspectives and Implications*

In this study, we observed different post-ingestive dynamics following two breakfast meals either high in protein or carbohydrate. The present study shows that consuming food not only affects appetite sensations such as hunger and satiety. It also affects other post-ingestive sensations like feeling Energized, Sleepiness, Desires and more holistic sensations such as overall wellbeing, Food joy and Satisfaction. The presented results can be considered a contribute to supporting the development of a general description of post-ingestive sensation dynamics. Yet, studies on broader food categories would provide more knowledge and data to support a general description of subjective post-ingestive sensation dynamics.

Subjective measurements of post-ingestive sensations are not widely used within sensory and consumer science, and with this study, we contribute with new knowledge about the perceived sensations in the time after food intake. The present study showed that the post-ingestive sensations evolve differently over time and depending on product characteristics, in this case protein or carbohydrate. One thing is product factors and their effect on post-ingestive sensations as researched in this study. Another thing is post-ingestive sensations and their effect on food choice and eating behavior, for example post meal snacking providing additional calories. Consumers could learn from their own post-ingestive sensations and use them in their daily food decisions on a self-conscious level. Different post-ingestive sensations such as post meal desires or post meal wellbeing might influence future decisions on food and snack intake. For instance, different sensations might alter the specific food we choose to consume. Post-ingestive sensations could also modify the time intervals between meals or snacking or cause reduced or increased next intake. The contribution of the different sensations to food intake is therefore important, with the overall purpose to explore how post-ingestive sensations contribute to human eating behavior and food appreciation [12,18]. Further research on post-ingestive sensations' influence on eating behavior is thus warranted.

The concept of post-ingestive sensations provides both research and industry with new opportunities to explore consumer experiences around eating, and in particularly this new area can facilitate development of new products designed to address and perhaps provoke/induce certain wanted/desired sensations to help regulate body weight, manage calorie intake, and perhaps avoid eating when not hungry. This approach focusing on consumers' subjective post-ingestive sensations offers new promises to predict consumer food choices and can give further insight into (over)eating. Findings can therefore have impact in a broader context, such as implications for public health issues and obesity, which is one of our grand challenges in the world, evolving into a global epidemic [1,64]. The importance of post-ingestive sensations relative to prediction of longer-term food intake is so far unstudied. More and longer-term research is needed to support our findings and to gain more in-depth knowledge about post-ingestive sensations and their importance to consumer product acceptance and to their relevance in food choice and eating behavior.

### 4.7. Limitations

The present study showed that the post-ingestive sensations evolve differently over time depending on product characteristics, in this case protein or carbohydrate. A study limitation was, the varying sensory characteristics of the two test meals, inducing significantly higher liking and desire-to-eat for the HighCHO meal. The high palatability possibly affected the evaluation of other variables like Satisfaction and Food joy, as discussed under Section 4.1: *Implications of Hedonic Parameters*. Another aspect to consider includes the duration of the experiment. The study showed the development of post-ingestive sensations in the short term over a three h period. The effect of product factors and dynamics of post-ingestive sensations long term was not studied. Thus, we cannot exclude different effects long term. Furthermore, it is important to consider the nature and composition of the sample. The number of participants was $n = 48$ and all participants were young Danish students at a sports academy. In terms of this as a limitation, we consider it a representative number linked to precedent in the research area in previous papers, e.g., Martini et al. (2018), Boelsma et al. (2010) and Karalus and Vickers (2016) whom had 20, 21 and 30 participants respectively [12,13,20]. Our results however, whilst representative for active young consumers may vary when considered in relation to the general population. Thus, further confirmatory studies are required with differing population groups to determine the generalisability of the results found in the present study.

## 5. Conclusions

This quantitative research study contributed with knowledge on how different post-ingestive sensations develop after food intake. The study further elucidated differences between two breakfast meals either high in protein or carbohydrate. The work involved development of a questionnaire as well as conducting of a consumer study measuring product and time effects for subjective evaluations of post-ingestive sensations. The present study indicates that consuming foods not only affects satiety but also other post-ingestive sensations such as Energy, Desires, Food joy, Food-related Wellbeing and Satisfaction. Results showed a main effect of time for all measured post-ingestive response variables, validating that these sensations do evolve and change over a period of three h after food intake. The HighCHO induced higher hedonic responses compared to the HighPRO breakfast meal, as well as higher ratings for post-ingestive responses such as Satisfaction, Food joy, Overall wellbeing and Fullness. HighPRO, on the other hand, induced higher ratings for Sweet desire post intake.

This research study indicates applicability of post-ingestive sensations in food studies that goes beyond mere satiety measurements. Post-ingestive sensations provide information important for consumers' overall appreciation of products and including post-ingestive sensations and their dynamics when addressing the eating experience is therefore highly relevant. The development of sensations after a meal might be important for consumers' following food choices and for extra calorie intake. Consumers could learn from their subjective sensations and this could influence their future decisions around eating. More detailed knowledge in this area might elucidate aspects of overeating and obesity.

**Author Contributions:** Conceptualization, M.D., B.V.A., and D.V.B.; methodology, M.D. and B.V.A.; software, M.D.; validation, M.D. and B.V.A.; formal analysis, M.D. and B.V.A.; investigation, M.D.; resources, M.D.; data curation, M.D.; writing—original draft preparation, M.D.; writing—review and editing, M.D., B.V.A., and D.V.B.; visualization, M.D.; supervision, B.V.A. and D.V.B.; project administration, M.D., B.V.A., and D.V.B.; funding acquisition, B.V.A., and D.V.B.

**Acknowledgments:** We would like to thank Britta Hansen, Birgitte Foged and Kamilla Hall Kragelund for practical help with conducting of the study. We would also like to thank Hanne Gyntzel, matron at Ollerup Sports Academy, for her help and collaboration.

## Appendix A Phrasing of Questions

**Table A1.** Response variables with their original Danish phrasing a well as the English translation.

| Variable | Original Danish Phrasing | Translated English Phrasing |
|---|---|---|
| Energy | "Hvor energisk er du lige nu?" | "How energetic are you right now?" |
| Relaxation | "Hvor afslappet er du lige nu?" | "How relaxed are you right now?" |
| Concentration | "Hvor koncentreret er du lige nu?" | "How is your concentration right now?" |
| Sleepiness | "Hvor træt er du lige nu?" | "How sleepy are you right now?" |
| Fullness | "Hvor mæt er du lige nu?" | "How full are you right now?" |
| Hunger | "Hvor sulten er du lige nu?" | "How hungry are you right now?" |
| Overall wellbeing | "I hvor høj grad fornemmer du en generel velvære lige nu?" | "Please rate your overall wellbeing as you feel it right now" |
| Physical wellbeing | "Hvor fysisk veltilpas er du lige nu?" | "Please rate your physical wellbeing as you feel it right now" |
| Mental wellbeing | "Hvor mentalt veltilpas er du lige nu maden?" | "Please rate your mental wellbeing as you feel it right now" |
| Desire-to-eat | "Hvor stor er din lyst til noget at spise lige nu?" | "How much do you desire to eat something right now?" |
| Sweet desire | "I hvor høj grad har du lyst til noget sødt lige nu?" | "How much do you desire to eat something sweet right now?" |

**Table A1.** *Cont.*

| Variable | Original Danish Phrasing | Translated English Phrasing |
|---|---|---|
| Salty desire | "I hvor høj grad har du lyst til noget salt lige nu?" | "How much do you desire to eat something salty right now?" |
| Fatty desire | "I hvor høj grad har du lyst til noget fedt lige nu?" | "How much do you desire to eat something fatty right now?" |
| In-need-of-food | "I hvor høj grad mangler du noget mad lige nu?" | "How much do you need food right now?" |
| Food joy | "I hvor høj grad fornemmer du en glæde ved den mad du har spist i dag?" | "Please rate your sense of joy when thinking of the meal you ate today". |
| Satisfaction | "Hvor tilfreds er du med det måltid du har spist i dag?" | "How satisfied are you with the breakfast meal you ate today?" |
| Desire to eat the rest | "I hvor høj grad har du lyst til at spise resten af portionen?" | "How much do you desire to eat the rest of the serving?" |
| Overall liking | "Hvor godt kan du lide morgenmaden?" | "How much do you like the breakfast?" |
| Sensory satisfaction | "Hvis du tænker på morgenmadens udseende, lugt, smag og konsistens samlet set hvor tilfredsstillet er du så?" | "If you consider the breakfast's appearance, smell, taste, and consistence as a whole, how satisfied are you?" |

## References


1. World Health Organization (WHO). European Food and Nutrition Action Plan 2015–2020. Available online: http://www.euro.who.int/en/publications/abstracts/european-food-and-nutrition-action-plan-20152020-2014 (accessed on 9 September 2019).
2. Yeomans, M.R. Measuring Appetite and Food Intake. Available online: https://doi.org/10.1016/B978-0-08-101743-2.00006-6 (accessed on 9 September 2019).
3. Butland, B.; Jebb, S.; Kopelman, P.; McPherson, K.; Thomas, S.; Mardell, J.; Parry, V. *Foresight—Tackling Obesities: Future Choices—Project Report*; Government Office for Science: London, UK, 2007.
4. Blundell, J.; Bellisle, F. Satiation, Satiety and the Control of Food Intake: Theory and Practice. Available online: https://www.elsevier.com/books/satiation-satiety-and-the-control-of-food-intake/blundell/978-0-85709-543-5 (accessed on 9 September 2019).
5. Blundell, J.; De Graaf, C.; Hulshof, T.; Jebb, S.; Livingstone, B.; Lluch, A.; Mela, D.; Salah, S.; Schuring, E.; Van Der Knaap, H.; et al. Appetite Control: Methodological Aspects of the Evaluation of Foods. *Obes. Rev.* **2010**, *11*, 251–270. [CrossRef] [PubMed]
6. Blundell, J. Pharmacological approaches to appetite suppression. *Trends Pharmacol. Sci.* **1991**, *12*, 147–157. [CrossRef]
7. Harrold, J.A.; Dovey, T.M.; Blundell, J.E.; Halford, J.C. CNS regulation of appetite. *Neuropharmacol.* **2012**, *63*, 3–17. [CrossRef] [PubMed]
8. Kringelbach, M.L.; Stein, A.; Van Hartevelt, T.J. The functional human neuroanatomy of food pleasure cycles. *Physiol. Behav.* **2012**, *106*, 307–316. [CrossRef] [PubMed]
9. Arentson-Lantz, E.; Clairmont, S.; Paddon-Jones, D.; Tremblay, A.; Elango, R. Protein: A nutrient in focus. *Appl. Physiol. Nutr. Metab.* **2015**, *40*, 755–761. [CrossRef]
10. Forde, C.G. Measuring Satiation and Satiety. Available online: https://doi.org/10.1016/B978-0-08-101743-2.00007-8 (accessed on 9 September 2019).
11. Morell, P.; Fiszman, S. Revisiting the role of protein-induced satiation and satiety. *Food Hydrocoll.* **2017**, *68*, 199–210. [CrossRef]
12. Boelsma, E.; Brink, E.J.; Stafleu, A.; Hendriks, H.F. Measures of postprandial wellness after single intake of two protein–carbohydrate meals. *Appetite* **2010**, *54*, 456–464. [CrossRef]
13. Martini, D.; Brusamolino, A.; Del Bo', C.; Laureati, M.; Porrini, M.; Riso, P. Effect of fiber and protein-enriched pasta formulations on satiety-related sensations and afternoon snacking in Italian healthy female subjects. *Physiol. Behav.* **2018**, *185*, 61–69. [CrossRef]
14. Halton, T.L.; Hu, F.B. The effects of high protein diets on thermogenesis, satiety and weight loss: A critical review. *J. Am. Coll. Nutr.* **2004**, *23*, 373–385. [CrossRef]

15. Douglas, S.M.; Ortinau, L.C.; Hoertel, H.A.; Leidy, H.J. Low, moderate, or high protein yogurt snacks on appetite control and subsequent eating in healthy women. *Appetite* **2013**, *60*, 117–122. [CrossRef]
16. Anderson, G.H.; Moore, S.E. The Emerging Role of Dairy Proteins and Bioactive Peptides in Nutrition and Health. Dietary Proteins in the Regulation of Food Intake and Body Weight in Humans. *J. Nutr.* **2004**, *134*, 974–979. [CrossRef] [PubMed]
17. Stubbs, R.J.; Van Wyk, M.C.; Johnstone, A.M.; Harbron, C.G. Breakfasts high in protein, fat or carbohydrate: Effect on within-day appetite and energy balance. *Eur. J. Clin. Nutr.* **1996**, *50*, 409–417. [PubMed]
18. Andersen, B.V.; Mielby, L.H.; Viemose, I.; Bredie, W.L.; Hyldig, G. Integration of the sensory experience and post-ingestive measures for understanding food satisfaction. A case study on sucrose replacement by Stevia rebaudiana and addition of beta glucan in fruit drinks. *Food Qual. Preference* **2017**, *58*, 76–84. [CrossRef]
19. Andersen, B.V.; Hyldig, G. Food satisfaction: Integrating feelings before, during and after food intake. *Food Qual. Preference* **2015**, *43*, 126–134. [CrossRef]
20. Karalus, M.; Vickers, Z. Satiation and satiety sensations produced by eating oatmeal vs. oranges. a comparison of different scales. *Appetite* **2016**, *99*, 168–176. [CrossRef] [PubMed]
21. Lowe, M.R.; Butryn, M.L. Hedonic hunger: A new dimension of appetite? *Physiol. Behav.* **2007**, *91*, 432–439. [CrossRef] [PubMed]
22. Andersen, B.; Byrne, D.; Bredie, W.; Møller, P. Cayenne pepper in a meal: Effect of oral heat on feelings of appetite, sensory specific desires and well-being. *Food Qual. Preference* **2017**, *60*, 1–8. [CrossRef]
23. Duerlund, M.; Andersen, B.V.; Grønbeck, M.S.; Byrne, D.V. Consumer Reflections on Post-Ingestive Sensations. A Qualitative Approach by Means of Focus Group Interviews. *Appetite* **2019**, *142*, 104350. [CrossRef]
24. Andersen, B.V.; Hyldig, G. Consumers' View on Determinants to Food Satisfaction. A Qualitative Approach. *Appetite* **2015**, *95*, 9–16. [CrossRef]
25. Meiselman, H.L. The Future in Sensory/Consumer Research: ... ... Evolving to a Better Science. *Food Qual. Prefer.* **2013**, *27*, 208–214. [CrossRef]
26. Meiselman, H.L. Quality of Life, Well-Being and Wellness: Measuring Subjective Health for Foods and Other Products. *Food Qual. Prefer.* **2016**, *54*, 101–109. [CrossRef]
27. Craig, A.D. How Do You Feel? Interoception: The Sense of the Physiological Condition of the Body. *Nat. Rev. Neurosci.* **2002**, *3*, 655–666. [CrossRef] [PubMed]
28. Simmons, W.K.; Deville, D.C. Interoceptive Contributions to Healthy Eating and Obesity. *Curr. Opin. Psychol.* **2017**, *17*, 106–112. [CrossRef] [PubMed]
29. Stevenson, R.J.; Mahmut, M.; Rooney, K. Individual differences in the interoceptive states of hunger, fullness and thirst. *Appetite* **2015**, *95*, 44–57. [CrossRef] [PubMed]
30. Herbert, B.M.; Pollatos, O. Attenuated interoceptive sensitivity in overweight and obese individuals. *Eat. Behav.* **2014**, *15*, 445–448. [CrossRef]
31. Møller, P. Satisfaction, Satiation and Food Behaviour. *Curr. Opin. Food Sci.* **2015**, *3*, 59–64. [CrossRef]
32. Ares, G.; De Saldamando, L.; Giménez, A.; Claret, A.; Cunha, L.M.; Guerrero, L.; De Moura, A.P.; Oliveira, D.C.; Symoneaux, R.; Deliza, R. Consumers' associations with wellbeing in a food-related context: A cross-cultural study. *Food Qual. Preference* **2015**, *40*, 304–315. [CrossRef]
33. Brener, J.; Ring, C. Towards a psychophysics of interoceptive processes: The measurement of heartbeat detection. *Philos. Trans. R. Soc. B Boil. Sci.* **2016**, *371*, 20160015. [CrossRef]
34. King, S.C.; Snow, J.; Meiselman, H.L.; Sainsbury, J.; Carr, B.T.; McCafferty, D.; Serrano, D.; Gillette, M.; Millard, L.; Li, Q. Development of a questionnaire to measure consumer wellness associated with foods: The WellSense Profile™. *Food Qual. Preference* **2015**, *39*, 82–94. [CrossRef]
35. Ares, G.; Giménez, A.; Vidal, L.; Zhou, Y.; Krystallis, A.; Tsalis, G.; Symoneaux, R.; Cunha, L.M.; De Moura, A.P.; Claret, A.; et al. Do we all perceive food-related wellbeing in the same way? Results from an exploratory cross-cultural study. *Food Qual. Preference* **2016**, *52*, 62–73. [CrossRef]
36. Act on Research Ethics Review of Health Research Projects. Available online: https://www.retsinformation.dk/Forms/r0710.aspx?id~=~192671 (accessed on 9 September 2019).
37. Cohen, J.E. Statistical Power Analysis for the Behavioral Sciences. Available online: https://www.elsevier.com/books/statistical-power-analysis-for-the-behavioral-sciences/cohen/978-0-12-179060-8 (accessed on 9 September 2019).

38. Sawilowsky, S.S. New Effect Size Rules of Thumb. *J. Mod. Appl. Stat. Methods* **2009**, *8*, 597–599. [CrossRef]
39. Blatt, A.D.; Roe, L.S.; Rolls, B.J. Increasing the protein content of meals and its effect on daily energy intake. *J. Am. Diet. Assoc.* **2011**, *111*, 290–294. [CrossRef] [PubMed]
40. Bull, S.P.; Hong, Y.; Khutoryanskiy, V.V.; Parker, J.K.; Faka, M.; Methven, L. Whey protein mouth drying influenced by thermal denaturation. *Food Qual. Preference* **2017**, *56*, 233–240. [CrossRef] [PubMed]
41. Carter, B.; Drake, M. Invited review: The effects of processing parameters on the flavor of whey protein ingredients. *J. Dairy Sci.* **2018**, *101*, 6691–6702. [CrossRef]
42. Drewnowski, A.; Mennella, J.A.; Johnson, S.L.; Bellisle, F. Sweetness and Food Preference. *J. Nutr.* **2012**, *142*, 1142S–1148S. [CrossRef] [PubMed]
43. Bolhuis, D.P.; Costanzo, A.; Keast, R.S. Preference and perception of fat in salty and sweet foods. *Food Qual. Preference* **2018**, *64*, 131–137. [CrossRef]
44. Mattes, M.Z.; Vickers, Z.M. Better-liked foods can produce more satiety. *Food Qual. Preference* **2018**, *64*, 94–102. [CrossRef]
45. Cardello, A.V.; Schutz, H.; Snow, C.; Lesher, L. Predictors of food acceptance, consumption and satisfaction in specific eating situations. *Food Qual. Preference* **2000**, *11*, 201–216. [CrossRef]
46. Andersen, B.V. Sensory Factors in Food Satisfaction. An Understanding of the Satisfaction Term and a Measurement of Factors Involved in Sensory and Food Satisfaction. Ph.D. Thesis, Technical University of Denmark, Lyngby, Denmark, 2014.
47. Rolls, B.J.; Rolls, E.T.; Rowe, E.A.; Sweeney, K. Sensory specific satiety in man. *Physiol. Behav.* **1981**, *27*, 137–142. [CrossRef]
48. Hetherington, M.; Rolls, B.J.; Burley, V.J. The time course of sensory-specific satiety. *Appetite* **1989**, *12*, 57–68. [CrossRef]
49. Rolls, B.J. Sensory-Specific Satiety. *Nutr. Rev.* **1986**, *44*, 93–101. [CrossRef] [PubMed]
50. Havermans, R.C.; Janssen, T.; Giesen, J.C.A.H.; Roefs, A.; Jansen, A. Food Liking, Food Wanting, and Sensory-Specific Satiety. *Appetite* **2009**, *52*, 222–225. [CrossRef] [PubMed]
51. Olsen, A.; Ritz, C.; Hartvig, D.L.; Møller, P. Comparison of sensory specific satiety and sensory specific desires to eat in children and adults. *Appetite* **2011**, *57*, 6–13. [CrossRef] [PubMed]
52. Mela, D.J. Eating for pleasure or just wanting to eat? Reconsidering sensory hedonic responses as a driver of obesity. *Appetite* **2006**, *47*, 10–17. [CrossRef] [PubMed]
53. Harington, K.; Smeele, R.; Van Loon, F.; Yuan, J.; Haszard, J.J.; Drewer, A.; Venn, B.J. Desire for Sweet Taste Unchanged After Eating: Evidence of a Dessert Mentality? *J. Am. Coll. Nutr.* **2016**, *35*, 1–6. [CrossRef] [PubMed]
54. Murray, M.; Vickers, Z. Consumer Views of Hunger and Fullness. A Qualitative Approach. *Appetite* **2009**, *53*, 174–182. [CrossRef]
55. Reyner, L.; Wells, S.; Mortlock, V.; Horne, J. 'Post-lunch' sleepiness during prolonged, monotonous driving—Effects of meal size. *Physiol. Behav.* **2012**, *105*, 1088–1091. [CrossRef]
56. Smit, H.J.; Finnegan, Y.E.; Rogers, P.J. Post-Lunch Dip? Get out and Stay Out! Available online: https://doi.org/10.1016/j.appet.2006.07.003 (accessed on 9 September 2019).
57. Monk, T.H. The Post-Lunch Dip in Performance. *Clin. Sports Med.* **2005**, *24*, e15–e23. [CrossRef]
58. Wells, A.S.; Read, N.W.; Idzikowski, C.; Jones, J. Effects of meals on objective and subjective measures of daytime sleepiness. *J. Appl. Physiol.* **1998**, *84*, 507–515. [CrossRef]
59. Holt, S.H.A.; Brand-Miller, J.C.; Petocz, P.; Farmakalidis, E. A Satiety Index of Common Foods. *Eur. J. Clin. Nutr.* **1995**, *49*, 675–690.
60. Dodge, R.; Daly, A.P.; Huyton, J.; Sanders, L.D. The challenge of defining wellbeing. *Int. J. Wellbeing* **2012**, *2*, 222–235. [CrossRef]
61. McMahon, A.-T.; Williams, P.; Tapsell, L. Reviewing the meanings of wellness and well-being and their implications for food choice. *Perspect. Public Heal.* **2010**, *130*, 282–286. [CrossRef] [PubMed]
62. Miller, G.; Foster, L.T. Critical Synthesis of Wellness Literature. Available online: https://dspace.library.uvic.ca/handle/1828/2894 (accessed on 9 September 2019).
63. Rogers, P.J.; Hardman, C.A. Food Reward. What It Is and How to Measure It. *Appetite* **2015**, *90*, 1–15. [CrossRef] [PubMed]
64. Jones, S.E. The Global Problem of Obesity. Practical Guide to Obesity Medicine. Available online: https://doi.org/10.1016/B978-0-323-48559-3.00001-4 (accessed on 9 September 2019).

# 4

# Consumer Preference Heterogeneity Evaluation in Fruit and Vegetable Purchasing Decisions using the Best–Worst Approach

**Stefano Massaglia [1], Danielle Borra [1], Cristiana Peano [1], Francesco Sottile [2] and Valentina Maria Merlino [1,*]**

1 Department of Agricultural, Forest and Food Sciences, University of Turin, Largo Paolo Braccini 2, 10095 Grugliasco, Italy
2 Department of Architecture, University of Palermo, Viale delle Scienze 19, 90128 Palermo, Italy
* Correspondence: valentina.merlino@unito.it.

**Abstract:** This study assesses consumer preferences during fruit and vegetable (FV) sales, considering the sociodemographic variables of individuals together with their choice of point of purchase. A choice experiment was conducted in two metropolitan areas in Northwest Italy. A total of 1170 consumers were interviewed at different FV purchase points (mass retail chains and open-air markets) using a paper questionnaire. The relative importance assigned by consumers to 12 fruit and vegetable product attributes, including both intrinsic and extrinsic quality cues, was assessed by using the best–worst scaling (BWS) methodology. The BWS results showed that "origin", "seasonality", and "freshness" were the most preferred attributes that Italian consumers took into account for purchases, while no importance was given to "organic certification", "variety", or "brand". Additionally, a latent class analysis was employed to divide the total sample into five different clusters of consumers, characterized by the same preferences related to FV attributes. Each group of individuals is described on the basis of sociodemographic variables and by the declared fruit and vegetable point of purchase. This research demonstrates that age, average annual income, and families with children are all discriminating factors that influence consumer preference and behavior, in addition to affecting which point of purchase the consumer prefers to acquire FV products from.

**Keywords:** best–worst scaling; cluster analysis; consumer preferences; fruits and vegetables

---

## 1. Introduction

The fresh fruit and vegetable (FV) market is an extremely widespread, complex, and global network. It is heterogeneous in terms of variety of products, with specific dynamics involving several stakeholders (producer, processor, and distributor) [1]. Together with the technological, biological, and industrial improvements that impact the FV sector, the exploration and introduction of new and different products has provoked a shift in this market, making it more globalized. The fruit and vegetable supply chain is characterized by important and dynamic logistics and management systems [2], which must respond effectively and promptly to changes that are both dictated by the market and often determined by consumer-led demands and needs. The decision-making role of consumers in defining end-product characteristics and supply-chain processes is evolving and has become one of the main drivers that needs to be analyzed in order to create a successful product. Given the heterogeneity and the multiplicity of the qualitative aspects that can characterize FV products, several studies have investigated consumption and consumer behavior in this market, including a multitude of choice attributes related to the experience and pleasure aspects of consumption [3]. One of the most debated aspects among the quality attributes of fruits and vegetables is human health. The consumption of FV has been linked to the prevention of several diseases [4]. It has been reported that people are currently consuming fewer FV products than the global recommended amounts, and therefore several food educational projects have been set up by means of national health promotion programs in schools and at many fruit and vegetable points of purchase [4]. In addition to human health-related aspects, several authors have analyzed the extrinsic, hedonic, and taste-related/sensory attributes that are important in the purchasing process [5–9]. For example, in [10], extrinsic quality attributes, such as brand, price, and packaging, were revealed to not affect the process of quality perception by consumers. On the contrary, sensory factors (i.e., intrinsic attributes of FV), such as the visual appearance, taste, freshness, color, aroma, texture, shape, nutritional quality, and crispness (for some products) were important attributes for fruit and vegetable quality evaluation [3,11,12]. Visual detection of fruit defects also plays a primary role in consumer attitudes and is correlated to the rejection of future purchases [13]. The guarantee of quality and safety standards of imported products, which is compliant with the European food safety legislations [14,15], has become a prerequisite for consumers. Thus, environmentally friendly quality parameters (low environmental impact, free from pesticides, socially oriented, and sustainable agriculture) now represent important quality attributes for consumers [16,17]. Attributes such as seasonality, locality, origin, and organic certification play a role in the selection of products from non-European countries where, in some cases, the use of pesticides, herbicides, and fungicides exceeds the safety thresholds established by the parameters of the European Union [18]. Other quality attributes such as freshness, seasonality, and price have been considered to be discriminating factors by consumers during FV purchases [19,20]. Freshness includes sensory properties (appearance, taste, smell, chemical, and physical) [21–24] and elements linked to the product origin (place of production, conservation technologies and processing, and packaging) [22]. Among the other FV product quality attributes, organic certification has proven to be a discriminatory consumer choice [25]. Finally, the product price is another important attribute for consumer decision making applied to all food products, including fruit and vegetables. In addition to the intrinsic/extrinsic product characteristics, research in the literature has also demonstrated that the consumer sociodemographic variables significantly discriminate their decision-making process [26,27], including the type of purchase outlet, which has proven to be related to an individual's behavior and preference [28,29]. The aim of this paper was the exploration of consumer preferences and purchasing behaviors in Northern Italy, considering different fruit and vegetable sensory intrinsic/extrinsic attributes. The best–worst scaling methodology was applied to measure the relative importance of each attribute and a latent class cluster analysis was used for sample segmentation on the basis of consumer preferences. These methodologies have already been applied in several other studies in the agri-food sector [30,31] and enable the evaluation and comparison of the impact of the sociodemographic variables of the individual on the preferences and the selection of the fruit and vegetable point of purchase of the consumer.

## 2. Materials and Methods

### 2.1. Data Collection and Sample Characteristics

A choice experiment was conducted between 2017 and 2018 in two metropolitan areas of the Piedmont and Liguria regions in northwest Italy. Face-to-face interviews were conducted at six points of purchase (two mass retail chains and four open-air markets) in these geographical areas, alternating the day of the week (from Sunday to Saturday) and considering two time slots (from 9 a.m. to 1 p.m. and from 4 p.m. to 8 p.m.). During the intercept survey, individuals were randomly chosen, using the personal condition of "FV purchaser" as the only selection criterion. The first section included questions related to sociodemographic characteristics (gender, age, employment, education, and annual average income). The second section focused on fruit and vegetable purchasing and consumption behavior. In particular, this included examining the habitual points of purchase of fruit and vegetables (green grocer, mass retail channels, open-air market, food shop, organic food shop, purchasing groups, and producer). Consumer-declared preferences were collected in the third section, by means of the

BWS methodological design implementation, which aimed to measure the relative importance placed by consumers on the previously selected 12 FV intrinsic and extrinsic attributes.

### 2.2. Best–Worst Scaling Design

The best–worst scaling (BWS) methodology is a method for collecting declared preferences which consists of requiring the respondent to select, among a series of attributes, the best and worst alternative from the list of proposals [32]. In applying this methodology, 12 extrinsic and intrinsic attributes (selection criteria) describing fruits and vegetables were selected from an in-depth literature research, which are presented in Table 1.

**Table 1.** The 12 attributes of fruits and vegetables used for the best–worst scaling analysis and associated references.

| Fruit & Vegetable Quality Attributes | Attribute Description | References |
|---|---|---|
| Brand (or seller) | Brand allows the consumer to identify and discriminate a product. The evaluation of the brand associates "research", in relation to the "experience" of the characteristics of a product, together with information about the manufacturer. | [33–35] |
| Organic label | Organic certification has been recognized, in various studies, as an attribute that positively influences consumer choices at the time of purchase. | [36–38] |
| Quality certifications | This attribute is often related to greater product safety and wholesomeness that, in FV, often results in the reduction of pesticide risk. | [39,40] |
| Origin | Product origin is an intrinsic cue, linked to the consumer information acquisition of product/producer identification in addition to product quality assessment. Consumers believe in a higher quality of domestic food, in comparison to foreign products. | [41–43] |
| Price | Several studies have given evidence that the selection of fresh fruit and vegetables is often not influenced by price. Thus, considering a broader product category in our study, fruit and vegetables, price becomes relevant. | [16,44,45] |
| Offer | The evaluation of special offers (promotional prices) for fruit and vegetables is an important tool for this category of products, often characterized by medium–high prices. This factor often depends on the place of purchase and seasonality. | [46–48] |
| Appearance | The outward appearance of FV is one of the attributes that highly influences decisions at the time of purchase. | [46,49,50] |
| Local | Local production has a lower perception of risk, which helps to increase loyalty to local producers and to guarantee the sustainability of local companies. | [3,51–57] |
| Geographical indication label | The consumer attitude towards certified products (e.g., Protected Designation of Origin (PDO) and Protected Geographical Indication (PGI)) has been studied in various researches, evaluating different fruit and vegetable products and confirming the recognition by consumers of a higher quality compared to a conventional or commercial product, with higher organoleptic and taste properties. | [58–61] |
| Seasonality | The consumption of seasonal fruit and vegetables is associated with consumer choice behavior oriented towards an ecological product; one that avoids excessive packaging (such as tin and plastic) and waste, and which is considered to have a higher organoleptic quality and freshness. | [62–64] |
| Variety | Greater attention to this attribute differs by various types of consumers, especially in identifying targets that are more attentive to the variety (cultivars). | [65–67] |
| Freshness | Freshness is a very important quality criterion for FV acceptability. Consumer assessment of freshness of fruit and vegetables occurs through the analysis of sensory and visual aspects of the product appearance during purchase, but also during/after consumption. | [20,22,68,69] |

FV: fruit and vegetable.

Using the Sawtooth Software (v.2.0.0.2, Orem, UT, USA), the 12 attributes were organized into four different versions of the questionnaire (integrated into the third section of the final questionnaire) and the order of the presented attributes varied in the different sets (nine per version). Each set contained four attributes and the single item appeared three times in the questionnaire, according to [70] and [39]. For each set of attributes, the respondent was required to indicate what was the most important (BEST) and what was the least important (WORST) item during their FV purchase. The experimental design followed the rule that, if the number of selected attributes is k and they are positioned in the subset C, then there are k(k−1)/2 pairs of BW (best-worst) and k(k−1)/2 pairs of WB (worst-best) associated with each subset [71]. Therefore, each choice set contains k(k−1) possible choice options (pairs BW and WB). In this way, the respondent chooses two maximum difference of preferences for each attribute set by providing more information than for single-choice options. This method is also called "maximum difference scaling", because the chosen attributes should maximize the difference in utility made by a respondent on a preference scale. Finally, the BWS rating scales create an orderly ranking of items, explaining the level of relative importance appointed by each respondent for the single attribute using the average BW raw score (A-RS) or standard score [30,71,72]. This is calculated by dividing the BW score (best minus worst) [32] by the number of observations and the frequency that each attribute appeared in the four questionnaire versions. The A-RS is considered as the mean number of times each item was chosen as the most important or the least important by a single individual [73]. Among the main advantages, the BWS methodology allows measurement of the interview-declared preferences rather than, for example, the revealed preference which happens in traditional conjoint analysis. The BWS methodology allows for transformation of qualitative information into a real preference score (quantitative), avoiding the limitations, errors, and cognitive effort of the interviewed subjects, as in the case of, instead, being asked to classify (ranking) or declare a point (rating); for example, with the Likert scale [72,74,75].

The confidence limit applied in the estimation of the attribute scores was set at 95% and the standard deviation was used as a raw indicator of variability present within the sample. A two-tailed t-test was used to assess whether one attribute was preferred to another within the sample of respondents [76]. The rank of preferences derived from the best–worst scaling analysis and related to the expressed preferences of the whole sample allowed the application of latent class clustering (lClass) analysis, where the sample was divided into five clusters, according to the weight that the individuals assigned to the different attributes (average raw score values). The five selected groups of consumers were derived from the choices made between the four segmentations generated by the software. The most appropriate was reconciled by the one corresponding to the lowest BIC (Bayesian information criterion) value, in accordance with [42] (Table 2).

For cluster segmentation, the *p*-value for each attribute was calculated following a variance homogeneity test. The SPSS.21.0 software was used for the quantitative analysis.

**Table 2.** Latent class clustering (lClass) analysis results: Comparison between the BIC (Bayesian information criterion) for cluster segmentation choice.

| Groups | Replication [1] | BIC |
|---|---|---|
| 2 | 3 | 46257.960 |
| 3 | 4 | 45162.476 |
| 4 | 2 | 44564.793 |
| 5 | 4 | 44171.250 |

[1] Number of replications of data coupling performed by software, based on initial setting (replication set = 5).

## 3. Results and Discussions

### 3.1. Participant Description

A sample of 1170 respondents was collected for this study during the survey period: 684 were interviewed at large-scale retail stores and 486 were interviewed at open-air markets (Table 3).

**Table 3.** Sociodemographic characteristics of the interviewed sample population ($n$ = 1170).

| Sample ($n$ = 1170) | | | | | |
|---|---|---|---|---|---|
| Gender | male | 31% | Employment | housewife | 6% |
| | female | 69% | | unemployed | 6% |
| Age | ≤30 | 9% | | employed | 42% |
| | 31–45 | 24% | | self-employed | 9% |
| | 46–55 | 22% | | retired | 34% |
| | 56–65 | 22% | | student | 3% |
| | >65 | 23% | Annual average income (€/year) | <25,000 | 40% |
| Education | primary school | 6% | | 25,000–35,000 | 33% |
| | lower secondary school | 26% | | >35,000 | 8% |
| | upper secondary school | 49% | | n.d. | 18% |
| | master's degree | 19% | | | |

## 3.2. Consumer Preferences for Fruit and Vegetable Intrinsic and Extrinsic Choice Attributes

The best–worst scaling analysis results supported the identification of the most important FV attributes considered by all interviewed consumers during purchase decision (Table 4).

**Table 4.** Best–worst scaling count report (number of BEST and number of WORST), BW average raw score (A-RS), and standard deviation for each considered attribute.

| Rank | Attribute | Number of Best | Number of Worst | BW Average Raw Score [1] | Standard Deviation |
|---|---|---|---|---|---|
| 1 | Origin (Italian/foreign) | 1521 | 345 | 1.6840 | 1.673 |
| 2 | Seasonality | 1563 | 278 | 1.6790 | 1.394 |
| 3 | Freshness | 1489 | 305 | 1.6170 | 1.052 |
| 4 | Local | 842 | 559 | 0.4030 | 1.461 |
| 5 | Price | 902 | 685 | 0.3700 | 2.003 |
| 6 | Offer | 798 | 831 | −0.0260 | 1.913 |
| 7 | Appearance | 704 | 914 | −0.2960 | 1.520 |
| 8 | Geographical indication labels | 444 | 1145 | −0.8820 | 1.575 |
| 9 | Certification | 458 | 1102 | −1.0110 | 1.407 |
| 10 | Brand (or seller) | 400 | 1166 | −1.0240 | 1.227 |
| 11 | Variety | 319 | 1133 | −1.0770 | 1.027 |
| 12 | Organic | 406 | 1383 | −1.4380 | 1.918 |

[1] A negative BW (best-worst) average raw score value is due to the attribute not commonly chosen as the best factor.

Consumer choices during FV purchases were particularly influenced by the intrinsic factors of fruit and vegetable products: the origin having the highest average raw score (1.684), seasonality (average raw score 1.679), followed by the product freshness (average raw score 1.617). Other studies confirmed the importance given by consumers to the freshness attribute as a driver for purchasing fruit and vegetables [77,78]. Therefore, our study revealed that consumers, first and foremost, assess and pay attention to the product origin when buying FV, a quality attribute that focuses on the product identity and environmental and social sustainability, as well as the specific unique organoleptic characteristics [79–81]. Origin evaluation is closely linked to the product quality assessment, particularly since consumers generally consider a national product to be of better quality and safer than imported products [77]. The safety aspect also emerges as a discriminating factor linked to the origin of FV, characterized by a wide variety of products from different countries, including those outside the European Union. Often, in other similar studies, the analysis of the importance of local characteristics was carried out simultaneously with the seasonality attribute [63], together with sustainability, confirming the correlation between these quality attributes in FV products [59,82]. The price has also been shown to be an important attribute for consumers during FV purchases, evaluated as a quality indicator [13]. On the contrary, the worst values assigned by consumers for FV selection were attributed to extrinsic factors: organic certification (average raw score −1.438), variety (average raw score −1.077), and brand (average raw score −1.024). The negative evaluation of brand and special offers has also been confirmed by research in the existing literature [83,84]. However, our results regarding the non-important attributes for FV purchases were in contrast to other references in the literature [65,85].

### 3.3. Latent Class Analysis

The sample division of five clusters with different sizes is reported in Table 5. Each consumer group was named as a function of the most influential fruit and vegetable attributes considered during their purchase decision-making process (five groups). Only three FV quality attributes had positive values of average BW raw scores in all clusters (seasonality, freshness, and origin), while variety and brand were considered not important by consumers in all the five groups. The other seven attributes were differently considered among the different clusters.

**Table 5.** The average BW raw score for each fruit and vegetable attribute resulting from the five clusters of consumers: price sensitive, proposed loyalty, value for money, undecided consumer, and local sensitive.

| | Price Sensitive | Proposed Loyalty | Value for Money | Undecided Consumer | Local Sensitive | *p*-Value [1] |
|---|---|---|---|---|---|---|
| Cluster dimension | 17.50% | 25.50% | 15.24% | 16.50% | 25.26% | |
| *Attributes* | | | *Average BW raw scores* | | | |
| Offer | 50.21 | −20.27 | −8.80 | 20.99 | −45.83 | 0.966 |
| Geographical indication labels | −30.48 | 18.74 | −41.19 | −32.70 | 5.71 | 0.548 |
| Seasonality | 11.91 | 45.11 | 51.82 | 23.17 | 54.17 | **0.011** |
| Appearance | 3.22 | −11.70 | 39.14 | −10.54 | −39.42 | 0.777 |
| Origin (Italian/foreign) | 7.92 | 20.44 | 31.11 | 48.21 | 50.45 | **0.017** |
| Price | 64.25 | −16.92 | 3.75 | 24.52 | −36.37 | 0.675 |
| Organic | −35.75 | −39.25 | −37.70 | −51.79 | 6.03 | **0.032** |
| Certification | −26.40 | 10.66 | −33.82 | −37.71 | −6.84 | 0.107 |
| Variety | −14.47 | −38.02 | −7.86 | −30.22 | −30.50 | **0.012** |
| Local | −13.97 | −28.89 | −11.80 | 24.61 | 37.54 | 0.911 |
| Brand (or seller) | −26.20 | −0.65 | −42.33 | −6.48 | −32.80 | 0.051 |
| Freshness | 9.76 | 60.75 | 57.67 | 27.93 | 37.87 | **0.014** |

[1] The significant scores are highlighted in bold ($p$-value < 0.05).

#### 3.3.1. Preferences and Sociodemographic Variables of Different Consumer Clusters

##### Proposed Loyalty

Proposed loyalty cluster was the most highly represented group (25.5%). The sociodemographic variables of respondents belonging to this cluster are described in Figure 1. They were mainly women (67.6%) and were equally distributed among all age groups, with the exception of the youngest bracket. Meanwhile, men (32.37%) were well represented, mainly individuals 65 years and older. In accordance with the average age between individuals, this cluster mainly included employees and pensioners (the latter, presumably, men), and especially couples or families with a child. These respondents had a medium-high educational level, and a medium economic level. This cluster represents individuals who most likely entrust their selection of fruit and vegetables according to the market suggestions of what is in season, and who are attentive towards product freshness and origin. Variety, organic certification, local production, price, and special offers were not important at the time of purchase. These individuals expressed low importance towards the price and supply of the product, focusing their purchases on high-quality seasonal products. Moreover, the positive evaluation of the "certification" and "geographical indication labels" attributes, which occurred only in this cluster, further confirms that this attitude is oriented towards certified guarantees offered by the market.

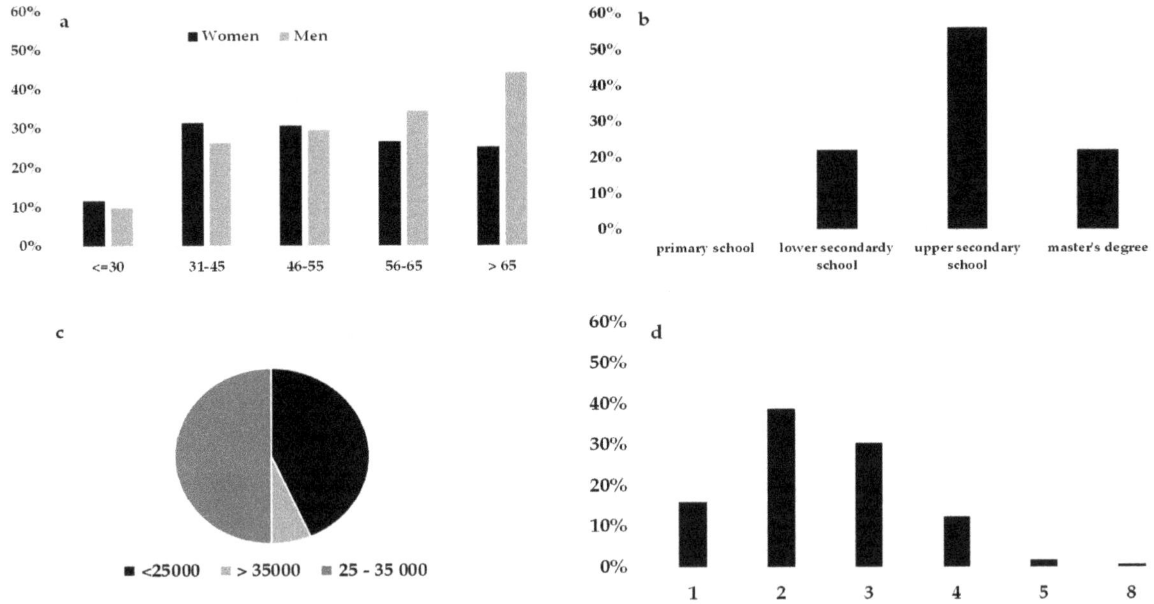


**Figure 1.** Sociodemographic characteristics of consumers belonging to the proposed loyalty cluster: (**a**) Gender and age group proportions, (**b**) educational level, (**c**) annual average income, and (**d**) number of family components.

Local Sensitive

This cluster was the second largest group (25.2% of the respondents) and had a clear majority of women, as compared to men; however, in this case, the women were concentrated in the middle-young age group, while the men were more mature individuals. Again, average income was over 35,000 euros/year, with an educational level distributed between lower school and master's degree (Figure 2).

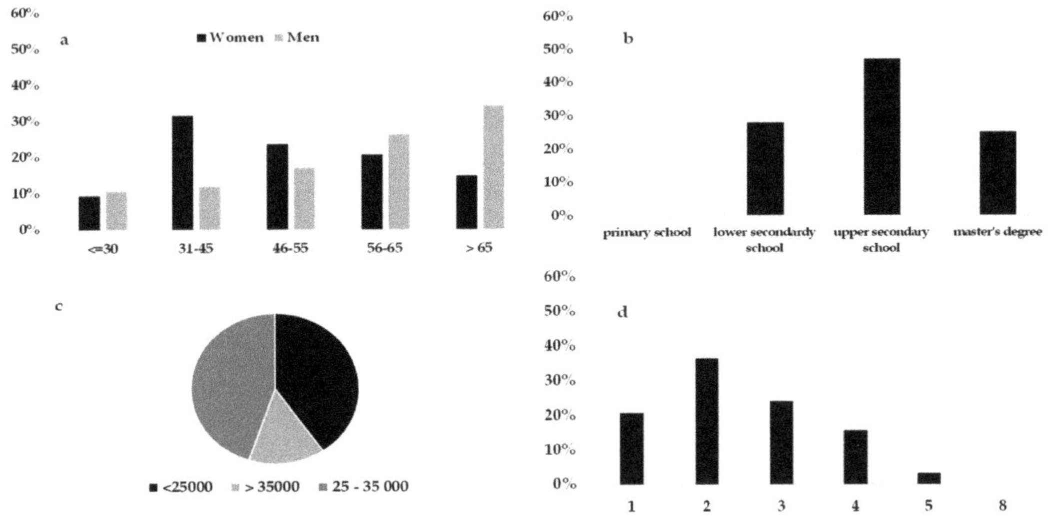


**Figure 2.** Sociodemographic characteristics of consumers belonging to the local sensitive cluster: (**a**) Gender and age groups proportion, (**b**) educational level, (**c**) yearly average income bracket, and (**d**) number of family components.

The local sensitive cluster had a majority percentage of three to four family members, but single individuals were also well-represented, in comparison to the other clusters. For these consumers, product price and promotion were not important during the FV purchasing process; however, seasonality linked to the proximity of production was fundamental in their choice. This was the only group where the consumers perceived the organic attribute positively, The interaction between "organic" and "locally grown" has also been studied and confirmed in [86], in addition to the preference of the local/organic combination by consumers [87], especially for young individuals. This was also confirmed in this cluster, since it was mostly represented by single women over 40 years old.

Price Sensitive

The consumer profile of the price sensitive cluster represented 17.5% of the sample. Price sensitive consumers were the most evenly distributed between the two genders and age groups, emerging as the youngest cluster. Among the sociodemographic characteristics of these subjects, the average income emerged as the lowest out of the five clusters and the educational status was also at a low level (Figure 3).

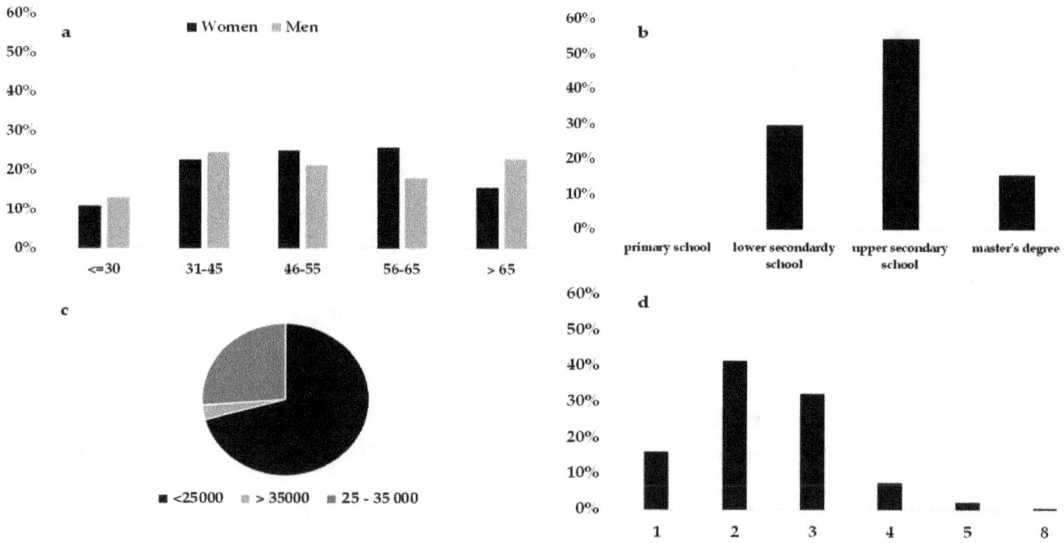


**Figure 3.** Sociodemographic characteristics of consumers belonging to price sensitive cluster: (**a**) Gender and age groups proportion, (**b**) educational level, (**c**) yearly average income bracket, and (**d**) number of family components.

These individuals focused their attention on product price and special offers/promotions during the FV purchasing process. The average annual income was a discriminating factor for defining the behavior of these individuals, confirming the correlation between economic availability and willingness to buy [88,89]. The lack of economic independence or an inexperienced spending model (youngest consumer) could define the individuals within this cluster. They buy only in relation to their bank account, minimally considering quality attributes such as freshness or appearance. Among the quality attributes of the product, these individuals are attentive to seasonality (in season products cost less), confirming the propensity to choose a convenient product.

Undecided

The undecided cluster (16.5% of the sample considered) was the cluster most represented by women (72%) who, most likely, did not devote much time to shopping, only purchasing generic FV products. The women were equally distributed among the different age groups, while the men were concentrated in the older age groups (Figure 4). The undecided individuals had a higher percentage of retirees among individuals (men) and of two-component families. Unlike other clusters, undecided individuals mainly represented two-component households with a low average annual income [89]. The preferences expressed by consumers were difficult to interpret in this group. In particular, these individuals were characterized by a widespread homogeneity of preference among the various factors of choice, buying what they found in the market at that particular time, looking at the origin (in some cases), but not exalting particular qualitative aspects of the product during their choice.

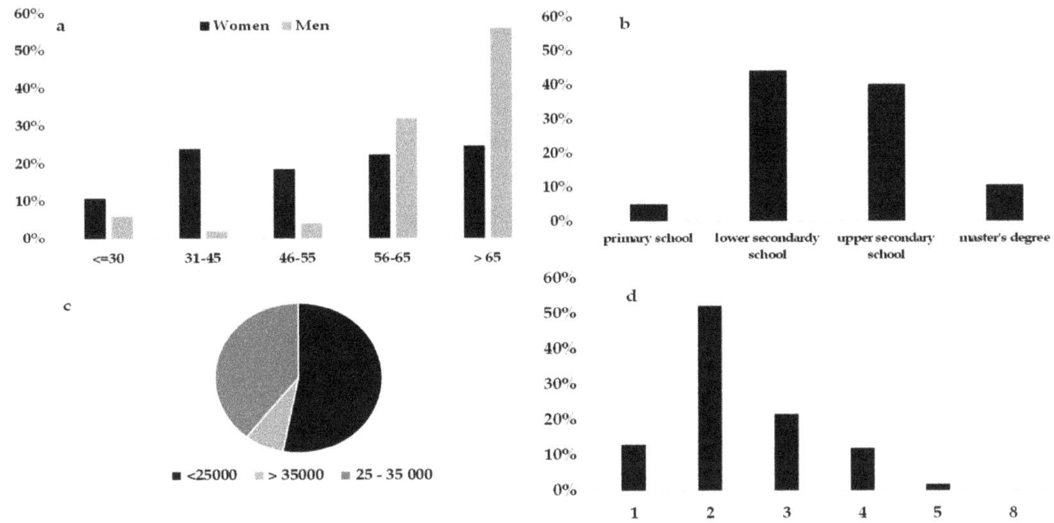


**Figure 4.** Sociodemographic characteristics of consumers belonging to the undecided cluster: (**a**) Gender and age groups proportion, (**b**) educational level, (**c**) yearly average income bracket, and (**d**) number of family components.

Value for Money

The value for money cluster was the smallest group (15.2%). It consisted mainly of women in different age groups (from 30 to 65 years old), and mature men (over 65 years). This group had the highest percentage of individuals with high incomes, although, on average, we defined them at an average income level. In this group, there were 20% singles, and almost 40% of the individuals had one or two children in the household (Figure 5).

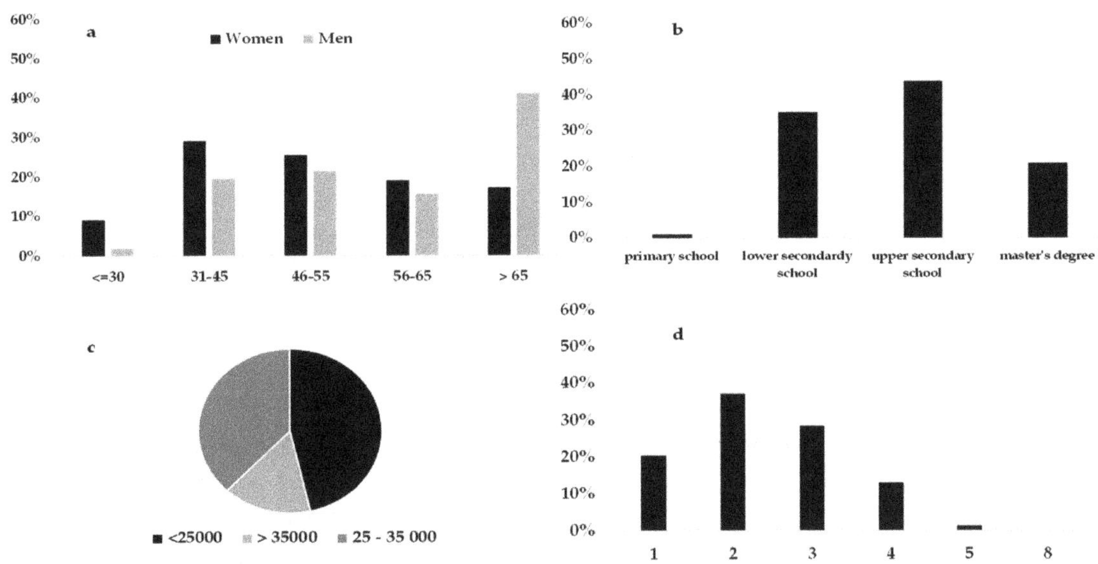


**Figure 5.** Sociodemographic characteristics of consumers belonging to value for money cluster: (**a**) Gender and age groups proportion, (**b**) educational level, (**c**) yearly average income bracket, and (**d**) number of family components.

These consumers expressed their preferences by interpreting the quality of the products provided by the experience, considering the freshness, appearance, and seasonality all as important, while not neglecting the price. The latter result is in line with the consumers in this cluster with the presence of children in households. In fact, it has been shown that special offers are often negatively evaluated in the fresh food products purchasing process, because it is perceived as a seller's willingness to dispose of products with a lower quality standard [46,90].

3.3.2. Fruit and Vegetable Point of Purchase Choice as a Function of Consumer-Declared Preferences

Differences between consumer choices regarding the fruit and vegetable points of purchase emerged from habit exploration of the different consumer clusters (Figure 6).

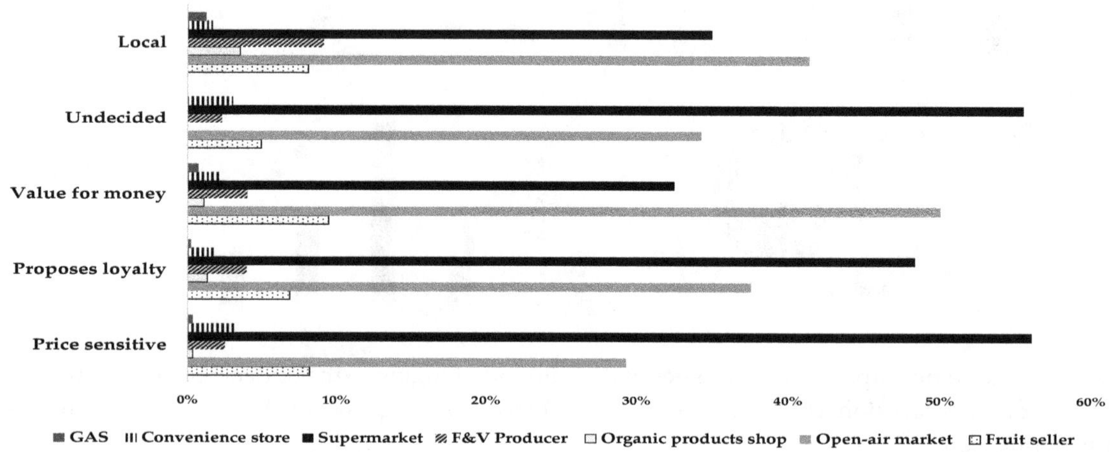


**Figure 6.** The points of purchase of fruit and vegetables chosen by consumers belonging to the five selected clusters.

Supermarkets were preferred by the price sensitive and undecided consumers. However, the sociodemographic factor that discriminated these two groups from the other clusters was the average low annual income. Therefore, it is possible to assume that the search for a product at a low price (price sensitive cluster) and the disinterested purchase (undecided cluster) can be satisfied by the offer from mass retail channels. The price sensitive group had the youngest respondents represented in the sample, confirming the attitudes of millennial consumers in preferring supermarkets for food purchases [91,92]. These consumers could coincide with a profile of buyers who are not attentive to locality or variety, in addition to brands and product certifications [48,83]. The choices of the undecided cluster coincided with the sociodemographic profile of active women, representing families composed of two members (husband and wife) who do not devote much time to shopping, but are also attentive towards the product origin, an attribute guaranteed by large and organized distribution. In the case of the proposed loyalty cluster, they bought FV mostly at the supermarkets, but also at open-air markets. The women from the proposed loyalty cluster represented families with children, highlighting their propensity to consider quality, pointing to extrinsic factors and seasonal products, relying on the proposals of the seller to protect the health of the family [93]. This also included the choice of supermarkets as a point of purchase for FV that guarantee information on the country of origin, the quality class, as well as higher prices which indicates greater guarantees of overall quality [94,95]. The research contained in [29] also pointed out that the need to satisfy consumer demand is reflected in an increase of shoppers at supermarkets. Therefore, characteristics such as appearance, sizing and packaging, and fast delivery, as well as permanent availability and a full range of products, are often requirements that cannot be met by local producers. A part of the proposed loyalty cluster is likely composed of more mature men who probably go to the open-air market or rely on the green grocer to buy products. In relation to the price sensitive and undecided clusters, a comparable sociodemographic composition is in the proposed loyalty cluster, which also has a medium to high average income as a discriminatory behavioral factor.

Local sensitive consumers mainly bought from open-air markets and from producers, confirming that their propensity for local products is linked to a short production chain [3,96,97], but not excluding supermarkets from their choices. Open-air markets satisfy a need for familiarity, in addition to safety and sustainability (environmental, social, and economic) for local production [97,98]. However, the percentage of local sensitive consumers who bought from supermarkets was significant, because local origin guarantees can also be provided by mass retail chain standards. In addition, these individuals expressed a preference towards the organic attribute, confirming that organic shoppers are part of this consumer group [99]. The value for money group was distinguished by the higher percentage of individuals who bought fruit and vegetables at open-air markets and at green grocers. The presence of children in the family composition was also evident in this cluster. These consumers were comparable to the proposed loyalty cluster, in the search for a quality and seasonal product, but differed in the place of purchase [100,101]. For the value for money cluster individuals, the product price was also an important factor in the FV choice. Our results are confirmed by [95], which showed that the number of children and full-time work of the consumers are variables directly related to the evaluation of a higher food price perceived as an indicator of guaranteed quality.

## 4. Conclusions

This research explored the expressed preferences of a sample of consumers in northwest Italy towards extrinsic and intrinsic quality attributes of fruits and vegetables. In general, when choosing fruit and vegetables, consumers were mainly influenced by intrinsic and sensory product attributes such as origin, freshness, and seasonality.

The extrinsic aspects of fruits and vegetables were not deemed important for the choice of this type of product. However, the importance appointed to each attribute was heterogeneous among the different clusters of individuals identified in this study. The consumer cluster analysis considered not only individual preferences, but also their sociodemographic variables, allowing for an understanding of the connection between preferences and attitude formation, the intrinsic characteristics of individuals, and the purchasing behavior expressed at the choice of fruit and vegetable point of sale. Our results describe the fruit and vegetable consumer as being mainly women and middle-aged men, over 55 years of age. Through analysis of the clusters, it emerged that consumer choices are greatly influenced by average income and age, but, above all, the presence of children in the family leads the buyer to make more reasoned choices aimed at the search for quality, safety, and seasonality found in the local products of outdoor markets. This last result has underlined that the guarantee of the quality of fruit and vegetable products can also be ensured by producers, highlighting the importance of sustainability linked to local production systems and short supply chains, as well as guarantee of freshness (an intrinsic aspect of the product among the most important for the choice). Attributes such as local origin, seasonality, and freshness describe a safe product, even for the youngest members of the family. On the contrary, the economic resources of the families have proven to be discriminatory during the purchase of FV, directing low income consumers towards the purchase of generic products with no particular characteristics. Family consumers were also the most likely to buy fruit and vegetables in supermarkets. The guarantee of a better quality of the product was, therefore, perceived as being linked to a higher cost, which can be found in outdoor markets and directly from the producer.

The limitations of this research may include the selection of attributes, which could be extended to a larger number. Although, this technique has made it possible to discriminate between the most important (origin and seasonality) and least important (organic and variety) attributes, consumer evaluations (credence and beliefs) could be deepened in future studies. In addition, the small geographical area and large metropolitan areas considered for research sampling represent the limits of our research; further research could be extended to other geographical areas to assess potential differences between the Italian regions and to compare consumer preferences of individuals belonging to small and large residential areas. Increased educational campaigns for school-age consumers could ensure a higher level of awareness of the benefits of consuming fruit and vegetables, in order to increase consumption even among younger consumers.

**Author Contributions:** Conceptualization, C.P., S.M., F.S., V.M.M., and D.B.; methodology, V.M.M. and S.M.; validation, V.M.M.; formal analysis, V.M.M.; investigation, S.M.; data curation, V.M.M.; writing—original draft preparation, V.M.M. and C.P.; writing—review and editing, C.P., F.S., S.M., V.M.M., and D.B.

## References

1. Yu, M.; Nagurney, A. Competitive food supply chain networks with application to fresh produce. *Eur. J. Oper. Res.* **2013**, *224*, 273–282. [CrossRef]
2. Camanzi, L.; Malorgio, G.; Azcárate, T.G. The role of Producer Organizations in supply concentration and marketing: A comparison between European Countries in the fruit and vegetables sector. *J. Food Prod. Market.* **2009**, *17*, 57990. [CrossRef]
3. Moser, R.; Raffaelli, R.; Thilmany-McFadden, D. Consumer preferences for fruit and vegetables with credence-based attributes: A review. *Int. Food Agribus. Manag. Rev.* **2011**, *14*, 121–142.
4. Food and Agricultural Organization; World Health Organization. *Fruit and Vegetables for Health Initiative*; FAO: Rome, Italy; WHO: Geneva, Switzerland, 2017.
5. Mueller, S.; Szolnoki, G. The relative influence of packaging, labelling, branding and sensory attributes on liking and purchase intent: Consumers differ in their responsiveness. *Food Qual. Prefer.* **2010**, *21*, 774–783. [CrossRef]
6. Enneking, U.; Neumann, C.; Henneberg, S. How important intrinsic and extrinsic product attributes affect purchase decision. *Food Qual. Prefer.* **2007**, *18*, 133–138. [CrossRef]
7. Motoki, K.; Sugiura, M. Disgust, sadness, and appraisal: Disgusted consumers dislike food more than sad ones. *Front. Psychol.* **2018**, *9*, 76. [CrossRef]
8. Motoki, K.; Saito, T.; Nouchi, R.; Kawashima, R.; Sugiura, M. A Sweet Voice: The Influence of Cross-Modal Correspondences Between Taste and Vocal Pitch on Advertising Effectiveness. *Multisens. Res.* **2019**, *32*, 401–427. [CrossRef]
9. Motoki, K.; Saito, T.; Nouchi, R.; Kawashima, R.; Sugiura, M. The paradox of warmth: Ambient warm temperature decreases preference for savory foods. *Food Qual. Prefer.* **2018**, *69*, 1–9. [CrossRef]
10. Chung, J.E.; Yu, J.P.; Pysarchik, D.T. Cue utilization to assess food product quality: A comparison of consumers and retailers in India. *Int. Rev. Retail Distrib. Consum. Res.* **2006**, *16*, 199–214. [CrossRef]
11. Schreiner, M.; Korn, M.; Stenger, M.; Holzgreve, L.; Altmann, M. Current understanding and use of quality characteristics of horticulture products. *Sci. Hortic.* **2013**, *163*, 63–69. [CrossRef]
12. Demattè, M.L.; Pojer, N.; Endrizzi, I.; Corollaro, M.L.; Betta, E.; Aprea, E.; Charles, M.; Biasioli, F.; Zampini, M.; Gasperi, F. Effects of the sound of the bite on apple perceived crispness and hardness. *Food Qual. Prefer.* **2014**, *38*, 58–64. [CrossRef]
13. Jaeger, S.R.; Antúnez, L.; Ares, G.; Johnston, J.W.; Hall, M.; Harker, F.R. Consumers' visual attention to fruit defects and disorders: A case study with apple images. *Postharvest Biol. Technol.* **2016**, *116*, 36–44. [CrossRef]
14. The European Parliament and the Council of the European Union. Regulation (EC) No 1107/2009 of the European Parliament and of the Council of 21 October 2009 concerning the placing of plant protection products on the market and repealing Council Directives 79/117/EEC and 91/414/EEC. *Off. J. Eur. Union* **2009**, *309*, 1–50.
15. The European Parliament and the Council of the European Union. Regulation (EC) No 396/2005 of the European Parliament and of the Council of 23 February 2005 on maximum residue levels of pesticides in or on food and feed of plant and animal origin and amending Council Directive 91/414/EECText with EEA relevance. *Off. J. Eur. Union* **2005**, *70*, 1–16.
16. Verain, M.C.; Sijtsema, S.J.; Antonides, G. Consumer segmentation based on food-category attribute importance: The relation with healthiness and sustainability perceptions. *Food Qual. Prefer.* **2016**, *48*, 99–106. [CrossRef]
17. Blanc, S.; Massaglia, S.; Brun, F.; Peano, C.; Mosso, A.; Giuggioli, N.R. Use of Bio-Based Plastics in the Fruit Supply Chain: An Integrated Approach to Assess Environmental, Economic, and Social Sustainability. *Sustainability* **2019**, *11*, 2475. [CrossRef]

18. Volpato, C. Consumi di frutta e verdura: Trend e prospettive internazionali. *Rivista di Frutticoltura e di Ortofloricoltura* **2017**, *81*, 8–11.
19. Péneau, S.; Hoehn, E.; Roth, H.R.; Escher, F.; Nuessli, J. Importance and consumer perception of freshness of apples. *Food Qual. Prefer.* **2006**, *17*, 9–19. [CrossRef]
20. Peparaio, M.; Sinesio, F.; Paoletti, F.; Moneta, E.; Saba, A.; Vassallo, M. Towards a multi-dimensional concept of vegetable freshness from the consumer's perspective. *Food Qual. Prefer.* **2017**, *66*, 1–12.
21. Péneau, S.; Brockhoff, P.B.; Escher, F.; Nuessli, J. A comprehensive approach to evaluate the freshness of strawberries and carrots. *Postharvest Biol. Technol.* **2007**, *45*, 20–29. [CrossRef]
22. Péneau, S.; Linke, A.; Escher, F.; Nuessli, J. Freshness of fruits and vegetables: Consumer language and perception. *Br. Food J.* **2009**, *111*, 243–256. [CrossRef]
23. Fillion, L.; Kilcast, D. Consumer perception of crispness and crunchiness in fruits and vegetables. *Food Qual. Prefer.* **2002**, *13*, 23–29. [CrossRef]
24. Fillion, L.; Kilcast, D. Concept and Measurement of Freshness of Fruits and Vegetables. Available online: https://www.researchgate.net/publication/223892341 (accessed on 20 June 2019).
25. Akgüngör, S.; Miran, B.; Abay, C. Consumer willingness to pay for organic products in urban Turkey. *J. Int. Food Agribus. Market.* **2010**, *22*, 299–313. [CrossRef]
26. Kyriacou, M.C.; Rouphael, Y. Towards a new definition of quality for fresh fruits and vegetables. *Sci. Hortic.* **2018**, *234*, 463–469. [CrossRef]
27. Grunert, K.G.; Sonntag, W.I.; Glanz-Chanos, V.; Forum, S. Consumer interest in environmental impact, safety, health and animal welfare aspects of modern pig production: Results of a cross-national choice experiment. *Meat Sci.* **2018**, *137*, 123–129. [CrossRef] [PubMed]
28. Panzone, L.; Hilton, D.; Sale, L.; Cohen, D. Socio-demographics, implicit attitudes, explicit attitudes, and sustainable consumption in supermarket shopping. *J. Econ. Psychol.* **2016**, *55*, 77–95. [CrossRef]
29. Umberger, W.J.; Reardon, T.; Stringer, R.; Mueller Loose, S. Market-Channel Choices of Indonesian Potato Farmers: A Best–Worst Scaling Experiment. *Bull. Indones. Econ. Stud.* **2015**, *51*, 461–477. [CrossRef]
30. Merlino, V.M.; Borra, D.; Girgenti, V.; Dal Vecchio, A.; Massaglia, S. Beef meat preferences of consumers from Northwest Italy: Analysis of choice attributes. *Meat Sci.* **2018**, *143*, 119–128. [CrossRef]
31. Nunes, F.; Madureira, T.; Oliveira, J.V.; Madureira, H. The consumer trail: Applying best-worst scaling to classical wine attributes. *Wine Econ. Policy* **2016**, *5*, 78–86. [CrossRef]
32. Louviere, J.J.; Flynn, T.N. Using best-worst scaling choice experiments to measure public perceptions and preferences for healthcare reform in Australia. *Patient* **2010**, *3*, 275–283. [CrossRef]
33. Louis, D.; Lombart, C. Retailers' communication on ugly fruits and vegetables: What are consumers' perceptions? *J. Retail. Consum. Serv.* **2018**, *41*, 256–271. [CrossRef]
34. Ryan, J.; Casidy, R. The role of brand reputation in organic food consumption: A behavioral reasoning perspective. *J. Retail. Consum. Serv.* **2018**, *41*, 239–247. [CrossRef]
35. De Wulf, K.; Odekerken-Schröder, G.; Goedertier, F.; Van Ossel, G. Consumer perceptions of store brands versus national brands. *J. Consum. Market.* **2005**, *22*, 223–232. [CrossRef]
36. Massaglia, S.; Merlino, V.M.; Borra, D.; Peano, C. Consumer perception of organic blueberry labeling in Italy. *Qual. Access Success* **2017**, *19*, 312–318.
37. Zhao, X.; Chambers, E.; Matta, Z.; Loughin, T.M.; Carey, E.E. Consumer Sensory Analysis of Organically and Conventionally Grown Vegetables. *J. Food Sci.* **2007**, *72*, S87–S91. [CrossRef]
38. Merlino, V.M.; Borra, D.; Lazzarino, L.L.; Blanc, S. Does the organic certification influence the purchasing decisions of milk consumers? *Qual. Access Success* **2019**, *20*, 382–387.
39. Girgenti, V.; Massaglia, S.; Mosso, A.; Peano, C.; Brun, F. Exploring perceptions of raspberries and blueberries by Italian consumers. *Sustainability* **2016**, *8*, 1027. [CrossRef]
40. Botonaki, A.; Polymeros, K.; Tsakiridou, E.; Mattas, K. The role of food quality certification on consumers' food choices. *Br. Food J.* **2006**, *108*, 77–90. [CrossRef]
41. Schjøll, A. Country-of-origin preferences for organic food. *Org. Agric.* **2017**, *7*, 315–327. [CrossRef]
42. Dekhili, S.; Sirieix, L.; Cohen, E. How consumers choose olive oil: The importance of origin cues. *Food Qual. Prefer.* **2011**, *22*, 757–762. [CrossRef]
43. Loureiro, M.L.; Hine, S. Discovering Niche Markets: A Comparison of Consumer Willingness to Pay for Local (Colorado Grown), Organic, and GMO-Free Products. *J. Agric. Appl. Econ.* **2002**, *34*, 477–487. [CrossRef]

44. Miller, S.; Tait, P.; Saunders, C.; Dalziel, P.; Rutherford, P.; Abell, W. Estimation of consumer willingness-to-pay for social responsibility in fruit and vegetable products: A cross-country comparison using a choice experiment. *J. Consum. Behav.* **2017**, *16*, e13–e25. [CrossRef]
45. Aschemann-Witzel, J.; De Hooge, I.; Amani, P.; Bech-Larsen, T.; Oostindjer, M. Consumer-Related Food Waste: Causes and Potential for Action MAPP—Centre for Research on Customer Relations in the Food Sector. *Sustainability* **2015**, *7*, 6457–6477. [CrossRef]
46. Aschemann-Witzel, J.; Jensen, J.H.; Jensen, M.H.; Kulikovskaja, V. Consumer behaviour towards price-reduced suboptimal foods in the supermarket and the relation to food waste in households. *Appetite* **2017**, *116*, 246–258. [CrossRef]
47. Gjerris, M.; Gaiani, S. Household food waste in Nordic countries: Estimations and ethical implications. *Etikk i Praksis* **2013**, *7*, 6–23. [CrossRef]
48. Glanz, K.; Yaroch, A.L. Strategies for increasing fruit and vegetable intake in grocery stores and communities: Policy, pricing, and environmental change. *Prev. Med.* **2004**, *39*, S75. [CrossRef]
49. Oliver, P.; Cicerale, S.; Pang, E.; Keast, R. Identifying Key Flavors in Strawberries Driving Liking via Internal and External Preference Mapping. *J. Food Sci.* **2018**, *83*, 1073–1083. [CrossRef]
50. Verhulst, A.; Normand, J.M.; Lombard, C.; Moreau, G. A study on the use of an immersive virtual reality store to investigate consumer perceptions and purchase behavior toward non-standard fruits and vegetables. In Proceedings of the EEE Virtual Reality, Los Angeles, CA, USA, 18–22 March 2017; pp. 55–63.
51. Mugera, A.; Burton, M.; Downsborough, E. Consumer Preference and Willingness to Pay for a Local Label Attribute in Western Australian Fresh and Processed Food Products. *J. Food Prod. Mark.* **2017**, *23*, 452–472. [CrossRef]
52. Peano, C.; Girgenti, V.; Baudino, C.; Giuggioli, N.R. Blueberry supply chain in Italy: Management, innovation and sustainability. *Sustainability* **2017**, *9*, 261. [CrossRef]
53. Blanc, S.; Accastello, C.; Girgenti, V.; Brun, F.; Mosso, A. Innovative strategies for the raspberry supply chain: An environmental and economic assessment. *Qual. Access Success* **2018**, *19*, 139–142.
54. Bulsara, H.P.; Trivedi, K.G. An Exploratory study of factors related to Consumer Behaviour towards purchase of Fruits and Vegetables from different Retail Formats. *J. Res. Market.* **2016**, *6*, 397. [CrossRef]
55. Midmore Peter, N.S.; Vairo Daniela, W.M.S.A.-M.; Raffaele, Z. Consumer Attitudes towards the Quality and Safety of Organic and Low Input Foods. Available online: http://orgprints.org/8181/ (accessed on 20 June 2019).
56. Rodríguez-Ibeas, R. Environmental Product Differentiation and Environmental Awareness. *Environ. Res. Econ.* **2007**, *36*, 237–254. [CrossRef]
57. Thilmany, D.; Bond, C.A.; Bond, J.K. Going Local: Exploring Consumer Behavior and Motivations for Direct Food Purchases. *Am. J. Agric. Econ.* **2008**, *90*, 1303–1309. [CrossRef]
58. Ingrassia, M.; Sgroi, F.; Tudisca, S.; Chironi, S. Study of Consumer Preferences in Regard to the Blonde Orange Cv. *Washington Navel* "Arancia Di Ribera PDO". *J. Food Prod. Mark.* **2017**, *23*, 799–816. [CrossRef]
59. Lazzarini, G.A.; Visschers, V.H.M.; Siegrist, M. Our own country is best: Factors influencing consumers' sustainability perceptions of plant-based foods. *Food Qual. Prefer.* **2017**, *60*, 165–177. [CrossRef]
60. Xie, J.; Gao, Z.; Swisher, M.; Zhao, X. Consumers' preferences for fresh broccolis: Interactive effects between country of origin and organic labels. *Agric. Econ.* **2016**, *47*, 181–191. [CrossRef]
61. Mascarello, G.; Pinto, A.; Parise, N.; Crovato, S.; Ravarotto, L. The perception of food quality. Profiling Italian consumers. *Appetite* **2015**, *89*, 175–182. [CrossRef]
62. Rekhy, R.; McConchie, R.R. Promoting consumption of fruit and vegetables for better health. Have campaigns delivered on the goals? *Appetite* **2014**, *79*, 113–123. [CrossRef]
63. Wilkins, J.L. Consumer perceptions of seasonal and local foods: A study in a U.S. community. *Ecol. Food Nutr.* **2002**, *41*, 415–439. [CrossRef]
64. Verbeke, W.; Vanhonacker, F.; Van Loo, E.J.; Gellynck, X. Flemish consumer attitudes towards more sustainable food choices. *Appetite* **2013**, *62*, 7–16.
65. Kelley, K.M.; Primrose, R.; Crassweller, R.; Hayes, J.E.; Marini, R. Consumer peach preferences and purchasing behavior: a mixed methods study. *J. Sci. Food Agric.* **2016**, *96*, 2451–2461. [CrossRef]
66. Bonany, J.; Buehler, A.; Carbó, J.; Codarin, S.; Donati, F.; Echeverria, G.; Egger, S.; Guerra, W.; Hilaire, C.; Höller, I.; et al. Consumer eating quality acceptance of new apple varieties in different European countries. *Food Qual. Prefer.* **2013**, *30*, 250–259. [CrossRef]

67. Konopacka, D.; Stehr, R.; Schoorl, F.; Buehler, A.; Jesionkowska, K.; Kruczyn, D.; Egger, S.; Codarin, S.; Hilaire, C.; Ho, I.; et al. Apple and peach consumption habits across European countries. *Appetite* **2010**, *55*, 478–483. [CrossRef]
68. Jung, Y.J.; Padmanabahn, A.; Hong, J.H.; Lim, J.; Kim, K.O. Consumer freshness perception of spinach samples exposed to different storage conditions. *Postharvest Biol. Technol.* **2012**, *73*, 115–121. [CrossRef]
69. Peano, C.; Girgenti, V.; Palma, A.; Fontanella, E.; Giuggioli, N.R. Film type and map on CV. Himbo top raspbery fruit quality, composition and volatiles. *Ital. J. Food Sci.* **2013**, *25*, 421–432.
70. Orme, B.K. Accuracy of HB Estimation in MaxDiff Experiments (Sawtooth Software Research Paper Series). Available online: www.sawtoothsoftware.com (accessed on 26 June 2019).
71. Goodman, S.; Lockshin, L.; Cohen, E. Best-Worst Scaling: A Simple Method to Determine Drinks and Wine Style Preferences. Available online: https://hekyll.services.adelaide.edu.au/dspace/handle/2440/33743 (accessed on 20 June 2019).
72. Liu, C.; Li, J.; Steele, W.; Fang, X. A study on Chinese consumer preferences for food traceability information using best-worst scaling. *PLoS ONE* **2018**, *13*, e0206793. [CrossRef]
73. Loose, S.M.; Lockshin, L. Testing the robustness of best worst scaling for cross-national segmentation with different numbers of choice sets. *Food Qual. Prefer.* **2013**, *27*, 230–242. [CrossRef]
74. Lee, J.A.; Soutar, G.N.; Louviere, J. Measuring values using best-worst scaling: The LOV example. *Psychol. Mark.* **2007**, *24*, 1043–1058. [CrossRef]
75. Lee, J.A.; Soutar, G.; Louviere, J. The best–worst scaling approach: an alternative to Schwartz's values survey. *J. Personal. Assess.* **2008**, *90*, 335–347. [CrossRef]
76. Jaeger, S.R.; Jørgensen, A.S.; Aaslyng, M.D.; Bredie, W.L.P. Best-worst scaling: An introduction and initial comparison with monadic rating for preference elicitation with food products. *Food Qual. Prefer.* **2008**, *19*, 579–588. [CrossRef]
77. Terano, R.; Mohamed, Z.; Rezai, G.; Hanum, Z. Preference for Locally Grown or Imported Fruit Among the Millennial Generation in Johor, Malaysia. *J. Food Prod. Mark.* **2016**, *22*, 891–904. [CrossRef]
78. Cardello, A.V.; Schutz, H.G. The Concept of Food Freshness: Uncovering Its Meaning and Importance to Consumers. In *Freshness and Shelf Life of Foods*; ACS Symposium Series; American Chemical Society: Washington, DC, USA, 2009; pp. 22–41.
79. Giacalone, G.; Peano, C.; Iacona, T.; Iacona, C. Consumer Testing on Local and New Cultivars of Peach in the Roero Area, Piedmont, Italy. In Proceedings of the VI International Peach Symposium, Santiago, Chile, 9 January 2005; Acta Horticulturae: Leuven, Belgium, 2006; Volume 713, pp. 457–460.
80. Chambers, S.; Lobb, A.; Butler, L.; Harvey, K.; Bruce Traill, W. Local, national and imported foods: A qualitative study. *Appetite* **2007**, *49*, 208–213. [CrossRef]
81. Memery, J.; Angell, R.; Megicks, P.; Lindgreen, A. Unpicking motives to purchase locally-produced food: Analysis of direct and moderation effects. *Eur. J. Mark.* **2015**, *49*, 1207–1233. [CrossRef]
82. Thorndike, A.N.; Franckle, R.L.; Polacsek, M.; Boulos, R.; Greene, J.C.; Moran, A.; Rimm, E.B.; Block, J.P.; Blue, D.J. A Supermarket Double-Dollar Incentive Program Increases Purchases of Fresh Fruits and Vegetables Among Low-Income Families with Children: The Healthy Double Study. *J. Nutr. Educ. Behav.* **2017**, *50*, 217–228.e1.
83. Pearson, D. Fresh fruits and vegetables: Why do so many of them remain unbranded? *Aust. Agribus. Rev.* **2003**, *11*, 1–9.
84. Hussin, S.; Yee, W.; Bojei, J. Essential Quality Attributes in Fresh Produce Purchase by Malaysian Consumers. *J. Agribus. Market.* **2010**, *3*, 1–19.
85. Asioli, D.; Canavari, M.; Malaguti, L.; Mignani, C. Fruit Branding: Exploring Factors Affecting Adoption of the New Pear Cultivar 'Angelys' in Italian Large Retail. *Int. J. Fruit Sci.* **2016**, *16*, 284–300. [CrossRef]
86. Dentoni, D.; Tonsor, G.T.; Calantone, R.J.; Peterson, H.C. The Direct and Indirect Effects of 'Locally Grown' on Consumers' Attitudes towards Agri-Food Products. *Agric. Resour. Econ. Rev.* **2009**, *38*, 384–396. [CrossRef]
87. Ureña, F.; Bernabéu, R.; Olmeda, M. Women, men and organic food: Differences in their attitudes and willingness to pay. A Spanish case study. *Int. J. Consum. Stud.* **2007**, *32*, 18–26. [CrossRef]
88. Andreyeva, T.; Long, M.W.; Brownell, K.D. The Impact of Food Prices on Consumption: A Systematic Review of Research on the Price Elasticity of Demand for Food. *Am. J. Public Health* **2010**, *100*, 216–222. [CrossRef]
89. Webber, C.B.; Sobal, J.; Dollahite, J.S. Shopping for fruits and vegetables. Food and retail qualities of importance to low-income households at the grocery store. *Appetite* **2010**, *54*, 297–303. [CrossRef]

90. Merlino, V.; Borra, D.; Verduna, T.; Massaglia, S. Household Behavior with Respect to Meat Consumption: Differences between Households with and without Children. *Vet. Sci.* **2017**, *4*, 53. [CrossRef]
91. Cavaliere, A.; Ventura, V. Mismatch between food sustainability and consumer acceptance toward innovation technologies among Millennial students: The case of Shelf Life Extension. *J. Clean. Prod.* **2018**, *175*, 641–650. [CrossRef]
92. Brown, S.; Maloney, K. Making Sense of New Apple Varieties, Trademarks and Clubs: Current Status. *N. Y. Fruit Q.* **2009**, *17*, 9–12.
93. Heyman, M.B.; Abrams, S.A. Fruit Juice in Infants, Children, and Adolescents: Current Recommendations. *Pediatrics* **2017**, *139*, e20170967. [CrossRef]
94. Yu, H.; Neal, J.A.; Sirsat, S.A. Consumers' food safety risk perceptions and willingness to pay for fresh-cut produce with lower risk of foodborne illness. *Food Control* **2018**, *86*, 83–89. [CrossRef]
95. Scholderer, J.; Grunert, K.G. Consumers, food and convenience: The long way from resource constraints to actual consumption patterns. *J. Econ. Psychol.* **2005**, *26*, 105–128. [CrossRef]
96. Torres, A.P.; Marshall, M.I.; Alexander, C.E.; Delgado, M.S. Are local market relationships undermining organic fruit and vegetable certification? A bivariate probit analysis. *Agric. Econ.* **2017**, *48*, 197–205. [CrossRef]
97. Bartolini, S.; Viti, R.; Ducci, E. Local fruit varieties for sustainable cultivations: Pomological, nutraceutical and sensory characterization. *Agrochimica* **2015**, *59*, 281–294.
98. Massaglia, S.; Merlino, V.M.; Borra, D. Marketing strategies for animal welfare meat identification: Comparison of preferences between millennial and conventional consumers. *Qual. Access Success* **2018**, *19*, 305–311.
99. Hughner, R.S.; McDonagh, P.; Prothero, A.; Shultz, C.J.; Stanton, J. Who are organic food consumers? A compilation and review of why people purchase organic food. *J. Consum. Behav.* **2007**, *6*, 94–110. [CrossRef]
100. Hunt, A.R. Consumer interactions and influences on farmers' market vendors. *Renew. Agric. Food Syst.* **2007**, *22*, 54–66. [CrossRef]
101. Septiari, E.D.; Kusuma, G.H. Understanding the Perception of Millennial Generation toward Traditional Market (A Study In Yogyakarta). *Rev. Integr. Bus. Econ. Res. Online* **2016**, *5*, 30–43.

# 5

# When a Combination of Nudges Decreases Sustainable Food Choices Out-of-Home—The Example of Food Decoys and Descriptive Name Labels

**Pascal Ohlhausen * and Nina Langen**

Department Education for Sustainable Nutrition and Food Science, Institute of Vocational Education and Work Studies, Technische Universität Berlin, 10587 Berlin, Germany; nina.langen@tu-berlin.de

* Correspondence: ohlhausen@tu-berlin.de

**Abstract:** This paper reports results from three consecutive studies focusing on the comparison of the effectiveness of different nudges and their combinations to increase sustainable food choices out of the home. The nudges compared are the use of descriptive name labels (DNLs) for the most sustainable dish of a choice set (menu) and the decoy effect (DE), created by adding a less attractive decoy dish to a more attractive target dish with the goal of increasing the choice frequency of the target dish. In the literature, both nudges have been found to influence consumers' choices. In the first study, six category names of sustainability indicators were deduced from a focus group. These were tested with 100 students to identify the most attractive DNLs. Study II, a randomized choice study ($n = 420$), tested the DE, the DNLs and a combination of the DNLs and the DE used on four different dishes in a university canteen. In study III, 820 guests of a business canteen voted during four weeks for the special meals of the following week (identical to the four choice sets displayed in study II). Results indicate that the combination of DNLs and the DE is not recommended for fostering sustainable food choices. Pure DNLs were more efficient in increasing the choice frequency of the more sustainable meal, whereas the decoy effect resulted in decreased choice frequencies. Regional and sustainable DNLs were favoured by consumers.

**Keywords:** descriptive name labels; out-of-home; catering; sustainable nutrition; food; nudge; decoy

---

## 1. Introduction

The out-of-home catering (OOHC) sector is a rapidly growing market in Germany with growth rates of about three to four percent per year [1–4]. Due to rising mobility and urbanization, increasing shares of single-person households, higher incomes and time pressure, the amount of meals consumed at home is constantly decreasing [5–7]. Despite consumers' interest in the topics of sustainability or health [8,9], a considerable share of their food choices remain unsustainable and unhealthy [10–13]. One of the problems for consumers is procrastination (i.e., having an intention but failing to realise it) [14–16]. One possible approach to make customers' food choices more sustainable is to nudge them, and thereby stimulate, facilitate and encourage the sector's transformation towards sustainable development. The concept of nudging was first defined by Thaler and Sunstein [17] (p. 6): "A nudge, [ ... ] is any aspect of the choice architecture that alters people's behaviour in a predictable way without forbidding any options or significantly changing their economic incentives. To count as a mere nudge, the intervention must be easy and cheap to avoid." From this point on, the literature has focused on implementing and assessing nudges in several disciplines, e.g., tax and health policy, old-age provision and environmental protection [18–23]. In the sector of nutrition, several systematic reviews and meta-analyses have been conducted [24–29]. A meta-analysis by Arno and Thomas [29], for example, included 42 studies and revealed an increase in healthy food choices or decrease in energy intake of 15.3% when respective nudges were present. Besides laws and guidelines, nudges in the sector of nutrition can be used to further improve sustainability in the food sector, combat overweight as well as obesity and support the healthcare system without limiting consumers' wish for freedom of choice [8,29,30]. To avoid being used as a fraudulent marketing instrument, nudges should be transparent, never misleading, easy to opt-out of, consistent with people's values, improve the welfare of those being nudged and not violate individual rights [31,32]. Nudges in the nutrition sector range from choice-architecture-based nudges (also called system-1 nudges), such as changes in menu design or counter position, to information-based nudges (also called system-2 nudges), e.g., labels or additional nutritional information [20,33,34].

Employing the concept of nudging, this paper looks at consumers' food choices in education and business catering settings with the goal of fostering sustainable OOHC transformation.

To date, very little research has aimed at combining and comparing different nudging interventions in OOHC and surveyed their effects in simultaneous use [24,26,27]. Therefore, we opted to close this gap by combining and comparing two nudging interventions with the aim of increasing choice frequencies of sustainable dishes in OOHC. The nudges we combine and compare are (i) descriptive name labels (DNL), commonly referred to as food names, and (ii) the attraction effect or decoy effect (DE). The decoy effect in general, e.g., [35–40], and particularly descriptive name labels, e.g., [34,41–45], have been found to influence consumers' perceptions and choices of goods.

Most recent labelling studies have focused on food labels such as nutrition, health or warning labels and not on food names [41,46–49]. Nevertheless, DNLs are already used in restaurants and canteens, and the literature states increased choice frequency, higher taste ratings and more positive attitudes towards labelled dishes [41–43]. For preadolescent children, DNLs were used to boost vegetable and fruit intake [44]. The spectrum of name categories ranges from geographic DNLs (e.g., Southwestern), sensory DNLs (e.g., savory), nostalgic DNLs (e.g., grandma's) to heroic DNLs (e.g., for superheroes) [41,42,44].

The decoy effect was first described as a concept of asymmetrically dominated alternatives by Huber et al. [50]. A decoy is an alternative added to a choice set in order to alter the relative attractiveness of the other alternatives in the choice set. The addition of asymmetrically dominated alternatives called decoys increases the proportion of choices in favour of the target. In supermarkets, for example, the decoy effect can be used to enhance chocolate bar sales by placing a similar but slightly inferior chocolate bar next to the target product [51]. "Inferior" relates to the price or other quality attributes of the target product. Hence, the decoy effect can be used to push customers in the direction of buying a certain target product by showing a slightly worse alternative to the target product [50,52]. In the food sector, a choice study found evidence for the attraction effect when choosing dishes that display their calorie ratings [53].

The research aim is to assess the effectiveness of a combination of the decoy effect (DE) and descriptive name labels to increase consumers' choice of sustainable dishes. Hence, it is important to compare whether using a combination of these two nudging approaches is more effective in fostering sustainable food choice than the isolated nudging approaches on their own. The base of the studies was formed by a list of commonly offered OOHC dishes in Germany. Building on these dishes, the objective of study I is to find the best DNL wordings for six different DNL categories, which were derived by a focus group and literature survey, via a pre-test with students. The best DNLs per category are used in the choice experiment in study II. The objective of study II is to assess the effectiveness of DNLs in combination with the DE versus their single nudging variants in a university canteen setting. To reveal possible differences between settings, study III tests the combination of DNLs and the DE in a business canteen. For an overview of these studies, see Table 1.

**Table 1.** Overview of the three studies.

| | Study I | Study II | Study III |
|---|---|---|---|
| Content | Determine the best Descriptive name label (DNL) wording | Test of the isolated nudges decoy effect (DE) and DNLs, as well as the combination DNLs and DE | Test of the nudge combination DNLs and DE |
| Method | Focus group; Choice experiment; Best choice | Choice experiment; Linear regression | Choice experiment; U-Test (Mann/Whitney) |
| Sample | Students; $n = 100$ | University canteen; $n = 420$ | Business canteen; $n = 820$ |

## 2. Study I

### 2.1. Study I—Design

The first step was to define the dishes used as sustainable target dishes in the study. We, therefore, applied a tool developed and described by Engelmann et al. [54]. The tool is able to determine and compare the environmental and health impacts of dishes based on various criteria describing the four dimensions of sustainability: ecology, society, economy and health [54]. These criteria are, for example, the material footprint, carbon footprint, water demand, floor space demand, fair trade standards, animal welfare considerations, energy content, fibre content, fat content, carbohydrate content, sugar and salt content. Based on this assessment tool, we selected eight different commonly offered and frequently eaten OOHC meals with similar sales data (for more information see Study II), but with slightly higher health and sustainability ratings for the target dishes in comparison to the competitor dishes (see Table 2).

**Table 2.** Target and Competitor Dishes.

| Target Dishes | Competitor Dishes |
|---|---|
| Vegetable lasagne | Soy strips with noodles in mushroom sauce |
| Breaded fish with fried potatoes | Escalope chasseur with French fries |
| Spaghetti with rocket pesto | Mustard eggs with mashed potatoes |
| Chicken steak with tagliatelle | Spaghetti Bolognese |

To get more information about OOHC customers' sustainability preferences, we conducted a focus group in February 2016. The seven selected customers were between 20 and 60 years old and ate several times per week in OOHC canteens, making them experienced and involved participants. The most interesting question for the study regarded their desired and undesired indicators for eating meals out-of-home. Therefore, the focus group brainstormed relevant aspects first and then discussed the desirability of each indicator for OOHC. To avoid biasing participants' brainstorming process, no list of possible aspects was provided. Participants were free to mention any aspects they considered relevant. The 18 mentioned aspects and their desirability for meal characteristics are displayed in Table 3.

**Table 3.** Desirable and undesirable sustainability aspects in out-of-home catering (OOHC) (focus group).

| Desirable | Undesirable | Ambivalent |
|---|---|---|
| • Free of ...<br>• Regional<br>• Seasonal<br>• Pesticide-free<br>• Freshness<br>• Animal Welfare<br>• Hygiene | • CO2/Carbon Footprint<br>• Ecological Footprint<br>• Additives<br>• Percentage of animal products<br>• Resource input | • Organic<br>• Sugar<br>• Nutritional information<br>• Fair trade<br>• Free of genetic modifications<br>• Portion size |

The table, based on [55].

To combine dishes with different sustainable DNLs, different DNL categories were distinguished. The 18 aspects derived and discussed by the focus group were grouped into five sustainability categories for OOHC: regional (included the following aspects: regional, freshness, carbon footprint, ecological footprint, resource input) [41,42,56–58], seasonal (included the following aspects: seasonal, freshness, carbon footprint, ecological footprint, resource input) [56], organic (included the following aspects: free of, pesticide-free, animal welfare, carbon footprint, ecological footprint, additives, resource input, organic, free of genetic modifications) [59], sustainable (included the following aspects: animal welfare, carbon footprint, ecological footprint, resource input, fair trade, free of genetic modifications) [59] and healthy (included the following aspects: free of, pesticide-free, freshness, hygiene, additives, percentage of animal products, sugar, nutritional information, free of genetic modifications, portion size) [48,60]. These five categories were used as descriptive name label categories in studies I–III, described below.

In a following step, we screened menus from restaurants, canteens, cafés, bistros or snack bars as well as the literature [41,42] for insights into the great variety of possible DNLs for each category. It became clear that many DNLs describe traditional product and process characteristics. These were not identified during the focus group. Therefore, we added a sixth category to be used in studies I-III: the traditional name category.

A quantitative study was conducted in spring 2016 with students at TU Berlin. Students ($n$ = 100) were asked to indicate their preferred name labels for several food name categories displayed in a list. The question posed in the questionnaire was 'Imagine: It's lunchtime and you'e standing hungrily in the canteen in front of the menu. Which of the following phrases appeals to you most?'. The idea was to select the best DNLs per category for later application in the choice experiment of study II and to sort out unappealing DNLs. This procedure was recommended in Wansink et al. [41] and ensures that the DNLs used in the following studies are of relevance for consumers [41,42].

Each respondent was asked to answer two of a total of five questionnaires. These five questionnaire variants included one of our four target dishes for study II or the neutral "dish" variant. Each questionnaire included all six DNL categories (traditional, regional, seasonal, organic, sustainable and healthy) with all available DNLs applied to the dish, as can be seen in Table 4 for the neutral variant.

We used five different questionnaire variants to minimize the possibility that one DNL is preferred when applied to one specific dish but disliked on other dishes by the OOHC guests. The respondents were only to answer two out of the five questionnaires due to the possibility of a priming effect. That is, it was necessary to avoid that students kept selecting the same DNLs in all five questionnaires without considering the specific combination of dishes and DNLs.

**Table 4.** Overview of the pre-tested descriptive name labels (DNL) of Study I ($n$ = 100).

| Overview of The Descriptive Name Labels (DNL) of Study I | | | |
|---|---|---|---|
| Traditional | [%] | Regional | [%] |
| (Dish), traditional style | 63.5 | (Dish) from regional agriculture | 28.5 |
| Grandma's (Dish) | 16.5 | (Dish) from the region | 21.0 |
| (Dish) according to grandma's secret recipe | 11.5 | (Dish) from region XYZ | 18.5 |
| (Dish), the ancient way | 5.0 | (Dish) from regional production | 17.5 |
| (Dish) according to Aunt Martha's secret recipe | 2.5 | Region XYZ (Dish) | 14.0 |
| Missing | 1.0 | Missing | 0.5 |
| Seasonal | [%] | Organic | [%] |
| (Dish) with seasonal ingredients | 64.0 | Organic (Dish) | 35.5 |
| (Dish) from seasonal agriculture | 22.0 | (Dish) from organic agriculture | 16.0 |
| (Dish) from seasonal production | 14.0 | (Dish) from organic production | 15.5 |
| Missing | 0.0 | (Dish) from ecological production | 10.5 |
| | | (Dish) produced according to ecological standards | 7.0 |
| | | (Dish) from ecological agriculture | 7.0 |
| | | (Dish) from eco-friendly agriculture | 5.0 |
| | | (Dish) from eco-friendly production | 3.5 |
| | | Missing | 0.0 |
| Sustainable | [%] | Healthy | [%] |
| (Dish) from sustainable agriculture | 22.5 | (Dish) low-energy prepared | 24.5 |
| (Dish) from sustainable production | 18.5 | Low energy (Dish) | 19.0 |
| Fair trade (Dish) | 17.0 | Light (Dish) | 17.0 |
| (Dish) from fair production | 8.5 | (Dish) for light pleasure | 17.0 |
| (Dish) from fair trade agriculture | 6.5 | (Dish) with few calories | 6.0 |
| Fairly traded (Dish) | 6.5 | Calorie-reduced (Dish) | 5.0 |
| (Dish) from fair trade production | 6.0 | (Dish) for light nutrition | 5.0 |
| (Dish) from fair agriculture | 5.0 | (Dish) with reduced calories | 1.0 |
| (Dish) produced according to social standards | 3.5 | Missing | 5.5 |
| (Dish) produced according to ethical standards | 1.0 | | |
| Missing | 5.0 | | |

Note: For simplicity, this table bears the neutral "Dish" designation. In this evaluation, however, all votes across all five dishes are included. DNLs translated from German.

## 2.2. *Study I—Results*

Table 4 displays the results of Study I. For the traditional DNL category, the descriptive name "(Dish), traditional style" prevailed, with 63.5 percent of all votes. The choice results in the regional DNL category showed the DNL "(Dish) from regional agriculture" as the winner, with 28.5 percent of all votes. This DNL was modified for fish dishes, where the label "(Dish) from regional fishery" was used. Within the seasonal DNL category, the respondents clearly preferred the descriptive name "(Dish) with seasonal ingredients", with 64 percent of all votes. As expected, the descriptive name "Organic (Dish)" earned 35.5 percent of all votes and was the clear winner in the organic DNL category. By a narrow margin, the descriptive name "(Dish) from sustainable agriculture" prevailed in the sustainable DNL category, with 22.5 percent of all votes. The equivalent "(Dish) from sustainable fisheries" was used for the fish dish. In the healthy DNL category, the descriptive names "(Dish), low-energy prepared" as well as "Low energy (Dish)" received the most approval, with 24.5 and 19 percent of the votes. The implementation of these DNLs in OOHC is problematic due to European health claim regulations [61]. Therefore, the option "(Dish) for light pleasure" with 17 percent of votes was used as a healthy DNL in the later studies. According to European health claim regulations, food may only be designated as low in energy if the product contains no more than 40 kcal per 100 g for solid goods and no more than 20 kcal per 100 ml for liquid goods [61]. These extremely low limit values cannot be complied with by the majority of OOHC meals. However, it is possible to use the

DNL "(Dish) for light pleasure" or "Light (Dish)" in the OOHC. Here, the regulations require the same conditions that are applied to "reduced" dishes. "Energy-reduced" dishes must have a reduced caloric value of at least 30 percent compared to the common preparation of the dish [61].

## 3. Study II

### 3.1. Study II—Hypotheses

The overall goal of study II was to use both the decoy effect and DNL nudges to evaluate their combined impact on food choices in canteens. Currently, no published articles or other scientific studies on the topic exist. To close this gap, we developed the following five hypotheses:

- H1: "Consumers prefer meals with DNLs."

To test this hypothesis, we assessed whether target dishes with the single use of descriptive name labels were chosen more often than target dishes without the use of this nudge. This was necessary to replicate the effect of DNLs, as stated in the literature [41–45], in our OOHC setting.

- H2: "Using the DE increases the choice frequency of the target dish."

For the second hypothesis, we tested the decoy effect on target dishes compared to target dishes without this nudge to illustrate the general impact of the decoy effect [35–38,50,62] and its applicability in the OOHC setting.

- H3: "Consumers prefer those dishes promoted via the combination of both nudges in comparison to their common dish counterparts."

Hypothesis 3 aimed to answer the general research question and illustrates the general impact of the combination of DNLs and the DE in OOHC settings. Therefore, target dishes with the applied nudge combination of DNLs and the DE were compared to common target dishes (target dishes without any nudges).

- H4: "The combination of DNLs and the DE leads to more sustainable choices than the isolated effects of the single nudges."

The fourth hypothesis tested the impact of the combination of the two effects on consumer choices in comparison to meals with only DNLs or only the DE. The goal of this hypothesis was to avoid the use of inefficient nudge variants in OOHC.

- H5: "There is a significant difference between the effectiveness of different DNL categories and commonly named dishes in fostering sustainable meal choices."

Recent studies showed a trend for regional-sourced, organic, as well as sustainable OOHC dishes in Germany [57]. The last hypothesis assessed which of the DNL categories are favoured over commonly named dishes. For this reason, we checked for significant differences between the DNLs and their common variants as well as between the single DNL categories.

From a theoretical point of view, we would expect both hypothesis 1 and hypothesis 2 to be supported. In this case, the literature of DNLs [41–45] and DE [35–38,40,50,62] would be consistent with the study results of this work on sustainable OOHC food. Based on the combination of the two nudges, we also expect to find support for hypothesis 3. Hypothesis 4 has never been tested before [24,26,27]. However, the sales and marketing effects stated in the literature on the pure DE [35–38,50,62] and the pure DNLs [41–45] could complement each other and further increase the sales of sustainable OOHC dishes. Therefore, we expect to find support for hypothesis 4. Concerning hypothesis 5, studies [57,63] reveal that, in OOHC, the buzzwords organic, regional and sustainable used for ingredients create profitable sales data when used in communications to end consumers. Studies show [64] that the majority of consumers do not distinguish between different sustainability labelling formats. Hence, it is an open question whether DNLs are accepted differently by consumers based on the sustainability dimension highlighted (regional, seasonal, etc.).

### 3.2. Study II—Design

Study II, a randomized choice experiment with factorial design, tested the best DNLs for each of the target dishes compared to the DE in a university canteen setting ($n = 420$). The analysis of this survey provided insights into the single effects of DNLs, single effects of the DE, and the combination of both.

The decoy dishes were intended to be less attractive to consumers. Attractiveness was assessed by using the proxy sales data as indicated by our experts (see Figure 1). The attractiveness had been assessed by nutrition experts with theoretical and practical backgrounds as well as using data from chefs and the kitchen directors at OOHC canteens.

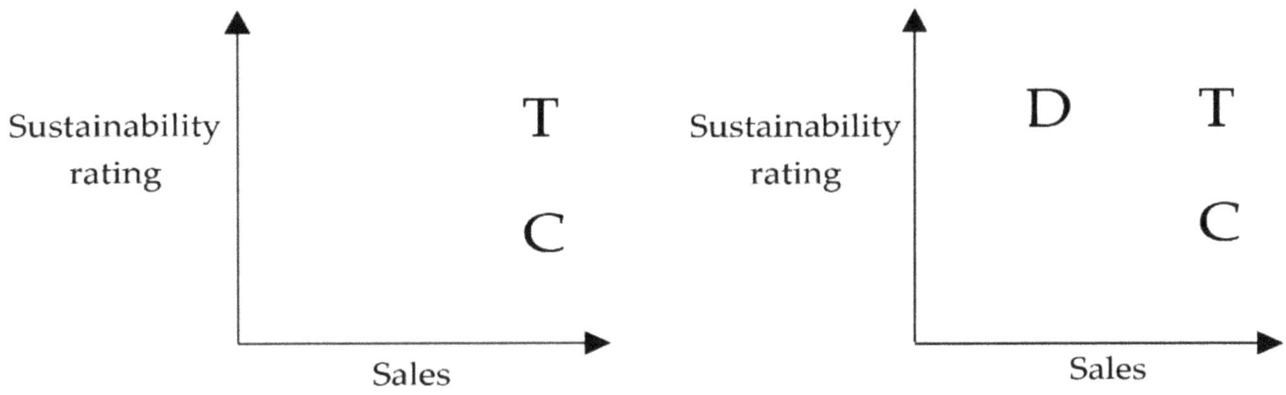


**Figure 1.** Exemplary illustration of the implementation of the decoy dishes. Note: Left: Without decoy dish/attraction effect, Right: With decoy dish/attraction effect. T = target dish, C = competitor dish, D = decoy dish. The figure based on [50].

The first of the decoy dishes mentioned in Table 5 was a carrot lasagne. This was inferior to the target vegetable lasagne, since the variety of ingredients was decreased to only one vegetable, the carrot. Experts have stated that this reduced variety negatively influences sales. The experts created the composition of the fish stew and noodles with pesto decoy dishes according to past sales data. Since the labels of these decoy dishes offer less information about the ingredients compared to the respective target dishes, sales data were lower. Chicken steak with celery puree was meant to be inferior due to the sales experience with celery puree in Germany (see Table 5).

**Table 5.** Target, Decoy and Competitor Dishes.

| Target Dishes | Decoy Dishes | Competitor Dishes |
|---|---|---|
| Vegetable lasagne | Carrot lasagne | Soy strips with noodles in mushroom sauce |
| Breaded fish with fried potatoes | Fish stew | Escalope chasseur with French fries |
| Spaghetti with rocket pesto | Noodles with pesto | Mustard eggs with mashed potatoes |
| Chicken steak with tagliatelle | Chicken steak with celery puree | Spaghetti Bolognese |

To test the five hypotheses, a complete factorial design to evaluate all variants and combinations of the nudges was used. For the DNLs, all six different name categories with their best descriptive names, determined in study I, were used, resulting in 14 choice sets for each of the four dishes and 56 choice sets overall (see Figure 2).

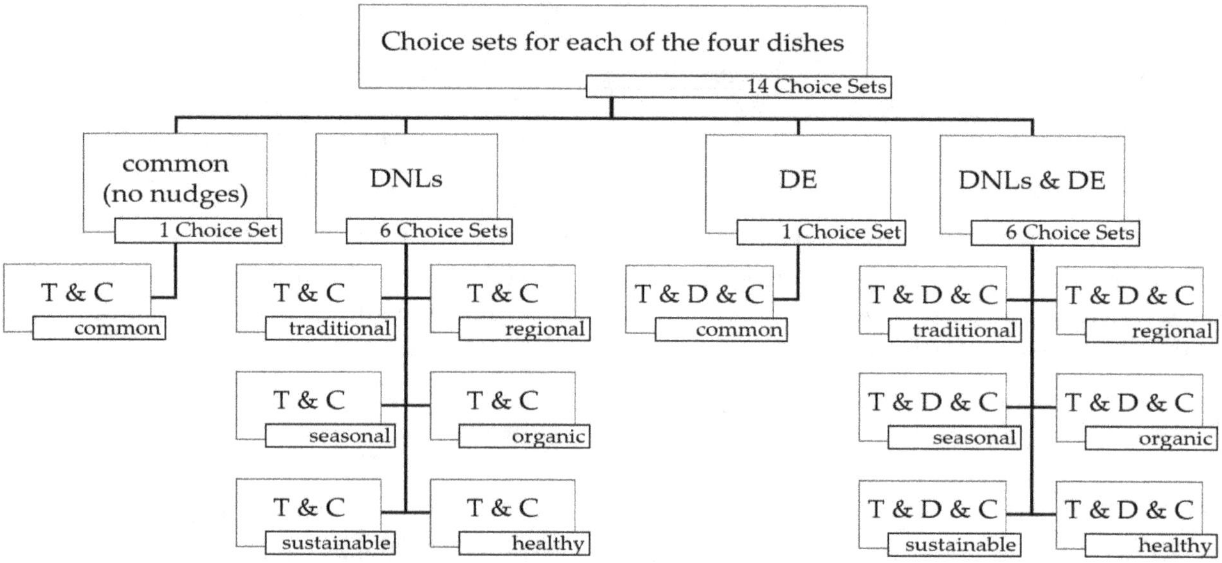


**Figure 2.** Factorial design to test the combination of DNLs and DE. Note: T = Target Dish (with DNLs), D = Decoy Dish, C = Competitor Dish.

To avoid priming effects within one dish variant with different nudges, each survey participant received only one choice set of the 14 available for each dish. This resulted in four randomized choice sets to answer. To secure compliance with the factorial design, the 14 choice sets were computer-generated and drawn out of urns, one urn for each dish, without putting them back into the urn. This procedure was repeated after every 14 respondents. In addition to the food choice variants offered in the questionnaire, the interviewees were given the possibility to choose the "opt-out option". For an exemplary choice set of the survey, see Figure A1 in the Appendix A.

Study II was carried out in several university canteens at two universities in Berlin via personal interview (on a tablet computer) in September and October 2016. A total of 420 people who had lunch in these canteens took part in the survey. These 420 people were personally interviewed inside the dining area after eating their lunch. The survey took about 2 min to complete. A small pre-test revealed that the guests were very reluctant to take part in the survey before eating their lunch and before visiting the university canteens due to time pressure, stress and hunger. This attempt was therefore omitted due to excessive time expenditure by interviewers and canteen visitors, some of whom felt harassed. All respondents completed the survey entirely.

To test the hypotheses, we used a linear regression in Stata. We used the voting scores of the target dishes (calculated in percentages) as the dependent variable. As independent factor variables (where each level of group is included as a separate covariate/dummy [65]), we used the dish variant (1 = vegetable lasagne, 2 = breaded fish with fried potatoes, 3 = spaghetti with rocket pesto, 4 = chicken steak with tagliatelle) and the nudging variant (1 = DNLs and DE, 2 = DNLs, 3 = DE, 4 = common) for hypotheses 1–4. For hypothesis 5, we used the dish variant and the descriptive name label variants (1 = traditional, 2 = regional, 3 = seasonal, 4 = organic, 5 = sustainable, 6 = healthy, 7 = common) to reveal possible differences between each of the single variants. Our results of the combination of DNLs and the DE (where the DE most likely had a strong negative influence) would influence the results used to answer hypothesis 5, if calculated in a joint model (14 choice sets per dish, see Figure 2). Therefore, we calculated a separate model for hypothesis 5 with a subset of data, where only pure DNLs and the base measurement without nudges were included (seven choice sets per dish, see Figure 2). The assumption checks of our two linear regression models were without conspicuities (no multicollinearity—mean variance inflation factors 2.39 and 1.64 [66]; normal distribution—skewness and kurtosis test (described by D'Agostino et al. [67] with the adjustment made by Royston [68]) $p$-value $p =$ 0.8950 and $p =$ 0.7817; no heteroscedasticity—White test $p$-values $p = 0.116$ and $p = 0.061$ [69]; no autocorrelation—Durbin–Watson-test d(7.56) = 2.08 and d(10.28) = 1.28 [70]).

### 3.3. Study II—Results

Over the entire survey, the common target dishes were chosen by participants about 40.0 percent of the time (see Figure 3). On average, target dishes with DNLs were about ten percent more favoured at 50.8 percent and were still ahead of their competitor dishes.

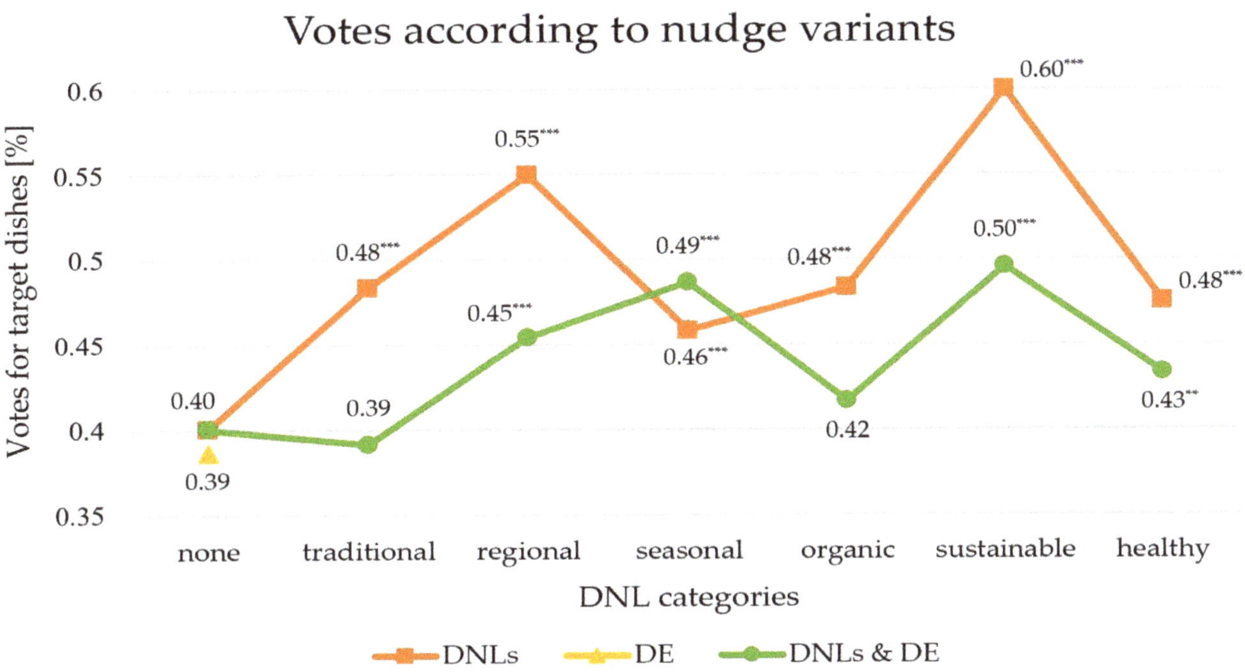


**Figure 3.** Descriptive results of Study II—DNLs, DE and DNLs and DE. Note: *** $p < 0.001$, ** $p < 0.01$, * $p < 0.05$ (pairwise $t$-test). The single decoy effect (DE) has only one data point in the "none" category.

Adding a decoy to the choice set decreased the choice frequency of the target dish. The target food with the DE was less popular than its common variants and led to a drop in overall votes. If the target dishes were combined with the DNLs and the DE, the result was a 44.6 percent choice frequency and an overall increase in votes compared to common target dishes. As seen in Figure 3, the strong effect of DNLs becomes clear in comparison to the other two nudging variants—the DE and the combination of both. By using the decoy effect, even the overall votes for target foods decreased and clearly improved in combination with DNLs, but this was solely due to the effect of the DNLs. From a descriptive point of view, the observations show that only the DNLs had a positive influence on choice behaviour, especially regional (55 percent) and sustainable (60 percent) DNLs. Figure 3 shows the difference in choice frequencies between pure DNLs, the DE and the combination of both for each DNL category.

Table 6 displays the results of the linear regression for hypotheses 1–4. The test for a positive significant difference in respondents' voting behaviour between DNLs and their common dish variants (β-coefficient $\beta = 0.108$; $p$-value $p \leq 0.001$), supports hypothesis 1. The use of descriptive names leads to significantly higher choices of the target dishes among the respondents.

Hypothesis 2, the test of the DE, must be rejected ($\beta = -0.012$; $p = 0.266$). Target dishes with a decoy were not chosen significantly more frequently than the target dishes without a decoy but even were "cannibalized" by their decoy dishes.

Supporting hypothesis 3, the combination of DNLs and the DE ($\beta = 0.046$; $p \leq 0.001$) did lead to significantly higher choice results for the target dishes with nudges in comparison to their common dish counterparts without.

**Table 6.** Linear Regression (dependent variable (DV): Target dish selection, independent variable (IV): Nudge variant, Dish variant).

| Target | Coefficient | Standard Error | $t$ | P > $t$ |
|---|---|---|---|---|
| Nudge | | | | |
| DNL | 0.108 | 0.009 | 12.69 | 0.000 |
| DNL and DE | 0.046 | 0.009 | 5.44 | 0.000 |
| DE | −0.012 | 0.011 | −1.11 | 0.266 |
| common | 0 | (base) | | |
| Dish | | | | |
| Spaghetti with rocket pesto | 0.198 | 0.006 | 33.18 | 0.000 |
| Vegetable lasagne | 0.169 | 0.006 | 28.41 | 0.000 |
| Chicken steak with tagliatelle | 0.067 | 0.006 | 11.21 | 0.000 |
| Breaded fish with fried potatoes | 0 | (base) | | |
| constant | 0.292 | 0.009 | 33.61 | 0.000 |
| Number of observations | | 1680 | | |
| $R$-squared | | 0.5175 | | |

As a result of the first three hypotheses, hypothesis 4 must also be rejected. The pure DE did not show any significant difference and the combination of DNLs and the DE had a lower impact on the target dishes than pure DNLs, as seen by the β-coefficients. The DNLs therefore had the highest positive influence on sustainable meal choices. Within study II, the combination of the two nudges is inferior to the use of pure DNLs.

Table 7 displays the results of the linear regression for hypothesis 5. The test of significant differences in the effectiveness of DNL categories supports hypothesis 5. All DNL categories had a significant positive influence on the meal choice of the target dishes compared to their common counterparts. Indeed, for the categories seasonal ($\beta = 0.058; p \leq 0.001$), healthy ($\beta = 0.075; p \leq 0.001$), organic ($\beta = 0.083; p \leq 0.001$) and traditional ($\beta = 0.083; p \leq 0.001$) there were significant differences in choice compared to target dishes without DNLs. However, for the regional name category ($\beta = 0.150;$ $p \leq 0.001$), as well as for the sustainable name category ($\beta = 0.200; p \leq 0.001$), these differences were considerably stronger. A pairwise $t$-test underlined the strong marketing effect of regional and sustainable DNLs, with their increase in choices of up to 20 percent compared to their common dish variants without the use of nudges. Table 8 illustrates the pairwise $t$-test, in which the use of regional and especially sustainable DNLs scored significantly better than the other DNL options.

**Table 7.** Linear Regression (DV: Target dish selection, IV: DNL variant, Dish variant).

| Target | Coef. | Std. Err. | $t$ | P > $t$ |
|---|---|---|---|---|
| DNL | | | | |
| sustainable | 0.200 | 0.009 | 21.70 | 0.000 |
| regional | 0.150 | 0.009 | 16.24 | 0.000 |
| traditional | 0.083 | 0.009 | 9.01 | 0.000 |
| organic | 0.083 | 0.009 | 9.01 | 0.000 |
| healthy | 0.075 | 0.009 | 8.12 | 0.000 |
| seasonal | 0.058 | 0.009 | 6.30 | 0.000 |
| common | 0 | (base) | | |
| Dish | | | | |
| Spaghetti with rocket pesto | 0.248 | 0.007 | 35.59 | 0.000 |
| Vegetable lasagne | 0.167 | 0.007 | 23.95 | 0.000 |
| Chicken steak with tagliatelle | 0.091 | 0.007 | 13.01 | 0.000 |
| Breaded fish with fried potatoes | 0 | (base) | | |
| constant | 0.274 | 0.008 | 35.21 | 0.000 |
| Number of observations | | | 840 | |
| $R$-squared | | | 0.7043 | |

**Table 8.** Pairwise *t*-test between the pure descriptive name labels (DNL) on choice behaviour.

| DNL | Contrast | Std. Err. | *t* | P > *t* |
|---|---|---|---|---|
| regional vs. traditional | 0.067 | 0.015 | 4.43 | 0.000 |
| seasonal vs. traditional | −0.025 | 0.015 | −1.66 | 0.096 |
| organic vs. traditional | 0.00 | 0.015 | 0.00 | 1.000 |
| sustainable vs. traditional | 0.117 | 0.015 | 7.77 | 0.000 |
| healthy vs. traditional | −0.008 | 0.015 | −0.55 | 0.583 |
| common vs. traditional | −0.083 | 0.015 | −5.53 | 0.000 |
| seasonal vs. regional | −0.092 | 0.015 | −6.09 | 0.000 |
| organic vs. regional | −0.067 | 0.015 | −4.43 | 0.000 |
| sustainable vs. regional | 0.050 | 0.015 | 3.35 | 0.001 |
| healthy vs. regional | −0.075 | 0.015 | −4.98 | 0.000 |
| common vs. regional | −0.150 | 0.015 | −9.95 | 0.000 |
| organic vs. seasonal | 0.025 | 0.015 | 1.66 | 0.096 |
| sustainable vs. seasonal | 0.142 | 0.015 | 9.44 | 0.000 |
| healthy vs. seasonal | 0.017 | 0.015 | 1.12 | 0.265 |
| common vs. seasonal | −0.058 | 0.015 | −3.86 | 0.000 |
| sustainable vs. organic | 0.117 | 0.015 | 7.77 | 0.000 |
| healthy vs. organic | −0.008 | 0.015 | −0.55 | 0.583 |
| common vs. organic | −0.083 | 0.015 | −5.53 | 0.000 |
| healthy vs. sustainable | −0.125 | 0.015 | −8.32 | 0.000 |
| common vs. sustainable | −0.200 | 0.015 | −13.30 | 0.000 |
| common vs. healthy | −0.075 | 0.015 | −4.98 | 0.000 |

Another important factor besides the evaluation of our five hypotheses was the influence of the respective target dishes on choice behaviour. As can be seen in both regression models (see Tables 6 and 7), lasagne and especially spaghetti worked better with their attached nudges compared to the chicken steak or the breaded fish. Therefore, it is important to not only check for nudging variants but also for the specific dish and nudge combination.

## 4. Study III

### 4.1. Study III—Design

During four weeks in October and November 2016, study III was conducted in a business canteen with four selected DNLs and DE choice sets from study II (see Table 9). One of these choice sets was put on tablet computers in the dining area of the business canteen each week. These tablet computers had already been used by the business canteen to select the special dish offered the following week, with no further questions asked about sociodemographic characteristics. With this tool, the respondents chose their preferred meal spontaneously without knowing about the ongoing study. During the four weeks, a total of 820 visitors took part in the tablet survey (for detailed respondent numbers per week, see Table 10). Due to delivery difficulties, in week 4 not all the ingredients of the dish but only tomatoes could be locally sourced for the business canteen. With a Mann–Whitney U Test [71] we checked for choice differences between the university canteen (study II) and the business canteen (study III) settings.

**Table 9.** Results of the DNLs and DE choice experiment in the business canteen.

| Results of the DNLs and DE Choice Experiment in the Business Canteen | | | |
|---|---|---|---|
| | **Target Dishes** | **Decoy Dishes** | **Competitor Dishes** |
| Week 1 | Vegetable lasagne traditional style | Carrot lasagne | Soy strips with noodles in mushroom sauce |
| | 51.1% | 30.1% | 18.8% |
| Week 2 | Breaded fish from sustainable fisheries with fried potatoes | Fish stew | Escalope chasseur with French fries |
| | 47.9% | 10.6% | 41.5% |
| Week 3 | Spaghetti with rocket pesto with seasonal ingredients | Noodles with pesto | Mustard eggs with mashed potatoes |
| | 34.1% | 22.4% | 43.5% |
| Week 4 | Chicken steak with tagliatelle and tomatoes from regional agriculture | Chicken steak with celery puree | Spaghetti Bolognese |
| | 46.1% | 19.7% | 34.2% |

**Table 10.** Mann–Whitney U Test for OOHC setting comparison of the combined DNLs and DE nudges.

| Target Dishes | OOHC Setting | Number of Observations | Rank Sum | z | Asymp. Sig. (2-Tailed) |
|---|---|---|---|---|---|
| Vegetable lasagne traditional style | University | 25 | 1935.5 | −0.286 | 0.775 |
| | Business | 133 | 10625.5 | | |
| Breaded fish from sustainable fisheries with fried potatoes | University | 21 | 3174.5 | 1.647 | 0.100 |
| | Business | 236 | 29978.5 | | |
| Spaghetti with rocket pesto with seasonal ingredients | University | 26 | 4046.0 | 2.740 | 0.006 |
| | Business | 223 | 27079.0 | | |
| Chicken steak with tagliatelle (and tomatoes) from regional agriculture | University | 21 | 3118.5 | 1.806 | 0.071 |
| | Business | 228 | 28006.5 | | |

### 4.2. *Study III—Results*

Table 9 displays the voting results of the business canteen guests. During week 1 (vegetable lasagne: 51.1%), week 2 (breaded fish: 47.9%) and week 4 (chicken steak: 46.1%) of our survey, the target dishes prevailed, while in week 3 the competitor dish was preferred (mustard eggs: 43.5%). Table 10 reveals that for vegetable lasagne (z = −0.286; $p$ = 0.775), there was no change in choice of the target food between settings. For breaded fish (z = 1.647; $p$ = 0.100) and chicken steak (z = 1.806; $p$ = 0.071) there are significant tendencies. Only for spaghetti (z = 2.740; $p$ = 0.006), significant differences between the two OOHC settings could be recorded. With only one significant change out of four target dishes, assumptions cannot be made as to the direct transferability of results from the university canteen in study II to the company canteen. As seen in previous studies, the results of nudging interventions are not only influenced by the type of nudge or the type of setting but also by other variables, e.g., the offered target dishes (see Study II), or the weather or the day of eating [34].

## 5. Discussion and Conclusions

The present research provides insights into the ability to increase the choice of sustainable dishes by combining nudges. Two nudges were examined: the use of different descriptive name labels (DNLs) that, according to the previous literature [41–45], increase choice frequency compared to respectively named dishes without DNLs, and the decoy effect [35–38,40,50,62], which in non–food contexts has been able to influence consumer choices in an intended direction. The feasibility and application of both nudges were evaluated in two different OOHC settings. Our studies reveal that descriptive name labels were able to influence choices positively insofar as they resulted in the choice of a more sustainable dish. Significant differences between the various tested descriptions indicate that the story told to guests/consumers matters. For example, a meal labelled as 'healthy' is less attractive than a 'regional' meal. Overall, it became obvious that the general term, 'sustainable', was most successful in influencing choices. Therefore, we conclude that consumers differentiate between the various sustainability dimensions but appreciate sustainable meals in general. Since we did not ask guests about their understanding of sustainability as a general concept or how it relates to food production and consumption, we cannot elaborate on the reasons for this finding.

On the basis of our results, we cannot recommend the joint use of DNLs and the DE as a tool for boosting sustainable food choices. We reject hypothesis 4, as the DE had no significant impact on choice frequency of target dishes and, in fact, the inclusion of the pure DE lowered the overall choice frequency of target dishes (hypothesis 2). It is clearly superior to apply single DNLs, especially the two most effective DNLs ('regional'; 'sustainable') as identified in our work (compare hypothesis 5). When comparing the best DNLs from the six name categories, all variants were able to significantly and positively influence respondents' choices. Study participants preferred the descriptions 'regional' (choice increase of around 15%) and 'sustainable' (choice increase of about 20%). It should be noted, however, that only one possible DNL per category was tested in this study. For the regional DNL category, this was the DNL "from regional agriculture/fisheries" and for the sustainable name category the tested DNL was "from sustainable agriculture/fisheries".

The negative effect of applying the DE to influence dish choice stands in contrast to the previous non-food literature [35–38,62]. Regarding the food sector and especially OOHC, a recently published online choice experiment study focused on the addition of a price decoy to a menu with the aim of promoting vegetarian target options. In this study by Attwood et al. [72], the vegetarian decoy was more expensive (up to 30%) than the less expensive vegetarian target dish. Results revealed that the addition of the decoy dish did not lead to a higher selection rate of the vegetarian target dish compared to the meat competitor option. One possible reason for the failure of the nudge in that study was the use of a price decoy strategy. Alternative decoy strategies which do not focus on price but on other attributes (e.g., menu description or caloric content) might work when displayed to consumers [72]. Even though sales data and sustainability ratings instead of price differences were used to define the decoy dishes in our study, the DE nudge did not work as expected. A possible reason is that consumers were not aware of e.g. the different sustainability ratings of the dishes as they were not indicated besides the dishes name. Another possible explanation for the non-significant finding of the decoy effect is that the meals assigned as competitor dishes, decoy dishes and target dishes by the project team and the experts might not have been perceived in these roles by canteen guests. For example, the experts had not expected canteen guests to prefer the decoy carrot lasagne over the target vegetable lasagne, which offered a larger variety of vegetables for the same price. The experts thought that variety seeking consumers would prefer a combination of vegetables over just one variety. Hereby, the experts might have overlooked that the complete information on vegetable ingredients in the case of the carrot lasagne simplified consumers' choice compared to a vegetable lasagne. The unknown combination of various vegetables might have contained some ingredients consumers might not like or hesitate to eat from, e.g., a health perspective. However, we did not employ questionnaires and so were unable to confirm why consumers' evaluations differed from the experts' expectations. Hence, the careful selection and grouping of the target and the competitor dishes is of fundamental relevance for the study results. The challenge researchers and practitioners face when designing study designs and menus is that, to date, meals and dishes have never been categorized as target or competitor dishes. Our decision to classify dishes in the competitor/target group was based on the experts' opinion and not on a systematic assessment of similarities and differences between dishes. Identifying consumers' perception of perceived similarity of various product characteristics and attribute levels is thus one important step future research has to take.

Based on these results, further research should also uncover and explore possible applications of the decoy effect in the field of food choice. In addition, the discussion of possible moderators of decoy effects and possible decoys should be continued. Such debates should take place in specialized contexts, e.g., in the context of sustainable food consumption or OOHC settings, since preferences often differ between several choice environments [34,73], as seen in study III.

It is important that nudges be applied situation-specifically. It is also advisable to monitor changes in guest behaviour when using new or unknown nudges in a specific setting [42,74]. It may happen, as study II suggests, that nudges do not always work in the desired direction, especially when several nudges are combined, and varying target dishes are used. Future research uncovering the interaction of nudges and different dish categories in OOHC is therefore recommended.

The preferences of OOHC consumers towards more regional and sustainable DNLs point to the wish for more regional food and overall sustainable ingredients in the product ranges of OOHC, which also corresponds to today's trend in Germany [57]. Needless to say, the respective food quality of products advertised with DNLs should be maintained, as a decline in quality would cause consumer expectations to inflate [41,42].

Catering managers, canteen and restaurant owners should follow this regional and sustainable trend, with the use of behavioural insights, nudges and further interventional approaches to promote relevant products and contribute to the transformation towards a more sustainable economy.

**Author Contributions:** Conceptualization, P.O. and N.L.; Data curation, P.O.; Formal analysis, P.O.; Funding acquisition, N.L.; Investigation, P.O. and N.L.; Methodology, P.O. and N.L.; Project administration, N.L.; Resources, N.L.; Software, P.O.; Supervision, N.L.; Writing—original draft, P.O.; Writing—review and editing, P.O. and N.L. All authors have read and agreed to the published version of the manuscript.

**Acknowledgments:** This research was carried out as part of the project "NAHGAST". The aim of the project NAHGAST is to initiate, support and disseminate transformation processes to sustainable management in OOHC. The authors would like to thank all researchers involved in the NAHGAST project, specifically Silke Friedrich, Petra Teitscheid, Christine Göbel, Holger Rohn, Tobias Engelmann, Melanie Speck and Katrin Bienge. We thank the nutrition experts and kitchen chefs for their support and advice in the design stage of the study. The authors gratefully acknowledge the advice and support given by the reviewers. We acknowledge support by the German Research Foundation and the Open Access Publication Fund of TU Berlin.

## Appendix A

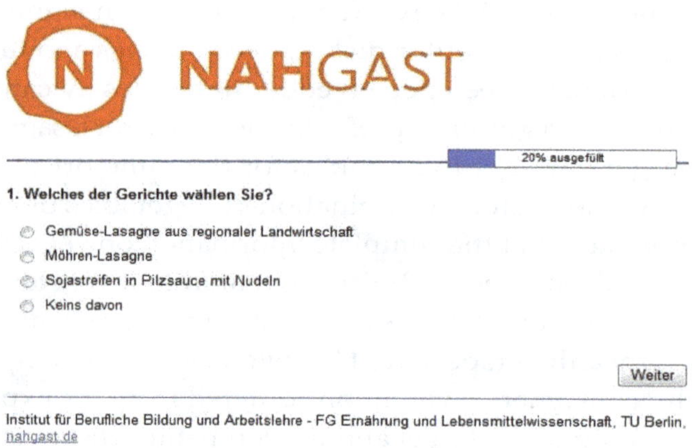


**Figure A1.** Exemplary choice set of Study II. Translation: Which of the dishes do you choose? Vegetable lasagne from regional agriculture vs. carrot lasagne vs. soy strips with noodles in mushroom sauce vs. none of them.

## References

1. BVE—Bundesvereinigung der Deutschen Ernährungsindustrie e.V. Jahresbericht 2015_2016. Available online: https://www.bve-online.de/presse/infothek/publikationen-jahresbericht/jahresbericht-2016 (accessed on 8 April 2020).
2. BVE—Bundesvereinigung der Deutschen Ernährungsindustrie e.V. Jahresbericht 2016_2017. Available online: https://www.bve-online.de/presse/infothek/publikationen-jahresbericht/jahresbericht-2017 (accessed on 8 April 2020).
3. BVE—Bundesvereinigung der Deutschen Ernährungsindustrie e.V. Jahresbericht 2017_2018. Available online: https://www.bve-online.de/presse/infothek/publikationen-jahresbericht/jahresbericht-2018 (accessed on 8 April 2020).
4. Rückert-John, J. Zukunftsfähigkeit der Ernährung außer Haus. In *Nachhaltigkeit und Ernährung. Produktion–Handel–Konsum*; Brunner, K.–M., Schönberger, G.U., Eds.; Campus–Verlag: Frankfurt am Main, Germany, 2005; pp. 240–262.
5. Mancino, L.; Newman, C. *Who has Time to Cook? How Family Resources Influence Food Preparation*; DIANE Publishing: Darby, PA, USA, 2007.
6. Statista Veränderung der Anzahl der Mahlzeiten, die zu Hause eingenommen werden, in Deutschland in den Jahren 2005 bis 2015 (in Millionen Mahlzeiten). Available online: https://de.statista.com/statistik/daten/studie/442026/umfrage/veraenderung$-$der$-$anzahl$-$der$-$mahlzeiten$-$in$-$deutschland/ (accessed on 8 April 2020).
7. United Nations. *2018 Revision of World Urbanization Prospects*; United Nations Department of Economic and Social Affairs: New York, NY, USA, 2018.
8. Bundesministerium für Ernährung und Landwirtschaft (BMEL) Deutschland, wie es isst–Der BMEL–Ernährungsreport 2019. Available online: https://www.bmel.de/SharedDocs/Downloads/Broschueren/Ernaehrungsreport2019.pdf?__blob=publicationFile (accessed on 9 April 2020).
9. Bundesministerium für Umwelt, Naturschutz, Bau und Reaktorsicherheit (BMUB) and Umweltbundesamt (UBA) Umweltbewusstsein in Deutschland 2016. Ergebnisse einer repräsentativen Bevölkerungsumfrage. Available online: https://www.umweltbundesamt.de/sites/default/files/medien/376/publikationen/umweltbewusstsein_deutschland_2016_bf.pdf (accessed on 8 April 2020).
10. Reisch, L.; Eberle, U.; Lorek, S. Sustainable food consumption: An overview of contemporary issues and policies. *Sustain. Sci. Pract. Policy* **2013**, *9*, 7–25. [CrossRef]
11. NCD Risk Factor Collaboration. Trends in adult body–mass index in 200 countries from 1975 to 2014: A pooled analysis of 1698 population–based measurement studies with 19 2 million participants. *Lancet* **2016**, *387*, 1377–1396. [CrossRef]
12. Organisation for Economic Co–operation and Development (OECD). Obesity Update 2017. Available online: http://www.oecd.org/health/health$-$systems/Obesity$-$Update$-$2017.pdf (accessed on 8 April 2020).
13. World Health Organization (WHO) Better Food and Nutrition in Europe: A Progress Report Monitoring Policy Implementation in the WHO European Region. Available online: www.euro.who.int/__data/assets/pdf_file/0005/355973/ENP_eng.pdf?ua=1 (accessed on 7 April 2020).
14. Cutler, D.M.; Glaeser, E.L.; Shapiro, J.M. Why have Americans become more obese? *J. Econ. Perspect.* **2003**, *17*, 93–118. [CrossRef]
15. DellaVigna, S.; Malmendier, U. Paying not to go to the gym. *Am. Econ. Rev.* **2006**, *96*, 694–719. [CrossRef]
16. Friedrichsen, J.; Hagen, K.; Wagner, L. Stupsen und Schubsen (Nudging): Ein neues verhaltensbasiertes Regulierungskonzept? *Vierteljahrshefte zur Wirtschaftsforschung/Q. J. Econ. Res.* **2018**, *87*, 5–13. [CrossRef]
17. Thaler, R.H.; Sunstein, C.R. *Nudge: Improving Decisions about Health, Wealth, and Happiness*; Penguin: London, UK, 2009.
18. Hallsworth, M.; List, J.A.; Metcalfe, R.D.; Vlaev, I. The behavioralist as tax collector: Using natural field experiments to enhance tax compliance. *J. Public Econ.* **2017**, *148*, 14–31. [CrossRef]
19. Johnson, E.J.; Goldstein, D. *Do Defaults Save Lives?* American Association for the Advancement of Science: Washington, DC, USA, 2003.
20. Lehner, M.; Mont, O.; Heiskanen, E. Nudging–A promising tool for sustainable consumption behaviour? *J. Clean. Prod.* **2016**, *134*, 166–177. [CrossRef]

21. Pichert, D.; Katsikopoulos, K.V. Green defaults: Information presentation and pro−environmental behaviour. *J. Environ. Psychol.* **2008**, *28*, 63–73. [CrossRef]
22. Schubert, C. Green nudges: Do they work? Are they ethical? *Ecol. Econ.* **2017**, *132*, 329–342. [CrossRef]
23. Camilleri, A.R.; Larrick, R.P. *Choice architecture. Emerging Trends in the Social and Behavioral Sciences: An Interdisciplinary, Searchable, and Linkable Resource*; John Wiley & Sons: Hoboken, NJ, USA, 2015; pp. 1–15.
24. Skov, L.R.; Lourenco, S.; Hansen, G.L.; Mikkelsen, B.E.; Schofield, C. Choice architecture as a means to change eating behaviour in self−service settings: A systematic review. *Obes. Rev.* **2013**, *14*, 187–196. [CrossRef]
25. Nørnberg, T.R.; Houlby, L.; Skov, L.R.; Peréz−Cueto, F.J.A. Choice architecture interventions for increased vegetable intake and behaviour change in a school setting: A systematic review. *Perspect. Public Health* **2016**, *136*, 132–142. [CrossRef] [PubMed]
26. Wilson, A.L.; Buckley, E.; Buckley, J.D.; Bogomolova, S. Nudging healthier food and beverage choices through salience and priming. Evidence from a systematic review. *Food Qual. Prefer.* **2016**, *51*, 47–64. [CrossRef]
27. Bucher, T.; Collins, C.; Rollo, M.E.; McCaffrey, T.A.; De Vlieger, N.; Van der Bend, D.; Truby, H.; Perez−Cueto, F.J. Nudging consumers towards healthier choices: A systematic review of positional influences on food choice. *Br. J. Nutr.* **2016**, *115*, 2252–2263. [CrossRef] [PubMed]
28. Broers, V.J.; De Breucker, C.; Van den Broucke, S.; Luminet, O. A systematic review and meta−analysis of the effectiveness of nudging to increase fruit and vegetable choice. *Eur. J. Public Health* **2017**, *27*, 912–920. [CrossRef]
29. Arno, A.; Thomas, S. The efficacy of nudge theory strategies in influencing adult dietary behaviour: A systematic review and meta−analysis. *BMC Public Health* **2016**, *16*, 676. [CrossRef]
30. Sunstein, C.R. The ethics of choice architecture. In *Choice Architecture in Democracies*; Nomos Verlagsgesellschaft mbH & Co. KG: Munich, Germany, 2016; pp. 19–74.
31. Sunstein, C.R.; Reisch, L.A. *Trusting Nudges: Toward a Bill of Rights for Nudging*; Routledge: Abingdon, UK, 2019.
32. Thaler, R.H. The power of nudges, for good and bad. *New York Times* **2015**, *31*, 2015.
33. Lorenz, B.A.; Langen, N. Determinants of how individuals choose, eat and waste: Providing common ground to enhance sustainable food consumption out−of−home. *Int. J. Consum. Stud.* **2018**, *42*, 35–75. [CrossRef]
34. Ohlhausen, P.; Langen, N.; Friedrich, S.; Speck, M.; Bienge, K.; Engelmann, T.; Rohn, H.; Teitscheid, P. Auf der Suche nach dem wirksamsten Nudge zur Absatzsteigerung nachhaltiger Speisen in der Außer−Haus−Gastronomie. *Vierteljahrshefte zur Wirtschaftsforschung* **2018**, *87*, 95–108. [CrossRef]
35. Simonson, I. Choice based on reasons: The case of attraction and compromise effects. *J. Consum. Res.* **1989**, *16*, 158–174. [CrossRef]
36. Simonson, I.; Tversky, A. Choice in context: Tradeoff contrast and extremeness aversion. *J. Mark. Res.* **1992**, *29*, 281–295. [CrossRef]
37. Doyle, J.R.; O'Connor, D.J.; Reynolds, G.M.; Bottomley, P.A. The robustness of the asymmetrically dominated effect: Buying frames, phantom alternatives, and in−store purchases. *Psychol. Mark.* **1999**, *16*, 225–243. [CrossRef]
38. Masicampo, E.J.; Baumeister, R.F. Toward a physiology of dual−process reasoning and judgment: Lemonade, willpower, and expensive rule−based analysis. *Psychol. Sci.* **2008**, *19*, 255–260. [CrossRef] [PubMed]
39. Momsen, K.; Stoerk, T. From intention to action: Can nudges help consumers to choose renewable energy? *Energy Policy* **2014**, *74*, 376–382. [CrossRef]
40. Lichters, M.; Bengart, P.; Sarstedt, M.; Vogt, B. What really matters in attraction effect research: When choices have economic consequences. *Mark. Lett.* **2017**, *28*, 127–138. [CrossRef]
41. Wansink, B.; Painter, J.; van Ittersum, K. How descriptive menu labels influence attitudes and repatronage. *ACR N. Am. Adv.* **2002**, *29*, 1.
42. Wansink, B.; Painter, J.; Ittersum, K.V. Descriptive menu labels' effect on sales. *Cornell Hotel Restaur. Adm. Q.* **2001**, *42*, 68–72.
43. Wansink, B.; Van Ittersum, K.; Painter, J.E. How descriptive food names bias sensory perceptions in restaurants. *Food Qual. Prefer.* **2005**, *16*, 393–400. [CrossRef]
44. Morizet, D.; Depezay, L.; Combris, P.; Picard, D.; Giboreau, A. Effect of labeling on new vegetable dish acceptance in preadolescent children. *Appetite* **2012**, *59*, 399–402. [CrossRef]
45. Okamoto, M.; Wada, Y.; Yamaguchi, Y.; Kimura, A.; Dan, H.; Masuda, T.; Singh, A.K.; Clowney, L.; Dan, I. Influences of food−name labels on perceived tastes. *Chem. Senses* **2008**, *34*, 187–194. [CrossRef]

46. Miller, D.L.; Castellanos, V.H.; Shide, D.J.; Peters, J.C.; Rolls, B.J. Effect of fat−free potato chips with and without nutrition labels on fat and energy intakes. *Am. J. Clin. Nutr.* **1998**, *68*, 282–290. [CrossRef]
47. Miller, L.M.S.; Cassady, D.L. The effects of nutrition knowledge on food label use. A review of the literature. *Appetite* **2015**, *92*, 207–216. [CrossRef] [PubMed]
48. Grunert, K.G.; Wills, J.M. A review of European research on consumer response to nutrition information on food labels. *J. Public Health* **2007**, *15*, 385–399. [CrossRef]
49. Cowburn, G.; Stockley, L. Consumer understanding and use of nutrition labelling: A systematic review. *Public Health Nutr.* **2005**, *8*, 21–28. [CrossRef] [PubMed]
50. Huber, J.; Payne, J.W.; Puto, C. Adding asymmetrically dominated alternatives: Violations of regularity and the similarity hypothesis. *J. Consum. Res.* **1982**, *9*, 90–98. [CrossRef]
51. Devetag, M.G. From utilities to mental models: A critical survey on decision rules and cognition in consumer choice. *Ind. Corp. Chang.* **1999**, *8*, 289–351. [CrossRef]
52. Ratneshwar, S.; Shocker, A.D.; Stewart, D.W. Toward understanding the attraction effect: The implications of product stimulus meaningfulness and familiarity. *J. Consum. Res.* **1987**, *13*, 520–533. [CrossRef]
53. Carroll, R.; Vallen, B. Compromise and attraction effects in food choice. *Int. J. Consum. Stud.* **2014**, *38*, 636–641. [CrossRef]
54. Engelmann, T.; Speck, M.; Rohn, H.; Bienge, K.; Langen, N.; Howell, E.; Göbel, C.; Friedrich, S.; Teitscheid, P.; Bowry, J. Sustainability assessment of out−of−home meals: Potentials and challenges of applying the indicator sets NAHGAST meal−basic and NAHGAST meal−pro. *Sustainability* **2018**, *10*, 562. [CrossRef]
55. Langen, N.; Ohlhausen, P. Design und Bewertung von Interventionskonzepten zur Förderung einer nachhaltigeren Ernährung in der Außer−Haus−Gastronomie am Beispiel NAHGAST. In *Nachhaltig Außer−Haus Essen*; Teitscheid, P., Langen, N., Speck, M., Rohn, H., Eds.; Oekom Verlag: München, Germany, 2018; pp. 292–304.
56. Irrgang, W. Internorga GV−Barometer 2016. Das Innovations- und Investitionsklima in der Gemeinschaftsverpflegung. Available online: https://www.internorga.com/fileadmin/internorga/2016/pdf/in16_gv-barometer.pdf (accessed on 8 April 2020).
57. Irrgang, W. Internorga GV-Barometer 2018. Das Innovations- und Investitionsklima in der Gemeinschaftsgastronomie. Available online: https://www.internorga.com/fileadmin/internorga/2018/pdf/in18_gv-barometer.pdf (accessed on 8 April 2020).
58. Gremmer, P.; Hempel, C.; Hamm, U.; Busch, C. Zielkonflikt beim Lebensmitteleinkauf: Konventionell regional, ökologisch regional oder ökologisch aus entfernteren Regionen. Available online: https://www.orgprints.org/30487/ (accessed on 8 April 2020).
59. Grunert, K.G.; Hieke, S.; Wills, J. Sustainability labels on food products: Consumer motivation, understanding and use. *Food Policy* **2014**, *44*, 177–189. [CrossRef]
60. Drichoutis, A.C.; Lazaridis, P.; Nayga, R.M., Jr. Consumers' use of nutritional labels: A review of research studies and issues. *Acad. Mark. Sci. Rev.* **2006**, *2006*, 1.
61. Regulation (EC). No 1924/2006 of the European Parliament and of the Council of 20 December 2006 on nutrition and health claims made on foods. *Off. J. Eur. Union* **2007**, *12*, 3–18.
62. Ariely, D.; Wallsten, T.S. Seeking subjective dominance in multidimensional space: An explanation of the asymmetric dominance effect. *Organ. Behav. Hum. Decis. Process.* **1995**, *63*, 223–232. [CrossRef]
63. Fülles, M.; Roehl, R.; Strassner, C.; a'verdis; Hermann, A.; Teufel, J.; Ökoinstitut e.V. Mehr Bio in Kommunen. Ein Praxisleitfaden des Netzwerks deutscher Biostädte. Available online: https://www.biostaedte.de/images/pdf/leitfaden_V4_verlinkt.pdf (accessed on 8 April 2020).
64. Janßen, D.; Langen, N. The bunch of sustainability labels – Do consumers differentiate? *J. Clean. Prod.* **2017**, *143*, 1233–1245. [CrossRef]
65. StataCorp LP. *Stata User's Guide*; StataCorp LP: College Station, TX, USA, 2013; ISBN 978-1−59718−115−0.
66. O'brien, R.M. A caution regarding rules of thumb for variance inflation factors. *Qual. Quant.* **2007**, *41*, 673–690. [CrossRef]
67. D'agostino, R.B.; Belanger, A.; D'Agostino, R.B., Jr. A suggestion for using powerful and informative tests of normality. *Am. Stat.* **1990**, *44*, 316–321.
68. Royston, P. Tests for departure from normality. *Stata Tech. Bull.* **1992**, *1*, 2.
69. White, H. A heteroskedasticity−consistent covariance matrix estimator and a direct test for heteroskedasticity. *Econom. J. Econom. Soc.* **1980**, 817–838. [CrossRef]

70. Tillman, J.A. The power of the Durbin–Watson test. *Econom. J. Econom. Soc.* **1975**, 959–974. [CrossRef]
71. Nachar, N. The Mann–Whitney U: A test for assessing whether two independent samples come from the same distribution. *Tutor. Quant. Methods Psychol.* **2008**, *4*, 13–20. [CrossRef]
72. Attwood, S.; Chesworth, S.J.; Parkin, B.L. Menu engineering to encourage sustainable food choices when dining out: An online trial of priced–based decoys. *Appetite* **2020**, *149*, 104601. [CrossRef]
73. Dhar, R.; Novemsky, N. Beyond rationality: The content of preferences. *J. Consum. Psychol.* **2008**, *18*, 175–178. [CrossRef]
74. Winkler, G.; Berger, B.; Filipiak-Pittroff, B. Small changes in choice architecture in self–service cafeterias. Do they nudge consumers towards healthier food choices? *Ernährungs Umsch.* **2018**, 170–178. [CrossRef]

6

# Individual Differences in Sweetness Ratings and Cross-Modal Aroma-Taste Interactions

**Anne Sjoerup Bertelsen, Line Ahm Mielby, Niki Alexi, Derek Victor Byrne and Ulla Kidmose** *

Department of Food Science, Faculty of Technical Sciences, Aarhus University, 8200 Aarhus N, Denmark; annesbertelsen@food.au.dk (A.S.B.); lineh.mielby@food.au.dk (L.A.M.); niki.alexi@food.au.dk (N.A.); derekv.byrne@food.au.dk (D.V.B.)

* Correspondence: ulla.kidmose@food.au.dk.

**Abstract:** Aroma-taste interactions, which are believed to occur due to previous coexposure (concurrent presence of aroma and taste), have been suggested as a strategy to aid sugar reduction in food and beverages. However, coexposures might be influenced by individual differences. We therefore hypothesized that aroma-taste interactions vary across individuals. The present study investigated how individual differences (gender, age, and sweet liker status) influenced the effect of aroma on sweetness intensity among young adults. An initial screening of five aromas, all congruent with sweet taste, for their sweetness enhancing effect was carried out using descriptive analysis. Among the aromas tested, vanilla was found most promising for its sweet enhancing effects and was therefore tested across three sucrose concentrations by 129 young adults. Among the subjects tested, females were found to be more susceptible to the sweetness enhancing effect of vanilla aroma than males. For males, the addition of vanilla aroma increased the sweet taste ratings significantly for the 22–25-year-olds, but not the 19–21-year-olds. Consumers were clustered according to their sweet liker status based on their liking for the samples. Although sweet taste ratings were found to vary with the sweet liker status, aroma enhanced the sweetness ratings similarly across clusters. These results call for more targeted product development in order to aid sugar reduction.

**Keywords:** sugar reduction; sweet; vanilla; consumers; age; gender; sweet liker status; young adults

## 1. Introduction

Excessive sugar intake has contributed to the prevalence of obesity and associated lifestyle diseases, such as type 2 diabetes [1,2]. Reduction of the content of sugar in food and beverages is therefore of high importance for both society and industry. Previously, non-nutritive sweeteners have been used extensively for sugar reduction purposes in the food and beverage industries [3]. However, many non-nutritive sweeteners have a negative effect on consumers' acceptability of the products, as they can have off-flavors and a slow onset or a lingering of the sweet taste [3,4]. As eating and drinking are multisensory experiences, cross-modal interactions have been suggested as an alternative strategy in the reformulation of products with the aim of reducing the sugar content [5–8]. The addition of aroma has, for example, been shown to increase sweetness intensity in many studies [5,9–15] and has therefore been suggested as a tool to aid sugar reduction. However, the effect of aroma on sweetness intensity is dependent on several factors, such as tastant concentration [14,16–19], aroma concentration [5,14,16,20,21], and the food matrix [5,7]. In addition, the type of aroma used has also been found to affect the ability to enhance sweetness intensity [5,9,13,15,22,23]. A careful screening of aromas is therefore of high importance in the development of sugar reduced products.

The extent to which an aroma is able to enhance sweetness intensity has been suggested to depend on whether the pairing of stimuli is congruent or not. Schifferstein and Verlegh [16] defined congruency as "the extent to which two stimuli are appropriate for combination in a food product". While some authors have found that congruency is necessary for aromas to enhance sweetness intensity [16,17,22], Wang et al. [15] found that both a congruent (vanilla) and an incongruent (beef) aroma could enhance sweetness intensity in sweetened milk. However, the sweetness enhancement effect was shown to be higher for the congruent compared to the incongruent aroma [15]. The majority of aromas that have been found to enhance sweet taste intensity are thus congruent with food products high in sweetness intensity, such as fruity or vanilla aromas. Sweetness enhancement by aroma is accordingly associated previous coexposures whereby specific combinations of aromas with sweet taste are learned [21,24,25].

Individual differences, such as age or gender, might possibly affect the sweetness enhancement ability of aromas due to either physiological differences or differences in consumption behavior between consumer groups. Doty and Cameron [26] suggested that for at least some aromas, females are more sensitive than males regarding odor detection, identification, discrimination, and memory. Other studies suggest that sweetness sensitivity decreases with age [27–29]. Differences in sensitivity might possibly change the perception and preference for food products. Indeed, older consumers with lower sensitivity for various stimuli were found to prefer higher intensity of sweetness than consumers with higher sensitivity [30]. Males and females have also been found to differ when it comes to their preferences for sweet taste and related products [31,32]. This might lead to differences in consumption behavior. Indeed, individuals with a higher preference for sweet taste have been found to have a higher intake of sweet beverages than individuals with a lower preference for sweet taste [33,34]. Different consumption behaviors might affect coexposures, and as explained previously, therefore also affect the abilities of aromas to enhance sweetness intensity. However, few studies have actually investigated the effect of individual differences on cross-modal interactions. Proserpio et al. [35] recently investigated how BMI, gender, and butter aroma in two different concentrations affected sweet taste intensity in custard desserts. For obese individuals, they found that females experienced a cross-modal effect of aroma on sweetness intensity for both aroma concentrations tested, while males only experienced this for the highest aroma concentration tested. However, for the normal weight individuals, no differences in ratings of sweetness intensity between genders were observed. Therefore, for obese individuals, females appeared to be more susceptible to the cross-modal effect than males. Moreover, Philipsen et al. [36] investigated the effect of age and aroma on sweetness intensity. They did not find a difference in the effect of cherry flavor on sweetness intensity between the group aged 18–22 years versus the group aged 60–75 years. Lavin and Lawless [37] similarly investigated the effect of age and vanilla flavor on sweet taste intensity in milk, but among different age groups: 5–7, 8–10, 11–14, and 18–31-year-olds. The only significant difference between groups in relation to sweetness enhancement of vanilla flavor was a less pronounced enhancement in sweetness intensity for the 5–7-year-olds in comparison with the 8–10-year-olds. However, as mentioned above, cross-modal aroma-sweetness interactions depend on many factors such as the type of aroma, their concentrations, and the food matrix. As these factors might also interact with the individual differences, the results of the mentioned studies might therefore not be valid for aromas, concentrations, and matrices different from those tested in each study. To the authors' knowledge, no studies have investigated the effect of sweet liker status on the cross-modal effect of aroma on sweetness intensity. Further studies are therefore warranted.

In the present study, we hypothesized that individual differences can affect the cross-modal effect of aroma on sweetness intensity. The aim was therefore to investigate how individual differences (gender, age, and sweet liker status) influence the effect of a congruent aroma on the dependent variable sweetness intensity. Besides ratings of sweetness intensity, the intensity of other sensory attributes as well as liking were also measured in order to better understand the extent of smell-taste cross-modal interactions and to be able to categorize individuals according to sweet liker status. As particularly adolescents and young adults have been found to consume too much sugar and sugar sweetened beverages [38–41], the consumer group used for this study consisted of young adults.

## 2. Experimental Design

In order to select an aroma for the consumer study, the first experiment was a screening of five aromas (vanilla, honey, banana, raspberry, and elderflower) which were all congruent with sweetness in a Danish cultural context. As descriptive sensory analysis (DA) previously has shown to be a successful method to identify aroma-related sensory interactions [42], DA was used for the screening of aromas. Based on the results of the DA, vanilla aroma was chosen for the consumer study (Experiment 2). Figure 1 provides an overview of the samples included in the DA as well as in the consumer study.

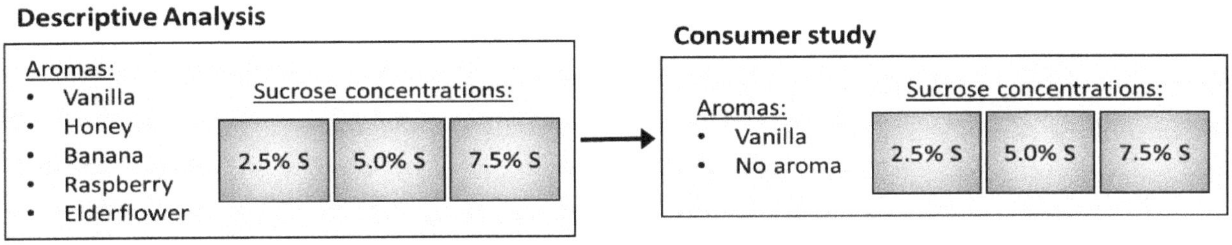


**Figure 1.** Overview of the samples in the descriptive analysis and the consumer study. % S = % *w/w* sucrose.

### 2.1. Samples

Fifteen samples were evaluated in the DA, while six samples were evaluated in the consumer study. Samples were prepared according to Table 1. All aromas were purchased from Bolsjehuset (Albertslund, Denmark) and the aroma concentrations (Table 1) were selected based on the recommendations from the supplier. As described in the introduction, the effect of aroma on sweetness intensity is dependent on the tastant concentration. The aromas were therefore combined with three different sucrose concentrations: low sucrose concentration (2.5% *w/w*), medium sucrose concentration (5% *w/w*), and high sucrose concentration (7.5% *w/w*) in aqueous blends based on local tap water. Sucrose was purchased from Merck KGaA (Darmstadt, DE) and due to the aim of sugar reduction, the sucrose concentrations were chosen to be lower than the 10% sugar, which is normally present in many beverages [43]. The sucrose concentrations were additionally pilot tested to give perceivable differences in sweetness intensity.

**Table 1.** Aroma concentrations in the samples and overview of what samples were included in each experiment. DA = descriptive analysis.

| Included in | Aroma Type | Sucrose Concentration (% *w/w*) | | |
|---|---|---|---|---|
| | | 2.5 | 5 | 7.5 |
| DA | Banana | 1.0 mL/kg | 1.0 mL/kg | 1.0 mL/kg |
| DA | Elderflower | 1.0 mL/kg | 1.0 mL/kg | 1.0 mL/kg |
| DA | Honey | 0.5 mL/kg | 0.5 mL/kg | 0.5 mL/kg |
| DA | Raspberry | 1.0 mL/kg | 1.0 mL/kg | 1.0 mL/kg |
| DA + consumer study | Vanilla | 1.0 mL/kg | 1.0 mL/kg | 1.0 mL/kg |
| Consumer study | No aroma | - | - | - |

Samples were prepared by first mixing sucrose with water. When the sucrose was dissolved, aroma was added according to Table 1. Samples were then poured into opaque sample tubes with red lids (Fisher Scientific, Roskilde, Denmark) with 20 mL in each. Prior to evaluation, samples were stored in 5 °C for 1–3 days. In both the DA and the consumer study, samples were coded with three-digit numbers and served at room temperature.

## 3. Experiment 1—Screening of Aromas

The objective of the first experiment was to screen aromas for their enhancing effect on sweetness intensity, in order to select the most effective one for the consumer study.

### 3.1. Method

The DA was done by an experienced trained sensory panel in the sensory lab at the Department of Food Science, Aarhus University, Årslev, Denmark. The panel consisted of nine women, with an age range of 22–60 years, who all gave informed verbal consent prior to participation. The DA began with an introductory discussion, followed by training and the final evaluation of the samples. In the introductory discussion, samples that were expected to span the relevant attributes were presented and discussed with the panel. During the discussion, the panel chose to continue with the following attributes (evaluated in the order shown): sweet aroma, sour aroma, intensity of aroma, sweet taste, sour taste, and intensity of flavor, as these attributes were able to describe the sensory differences within the samples presented. To allow for detection of differences in aroma and flavor attribute intensities, control samples without aroma were not included in the DA. In the training session, panelists evaluated a subgroup of the samples in triplicate.

The samples were evaluated in triplicates during the final evaluation, which took place over two consecutive days. For each panelist, samples were presented according to a Williams Latin Square design to limit carry over effects [44]. Attribute intensities were evaluated on 15 cm line scales with two anchor points placed at 1.5 and 13.5 cm of the scale, indicating low and high intensity, respectively. Data was recorded directly in the software Compusense (Compusense Inc., Ontario, Canada). A 30-s break was enforced between samples during which panelists were instructed to cleanse their palate using water, tepid very weak white tea, and/or crackers.

### 3.2. Statistical Analyses

For the DA, analysis of variance (ANOVA) and Tukey's HSD multiple comparison tests with $p \leq 0.05$ as significance level were performed for each attribute. This was done using XLSTAT (2018.6) (Addinsoft, NY, USA) [45]. Sucrose, aroma, and their interaction were treated as fixed effects, while panelists, replicates, and interactions were treated as random effects. None of the three-way interactions were significant and all models were therefore reduced to only include two-way interactions. To compare differences in sweetness intensity between aromas, ANOVAs were run with sucrose concentration as a fixed effect and panelists as a random effect. In order to visualize how the attributes and samples were correlated, Principal Component Analysis (PCA) of the mean centered average results from the DA was also performed with XLSTAT.

### 3.3. Results

A significant interaction between aroma and sucrose was found for the attributes sweet ($F_{(8,304)} = 2.02$, $p = 0.044$) and sour taste ($F_{(8,304)} = 2.71$, $p = 0.007$) (Table 2). This demonstrated that the effects of aroma and sucrose were interdependent for the rating of both sweet and sour taste intensity. Different cross-modal interactions were thus identified: aroma affected ratings of taste intensity, and sucrose affected the rating of retro-nasal aroma intensity (flavor intensity ($F_{(2,304)} = 17.52$, $p = < 0.0001$)). Samples with added vanilla aroma were in all sucrose concentrations rated the most sweet, while samples with added elderflower aroma were generally rated as the least sweet. Sucrose concentration significantly affected the enhancing effect of vanilla aroma compared to elderflower aroma on ratings of sweet taste intensity ($F_{(2,16)} = 6.17$, $p = 0.010$). The difference in sweet taste scores was significantly lower for the high sucrose concentration (difference = 4.5) than for the medium (difference = 6.6, $p$ = 0.011) and low sucrose concentrations (difference = 6.2, $p = 0.047$). These results indicate generally higher cross-modal effects in the medium and low sucrose concentrations when compared to the high sucrose concentration.

**Table 2.** Means ± standard (std.) errors and $p$-values for each design factor and attribute included in the DA. * indicates interaction. Significant effects are marked in bold. For sweet and sour taste, significant differences found with Tukey's comparison test are indicated with letters. Samples sharing a letter in each column are not significantly different.

| | | Sweet Aroma Ratings | Sour Aroma Ratings | Intensity of Aroma Ratings | Sweet Taste Ratings | | Sour Taste Ratings | | Intensity of Flavor Ratings |
|---|---|---|---|---|---|---|---|---|---|
| Sucrose | | $p = 0.825$ | $p = 0.920$ | $p = 0.425$ | **$p < 0.0001$** | | **$p < 0.0001$** | | **$p < 0.0001$** |
| Aroma | | **$p < 0.0001$** | **$p < 0.0001$** | **$p < 0.0001$** | **$p < 0.0001$** | | **$p < 0.0001$** | | **$p < 0.0001$** |
| Sucrose * Aroma | | $p = 0.969$ | $p = 0.488$ | $p = 0.715$ | **$p = 0.044$** | | **$p = 0.007$** | | $p = 0.246$ |
| 2.5% sucrose | Vanilla | 12.28 ± 0.61 | 1.48 ± 0.48 | 10.30 ± 0.97 | 8.46 ± 0.86 | de | 2.73 ± 0.80 | ab | 7.49 ± 0.88 |
| | Honey | 11.21 ± 0.69 | 1.66 ± 0.32 | 11.13 ± 0.64 | 7.04 ± 0.94 | cd | 3.21 ± 0.79 | ab | 7.56 ± 0.92 |
| | Banana | 10.75 ± 0.80 | 5.10 ± 0.86 | 11.35 ± 0.68 | 6.19 ± 0.89 | bc | 4.50 ± 0.80 | bc | 6.91 ± 0.81 |
| | Raspberry | 4.04 ± 0.62 | 11.67 ± 0.69 | 4.91 ± 0.72 | 2.93 ± 0.50 | a | 11.33 ± 0.62 | e | 2.56 ± 0.44 |
| | Elderflower | 2.38 ± 0.36 | 12.43 ± 0.44 | 3.61 ± 0.55 | 2.31 ± 0.43 | a | 10.65 ± 0.81 | de | 1.99 ± 0.33 |
| 5.0% sucrose | Vanilla | 12.58 ± 0.52 | 1.34 ± 0.28 | 11.30 ± 0.76 | 11.71 ± 0.69 | gh | 1.39 ± 0.27 | a | 10.88 ± 0.71 |
| | Honey | 11.15 ± 0.72 | 2.60 ± 0.66 | 11.42 ± 0.55 | 11.21 ± 0.65 | fgh | 2.55 ± 0.60 | ab | 10.25 ± 0.71 |
| | Banana | 10.59 ± 0.79 | 5.69 ± 0.86 | 11.33 ± 0.61 | 10.26 ± 0.73 | efg | 4.53 ± 0.75 | bc | 9.21 ± 0.68 |
| | Raspberry | 4.59 ± 0.71 | 10.56 ± 0.89 | 5.07 ± 0.76 | 5.81 ± 0.83 | bc | 10.22 ± 0.82 | de | 3.74 ± 0.51 |
| | Elderflower | 2.57 ± 0.51 | 12.42 ± 0.43 | 4.17 ± 0.72 | 5.10 ± 0.69 | b | 8.92 ± 0.90 | d | 3.57 ± 0.55 |
| 7.5% sucrose | Vanilla | 12.59 ± 0.63 | 1.80 ± 0.55 | 11.46 ± 0.80 | 13.88 ± 0.29 | i | 1.12 ± 0.26 | a | 12.73 ± 0.38 |
| | Honey | 11.58 ± 0.58 | 1.84 ± 0.45 | 10.95 ± 0.64 | 13.10 ± 0.20 | hi | 1.88 ± 0.37 | a | 12.00 ± 0.45 |
| | Banana | 10.61 ± 0.83 | 5.60 ± 0.91 | 10.90 ± 0.66 | 12.86 ± 0.43 | hi | 4.67 ± 0.82 | bc | 11.24 ± 0.60 |
| | Raspberry | 3.99 ± 0.60 | 10.72 ± 0.71 | 5.46 ± 0.73 | 9.38 ± 0.79 | ef | 8.99 ± 0.88 | d | 6.98 ± 0.72 |
| | Elderflower | 2.17 ± 0.40 | 12.05 ± 0.62 | 3.16 ± 0.62 | 9.42 ± 0.81 | ef | 6.59 ± 0.96 | c | 5.32 ± 0.71 |

Figure 2 shows a PCA biplot of the results from the DA. In total, the first two Principle Components (PCs) explain 97.46% of the variance in the data. As shown, the attribute sweet aroma was negatively correlated to the attribute sour aroma (Pearson's correlation = −0.81, $p < 0.0001$); thereby creating a sweet-sour aroma dimension, that separated the samples into two groups. Samples with added vanilla, honey, and banana aromas formed one group that was characterized by an intense sweet taste, sweet aroma, high overall intensity of aroma and flavor. The other group was formed by the samples with added raspberry or elderflower aromas and was characterized by intense sour aroma and taste.

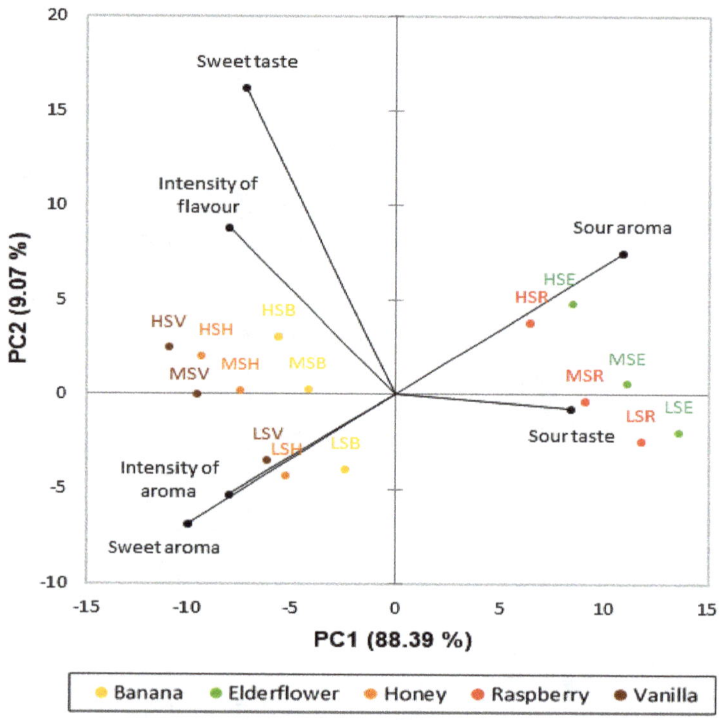


**Figure 2.** Principal Component Analysis biplot of the results from the DA. Attributes are marked in black while samples are colored according to aromas and coded according to sucrose concentration: LS = low sucrose concentration, MS = medium sucrose concentration, HS = high sucrose concentration, and aroma type: B = banana aroma, E = elderflower aroma, H = honey aroma, R = raspberry aroma, and V = vanilla aroma.

For each sucrose concentration, samples with added vanilla aroma were rated as having the highest intensities of sweet aroma (significant result, $p < 0.01$) and sweet taste (non-significant result, $p = 0.410$–$1.000$). Vanilla aroma was therefore found to be most promising in relation to sweet taste enhancement from a cross-modal effectiveness perspective. To follow up, the consumer study was set up to test the cross-modal effect of vanilla aroma across the same sucrose concentrations.

## 4. Experiment 2—Effect of Individual Differences in Consumers

The aim of the second experiment was to investigate the effect of individual differences (gender, age, and sweet liker status) on the enhancing effect of vanilla aroma on ratings of sweetness intensity. In order to better understand the extent of the aroma-taste interactions, results will also be presented for the attribute sweet aroma intensity.

### 4.1. Method

In order to recruit subjects from the relevant target group of young adults, the study was conducted in two different canteens at Aarhus University, Aarhus, Denmark. The study was carried out over two consecutive days and in total, 129 young adults (mean age 21.9 ± 1.5 years, 51 males) participated.

After the subjects had agreed to participate, they were served the six samples together with a paper questionnaire, a pen, and water for rinsing the palate between the samples. They were then given a short verbal introduction to the assignment and asked to match the three-digit numbers in the questionnaire with the three-digit numbers on the samples. For each sample, they rated the intensities of sweet aroma, sweet taste, and sour taste, as well as overall liking on nine-point scales anchored with "not at all" and "extremely" corresponding to point 1 and 9 on the scales, respectively. The liking scale also included an anchor in the middle named "neutral" corresponding to 5. To control for presentation bias, different versions of the questionnaires that differed in presentation orders were prepared and used. After participation, all subjects were offered small treats as compensation for their time. Prior to the actual consumer study, the procedure and questionnaire were pilot tested.

### *4.2. Statistical Analyses*

For the consumer study, consumers were divided into two age groups: 19–21-year-olds ($n = 56$, 20 males) and 22–25-year-olds ($n = 73$, 31 males). These groups were chosen as they gave the most equal and balanced groups in relation to total numbers of subjects as well as between genders. Using Minitab®19 (19.1.1) (Minitab, LLC, State College, PA, USA), mixed-model ANOVA and Tukey's HSD multiple comparison tests with $p \leq 0.05$ as the significance level, were carried out for the sensory attributes sweet aroma and sweet taste. Sucrose, aroma, age, gender, and their interactions were considered to be fixed effects, while location (canteen), subject nested within age and gender, and interactions involving these were treated as random effects. Location was found not to have any significant main or interaction effect. Thus, location as a factor was excluded from the reduced models. Gender was found to have a significant effect on several of the attributes. To examine the effect of gender, the dataset was thereafter split into two datasets: a "male" and a "female" dataset. The same ANOVA models as described above (excluding the gender factor) were then performed on the occurring datasets. As the three-way interaction between sugar, aroma, and age was not significant in any of the models, only two-way interactions were included in the final reduced models.

To identify groups of subjects with similar preferences, Agglomerative Hierarchical Clustering (AHC) analysis was performed on the liking data in XLSTAT by applying the Euclidean distances and Ward's agglomeration method. The clusters were compared in terms of gender and age group distribution using chi-square tests. Thereafter, ANOVA models were again carried out with sucrose, aroma, cluster, and their interactions as fixed effects, while subject nested in cluster was treated as a random effect, using the Minitab®19 software.

### *4.3. Results*

#### 4.3.1. Effect of Gender and Age among Young Adults on Ratings of Sweet Aroma and Sweet Taste Intensity

In Table 3, the results of the consumer study are shown separately for males and females. The addition of vanilla aroma increased the rating of sweet aroma intensity significantly for both genders (males: $F_{(1,49)} = 461.42$, $p < 0.001$; females: $F_{(1,76)} = 866.52$, $p < 0.001$), but for males, a significant interaction effect between sucrose and aroma was also identified ($F_{(2,100)} = 4.82$, $p = 0.010$). For the samples with added vanilla aroma, the sample with the medium sucrose concentration was rated significantly higher in sweet aroma intensity than the sample with the high sucrose concentration ($p = 0.001$). The sweet aroma intensity of the sample with low sucrose concentration was rated lower than the sample with medium sucrose concentration and higher than the sample with the high sucrose concentration, neither however significantly ($p = 0.185$ and $p = 0.433$, respectively).

**Table 3.** Means ± std. errors for the ratings of sweet aroma and sweet taste intensity for males and females, respectively, as well as *p*-values for the factors in the models. Significant effects are marked in bold.

| | | Males | | Females | |
|---|---|---|---|---|---|
| | | Sweet Aroma Ratings | Sweet Taste Ratings | Sweet Aroma Ratings | Sweet Taste Ratings |
| Age | | $p = 0.530$ | $p = 0.996$ | $p = 0.244$ | $\mathbf{p = 0.009}$ |
| Sucrose | | $\mathbf{p = 0.012}$ | $\mathbf{p < 0.001}$ | $p = 0.871$ | $\mathbf{p < 0.001}$ |
| Aroma | | $\mathbf{p < 0.001}$ | $\mathbf{p = 0.004}$ | $\mathbf{p < 0.001}$ | $\mathbf{p < 0.001}$ |
| Age*sucrose | | $p = 0.117$ | $p = 0.378$ | $p = 0.372$ | $p = 0.869$ |
| Age*aroma | | $p = 0.213$ | $\mathbf{p = 0.021}$ | $p = 0.566$ | $p = 0.472$ |
| Sucrose*aroma | | $\mathbf{p = 0.010}$ | $\mathbf{p = 0.001}$ | $p = 0.811$ | $\mathbf{p < 0.001}$ |
| 2.5% sucrose | No aroma | 1.3 ± 0.1 | 2.9 ± 0.3 | 1.5 ± 0.1 | 3.1 ± 0.2 |
| | Vanilla | 6.2 ± 0.3 | 4.4 ± 0.3 | 7.2 ± 0.2 | 5.4 ± 0.2 |
| 5.0% sucrose | No aroma | 1.4 ± 0.1 | 5.9 ± 0.2 | 1.6 ± 0.2 | 6.6 ± 0.2 |
| | Vanilla | 6.8 ± 0.3 | 6.1 ± 0.2 | 7.2 ± 0.2 | 6.7 ± 0.2 |
| 7.5% sucrose | No aroma | 1.4 ± 0.1 | 7.0 ± 0.3 | 1.5 ± 0.1 | 7.2 ± 0.2 |
| | Vanilla | 5.6 ± 0.3 | 7.5 ± 0.2 | 7.2 ± 0.2 | 8.4 ± 0.1 |

A significant sucrose-aroma interaction for sweet taste intensity was found for both genders (males: $F_{(2,100)} = 7.22$, $p = 0.001$; females: $F_{(2,154)} = 24.59$, $p < 0.001$). For males, just one significant cross-modal effect of vanilla aroma on sweet taste intensity was found, which was an enhancement of sweetness for the low sucrose concentration ($p < 0.001$). For females, a significant cross-modal effect of vanilla aroma on sweet taste intensity was found at both the low ($p < 0.001$) and high ($p < 0.001$) sucrose concentrations.

For males, a significant interaction effect was also seen between age and aroma affecting sweet taste intensity ($F_{(1,49)} = 5.69$, $p = 0.021$). The younger group of males rated the samples equally sweet independent of whether the samples were added vanilla aroma or not ($p = 0.972$). However, the older group of males rated the samples with added aroma significantly sweeter (mean 6.2) than the samples without aroma (mean 5.1) ($p < 0.001$). The cross-modal effect of vanilla aroma on sweet taste intensity thereby not only seems to be dependent on gender, but also on age for young males. For females, a significant main effect of age was seen on the rating of sweet taste intensity ($F_{(1,76)} = 7.22$, $p = 0.009$). For 19–21-year-old females, the mean sweet taste intensity score of all samples was 5.9, while the mean of the sweet taste intensity score for the 22–25-year-old females was 6.5. The older group of females thereby rated the samples sweeter than the younger group.

#### 4.3.2. Consumers' Acceptance and its Effect on Ratings of Sweet Aroma and Sweet Taste Intensity

For the consumer study, the subjects were divided into three clusters based on their liking of the samples. In general, samples acquired liking below the neutral point (grand mean ± SD = 4.2 ± 2) of the nine point scale. However, this was expected since the samples were aqueous blends and therefore very different from beverages normally consumed. In Table 4, gender and age characteristics for each of the clusters are shown together with the liking centroids.

**Table 4.** Cluster liking centroids for each sample together with the demographics for each cluster. Numbers in brackets shows the percentages from the total number of people, males, and 19–21-year-olds, respectively.

| Liking Clusters | | 1 Sweet Likers | 2 Neutrals | 3 Sweet Dislikers |
|---|---|---|---|---|
| 2.5% sucrose | No aroma | 3.00 | 5.38 | 4.00 |
| | Vanilla | 3.00 | 4.98 | 2.68 |
| 5.0% sucrose | No aroma | 4.48 | 5.40 | 3.09 |
| | Vanilla | 4.62 | 5.73 | 2.64 |
| 7.5% sucrose | No aroma | 5.57 | 5.40 | 2.51 |
| | Vanilla | 5.43 | 5.80 | 2.13 |
| Total number of people (%) | | 21 (16.3) | 55 (42.6) | 53 (41.1) |
| Number of males (%) | | 14 (27.5) | 28 (54.9) | 9 (17.6) |
| Number of 19-21-year-olds (%) | | 11 (19.6) | 25 (44.6) | 20 (35.7) |

When looking at the liking centroids for the clusters (Table 4), it seems the clusters can be characterized according to their preferences for sweetness intensity. The liking for the samples in Cluster 1 seems to increase along with the increase in sucrose concentration. No relation between either sucrose concentration or vanilla aroma and liking seems to exist for Cluster 2. For Cluster 3, liking seems to decrease with increasing sucrose concentration and the addition of vanilla aroma. According to this, the clusters can be described as sweet likers (1), neutrals (2), and sweet dislikers (3). No significant differences were found in the distribution of age groups between clusters ($p = 0.478$). The gender distribution was however skewed ($p < 0.0001$). Relative to the total number of males in the consumer study, there was a predominance of males among both the sweet likers and the neutrals. In the group of sweet dislikers, there was a predominance of females. To study if there was an effect of the clusters on the subjects' ratings of the sensory attributes, ANOVAs were performed with cluster included in the models as a factor (Table 5).

**Table 5.** Means ± std. errors for the ratings of sweet aroma and sweet taste intensity as well as *p*-values for the factors in the models. Significant effects are marked in bold.

| | | Sweet Aroma Ratings | Sweet Taste Ratings |
|---|---|---|---|
| Cluster | | $p = 0.542$ | **$p = 0.001$** |
| Sucrose | | $p = 0.212$ | **$p < 0.001$** |
| Aroma | | **$p < 0.001$** | **$p < 0.001$** |
| Cluster*sucrose | | $p = 0.307$ | $p = 0.864$ |
| Cluster*aroma | | $p = 0.959$ | $p = 0.783$ |
| Sucrose*aroma | | $p = 0.303$ | **$p < 0.001$** |
| 2.5% sucrose | No aroma | 1.4 ± 0.1 | 3.0 ± 0.2 |
| | Vanilla | 6.8 ± 0.2 | 5.0 ± 0.2 |
| 5.0% sucrose | No aroma | 1.5 ± 0.1 | 6.3 ± 0.2 |
| | Vanilla | 7.0 ± 0.2 | 6.5 ± 0.2 |
| 7.5% sucrose | No aroma | 1.4 ± 0.1 | 7.1 ± 0.2 |
| | Vanilla | 6.6 ± 0.2 | 8.0 ± 0.1 |

In accordance with the previous ANOVAs (Table 3), addition of vanilla aroma was found to increase the rating of sweet aroma intensity significantly ($F_{(1,126)} = 53.97$, $p < 0.001$). Furthermore, an interaction between sucrose and aroma significantly affected the rating of sweetness intensity ($F_{(2,256)} = 30.55$, $p < 0.001$). The addition of vanilla aroma only increased sweet taste intensity at the low and high sucrose concentrations. This is similar to the results previously described for females in Section 4.3.1.

Besides sucrose and aroma affecting sweetness intensity, the clusters were also found to significantly vary in their ratings of the sweetness intensity ($F_{(2,126)} = 7.67$, $p = 0.001$). As illustrated in Figure 3, sweet likers generally rated the sweet taste intensity of the samples significantly lower (mean 5.3) than both the neutrals (mean 5.9, $p = 0.045$) and the sweet dislikers (mean 6.3, $p < 0.001$) did. The neutrals and the sweet dislikers did not rate the sweet taste intensity significantly differently ($p = 0.123$).

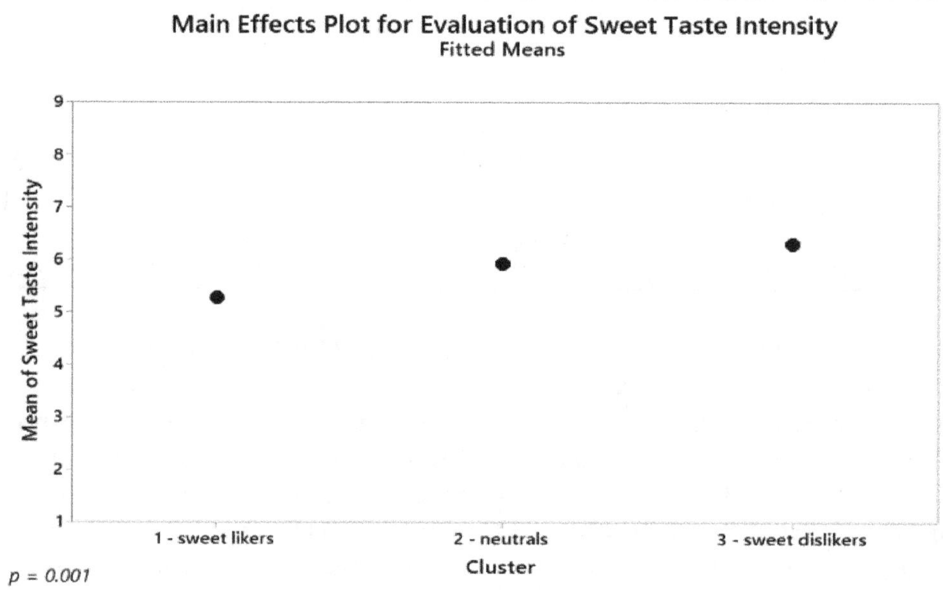


**Figure 3.** Fitted means of rated sweet taste intensity for the three clusters found in the consumer study.

## 5. Discussion

### *5.1. Screening of Aromas*

The DA revealed a sweet-sour aroma dimension (Figure 2), which is in accordance with Chifala and Polzella [46], who found a sweet-sour dimension when analyzing orthonasal evaluations of liqueurs using multidimensional scaling. The sweet-sour dimension in the present study was found to divide the aromas into two distinct perceptual groups: a "sweet" and a "sour" group. Aromas have previously been described with taste-like qualities such as sweet or sour even though they do not elicit tastes per se. This distinction of aromas being perceived either mainly as sweet or sour is likely created due to previous coexposures with sweet or sour taste qualities whereby a cognitive association of the aromas with the taste qualities is produced. Thus, aromas paired with either sucrose or citric acid have previously been found to subsequently smell sweeter and more sour than before pairing, respectively [47,48]. Raspberry and elderflower aromas therefore seem to be more associated with sour, rather than sweet taste. The aroma-induced sour taste of raspberry and elderflower samples, as seen in the DA, might therefore possibly have suppressed the sweet taste of these samples, as sour taste is known to suppress sweet taste due to taste-taste interactions [49]. The fact that the samples with added aromas from the "sweet" aroma group were consistently rated significantly higher in sweetness intensity than the samples with added aromas from the "sour" group is in accordance with the sweetness enhancing effect of aromas being dependent on how "sweet" they smell. Indeed, how "sweet" aromas smell has been shown to predict 60% of the sweet taste enhancement by aromas [21]. However, Barba et al. [23] found that just two out of nine aroma compounds associated with sweetness significantly increased the sweet taste intensity of 7% sucrose in water.

Among the aromas tested, the addition of vanilla aroma, which belonged to the "sweet" aroma group, led to the highest intensities for both sweet aroma and sweet taste (Table 2). Moreover, vanilla aroma has previously shown to increase sweetness intensity in several studies [5,7,14,15,18,37,50,51] and was thus found the most promising aroma with respect to sweet taste enhancement.

### 5.2. *Effect of Gender on the Cross-Modal Effect of Aroma on Sweet Taste Intensity*

In the consumer study, females seemed to be more susceptible for the cross-modal effect of vanilla aroma on sweetness intensity compared to males as the addition of vanilla aroma significantly increased sweet taste intensity scores at two sucrose concentrations for females, but just at one for males. To our knowledge, just one study has previously investigated the effect of gender on cross-modal aroma-sweetness interactions. Proserpio et al. [35] investigated the effect of BMI and gender on the evaluation of custard desserts added either no, 0.05%, or 0.1% butter aroma. They found a significant interaction effect between gender, BMI, and samples on the rating of sweet taste intensity. For obese subjects, females seemed to be most susceptible to the cross-modal effect of butter aroma on sweet taste intensity, while there for normal weight subjects were no differences between genders. In contrast, BMI was not investigated in the present study, and the current results can therefore not be directly compared with the study by Proserpio et al. [35].

Although conflicting results exist, most studies, according to Doty and Cameron's review on the subject, suggest that females on average are more sensitive than males when it comes to their ability to smell [26]. This might possibly explain why females in this study are more susceptible towards the cross-modal effect of vanilla aroma on sweetness intensity. A higher sensitivity might lead to females perceiving aromas as either stronger or more often than males. This in turn might cause females to experience coexposures of aromas with sweet taste either stronger or more frequently, and thus facilitating associative learning. Likewise, females perform better when it comes to odor memory [26]. Therefore, the gender differences seen in this study might also possibly be due to psychological or cognitive differences between males and females. For example, perhaps females are better at matching or remembering learned associations of tastes and aromas.

### 5.3. *Effect of age on Ratings of Sweet Aroma and Sweet Taste Intensity*

In the consumer study, a significant interaction between age and aroma was found affecting sweet taste intensity for males. For the younger group of males, no cross-modal effect of aroma was seen, while the older group of males rated samples with added vanilla aroma sweeter than samples without added aroma. Differences in cross-modal effects with age might possibly be due to differences in products normally consumed by each age group resulting in different associative learning. In 2017, 65.8% of the new students in the Danish universities were 21 years or younger [52]. As the consumer study was performed in the university canteens, it is therefore reasonable to assume that the majority of consumers in our study, especially in the younger group, were in the beginning of their university studies. Changes in consumption behavior with age in this group of consumers could thus occur due to changes in the students' life such as new friends or new living arrangements. As for the effect of gender on cross-modal aroma-sweetness interactions, just a few studies have investigated the effect of age on these interactions. The focus of these studies has primarily been on young adults versus senior adults (>60 years old) [30,36], and do therefore not compare relatively small age differences as seen in this study. One study however investigated the effect of adding vanilla flavor to milk among three different age groups of American children: 5–7, 8–10, and 11–14-year-olds, as well as for young adults aged 18–31 years old. All age groups found the sample with added vanilla flavor sweeter than the sample without. The only difference found was between the 5–7-year-olds and the 8–10-year-olds as the sweetness enhancement from vanilla flavor in the younger group was less pronounced than in the older group [37].

In this study, age did not affect the cross-modal effects for females, but it did have an effect on sweet taste intensity as the older group of females generally rated samples to be sweeter than the younger group of females. In contrast to cross-modal effects across ages, several studies have investigated taste perception across the life span. While some studies have found that neither the rating of sweet taste intensity nor the detection threshold of sucrose were affected by age in adults [53,54], others have found a decrease in sweetness sensitivity with age, especially after the late fifties [27–29]. As for the cross-modal interactions, few studies have focused on the effect of age in younger adults. Yamauchi et al. [55] studied different age groups of a Japanese population and found an interaction between age and gender on the detection threshold for sweet taste. Age (15–17 vs. 20–29-year-olds) affected sweet taste detection for men, but not for women. However, in the present study, supra-threshold intensities, and not detection thresholds were investigated, and these two parameters might not be correlated. Indeed, Mojet et al. [56] found that within 21 younger subjects between 19 and 33 years old, there was no relationship between threshold sensitivity and supra-threshold intensity in water or food products. Threshold sensitivity might therefore not be suitable to predict sensitivity of supra-threshold intensities in this group of consumers. Most recently, Dinnella et al. [57] among other things investigated the effect of individual differences on rated sweetness intensity of an aqueous 20% w/w sucrose solution among 1119 subjects. They found that neither fungiform papillae density, gender, nor age class (18–30, 31–45, and 46–60-year-olds) significantly affected the rated sweetness intensity of the aqueous solution in either 6-*n*-propylthiouracil (PROP) non-tasters or super-tasters. Besides differences in the age classes investigated, the different result from this study might also be due to the different sucrose concentrations used. Indeed, Dinnella et al. [57] wrote that intensity of sweetness was designed to be moderate/strong on a general Labeled Magnitude Scale, while the sucrose concentrations in this study were chosen to be lower than the sugar concentration normally found in beverages (10%) [43]. Finally, the discrepancy between the study of Dinnella et al. and our study might also be due to the population size, since our population size was much smaller than the population investigated by Dinnella et al. [57].

### *5.4. Effect of Sucrose on Release of Aroma*

In the consumer study, samples with added vanilla aroma were rated significantly higher in sweet aroma intensity compared to samples without added aroma. Interestingly though, an interaction between sucrose and aroma was found affecting the attribute sweet aroma intensity for males. The change in the aroma intensity noted between sucrose levels might be due to a chemical interaction between the sucrose molecules and the aroma compounds in the solution, changing the release of the aroma compounds. Indeed, the addition of sucrose (20% *w/w*) has previously shown to change the release of various aroma compounds compared to pure water [58]. In this way, sucrose seems to inhibit the release of the vanilla aroma. However, the sample with the low sucrose concentration was evaluated to be between the samples with medium and high sucrose concentrations in relation to the rating of sweet aroma intensity. The order of the samples with added vanilla aroma does not therefore follow the sucrose levels according to the rating of sweet aroma intensity, indicating that this might also be interpreted as a random effect. Another aspect supporting this view is that the sucrose-aroma interaction was only seen for males and not females. This might be explained by the fact that males are more sensitive to the aroma than women, but as mentioned earlier the opposite has been found more often [26]. This sucrose-aroma interaction affecting the rating of sweet aroma intensity for males does therefore not seem likely, but cannot be ruled out.

### *5.5. Classification of Sweet Liker Status and the Effect on Ratings of Sweet Taste Intensity*

The description of the clusters in the consumer study as sweet likers, neutrals, and sweet dislikers is in accordance with previous classifications of consumers according to their preferences for sweet taste [34,59]. In this study, 16.3% of consumers were classified as sweet likers, 42.6% as neutrals, and 41.1% as sweet dislikers. Iatridi et al. [59] reviewed different classification methods with aqueous sucrose solutions to determine sweet liker status and calculated the total weighed average proportions of the different sweet taste liker phenotypes. Across methodologies they found 48.5% sweet likers, 48.2% sweet dislikers, and 3.3% others/undefined phenotype. The biggest difference from Iatridi et al.'s results, therefore, seems to be a smaller proportion classified as sweet likers and a bigger proportion classified as neutrals in this study. Iatridi et al. [59] identified five overall classification methods used to determine sweet liker status: "Visual discrimination of hedonic responses", "Statistical discrimination of hedonic responses (algorithmic classification)", "Highest preference using ratings", "Average liking above mid-point"/"positive–negative average liking", and "Highest preference via paired comparisons". Compared to the other classification methods identified, the "Statistical discrimination of hedonic responses (algorithmic classification)", such as for example AHC analysis as used in this study, generally classified a higher proportion of the unclassified phenotypes [59]. Yet, even when accounting for this, the proportional difference of neutrals/unclassified phenotypes between the two studies is still big. The differences in distribution might therefore result from individual preferences for the vanilla aroma as well as a smaller range of sucrose concentrations tested in this study.

The distribution of the two age groups in the three clusters was not skewed, but the distribution of genders was. For the sweet likers and the group of neutrals, there was a predominance of males compared to the total number of males in the consumer study. For the sweet dislikers, there was instead a predominance of females. The skewed gender distribution for sweet dislikers (relatively more females than males) is in accordance with Yeomans et al. [31] who found that males consistently rated liking for sweetness higher than females. Similarly, Deglaire et al. [32] found that males rated liking for sweet foods and sugar higher than females across different BMI categories. Tuorila et al. [60] also found gender differences in the responses to sweetness. However, these referred to differences dependent on nationality and age, whereas others have found that there were no significant associations between gender and sweet liker/disliker status [4,61]. According to Deglaire et al. [32] a difference between males and females in their preference for sweet taste might possibly be due to different levels of health consciousness, as females have been shown to have higher health-related nutrition knowledge [62] and to find healthy eating more important than males [63–65].

As mentioned, ANOVAs were performed to evaluate whether the different clusters rated the samples similarly. The results showed that the clusters rated the sweet taste intensity differently from each other. Previous studies on the association between sweet liker status and sweetness intensity ratings are conflicting. As in this study, some studies have found that sweet dislikers generally rate sweetness intensity higher than sweet likers [61,66,67]. This might indicate that sweet dislikers perceive the same sucrose concentrations as more intense than sweet likers, and therefore do not like them. If this is actually the case, where the differences are due to differences in perception rather than hedonic preferences, then the sweet liker/disliker status might be an incorrect wording of the classification. However, others have found that sweet liker status is not associated with differences in ratings of sweetness intensity [4,68,69]. Differences might possibly be due to different methods used. Indeed, both Methven et al. [67] and Yang et al. [61] found their results to be dependent on matrix. Methven et al. [67] found that sweet dislikers rated sweetness intensity higher than sweet likers in jelly, but not in orange juice, while Yang et al. [61] found that low sweet likers (sweet dislikers) generally rated sweetness intensity of sucrose solutions higher than other groups, while this was not the case in iced tea.

### *5.6. Limitations and Future Research*

When analyzing the effects of individual differences such as age, gender, and sweet liker status, as done in this paper, it is important to keep the size of the population investigated in mind. The population sizes became relatively small when the dataset was split into both age and genders in the present study. The results seen, especially for the interaction between age and gender, could thus be random. Future studies should consequently aim to confirm our results with a larger population. It would also be interesting to reveal whether these results can be generalized to other populations such as other cultures, or other age groups regarding the gender effect.

As gender differences were found for the cross-modal effect of vanilla aroma on sweet taste intensity in the consumer study, it would have been preferable to have had a gender-balanced panel for the DA. Nevertheless, women were chosen for the panel because they performed better during the recruitment and annual retests of panelists. In the future, however, the results on individual differences from the present study should be considered when recruiting trained panelists for work on cross-modal interactions.

Another important factor to keep in mind is the food matrix. As described previously, the matrix for example affects whether sweet dislikers rate sweetness intensity higher than sweet likers or not [61,67]. In the present study, water was used as a model system. Additional studies with other more complex food matrices should be conducted in order to reveal if similar results can be obtained as in this study. On average, the samples in our study were not very liked. This could possibly be because water is not congruent with sweet taste and/or vanilla aroma. Indeed, congruency has been found to affect pleasantness of aroma-taste mixtures [16,70]. As aroma-taste interactions have also been suggested to be dependent on the matrix/product as well as previous food experience [5], future studies should investigate how congruency with the matrix affects aroma-taste interactions. Finally, it should be investigated whether the results also apply to other aromas than vanilla aroma.

## 6. Conclusions

To study individual differences in cross-modal effects among young adults, five aromas were first screened for their sweetness enhancing effect across three sucrose concentrations. Interestingly, the five aromas, all congruent with sweet taste, were found to belong to two distinct perceptual groups: a "sweet" and a "sour" group. Vanilla aroma belonged to the "sweet" group and was evaluated as the most promising aroma in relation to sweetness enhancement. Based on these results vanilla aroma was used for the consumer study that followed. Among young adult consumers, females were found more susceptible to the sweetness enhancing effect of vanilla aroma than males were. Furthermore, different age effects were found for males and females. For males, a significant cross-modal effect of vanilla aroma on sweet taste intensity was only found for the 22–25-year-olds, but not the 19–21-year-olds. While for females, the 22–25-year-olds were generally found to rate samples sweeter than the 19–21-year-olds. Finally, subjects were separated according to their sweet liker status. Interestingly, the groups identified were found to differ in their ratings of sweet taste intensity, but not in cross-modal effects. Sweet likers, a cluster with a predominance of men, rated sweet taste intensity significantly lower than the sweet dislikers, which were predominantly women. These results indicate that, even within a relatively small target group of young adults, individual differences in perception and in cross-modal interactions exist. These results could be used for more targeted product development and marketing in order to successfully aid sugar reduction and improve healthy eating. Furthermore, these results contribute to our understanding of cross-modal interactions and sensory integration. However, as results are restricted to the aroma and matrix tested, it would be interesting to further investigate this relevant target group with a more realistic matrix and more different aromas.

**Author Contributions:** Conceptualization, methodology, investigation, and resources, A.S.B., L.A.M., and U.K.; software, data curation, visualization, project administration, and writing—original draft preparation, A.S.B.; validation, and formal analysis A.S.B. and N.A.; writing—review and editing, A.S.B., L.A.M., N.A., D.V.B., and U.K.; supervision, and funding acquisition, L.A.M., D.V.B., and U.K. All authors have read and agreed to the published version of the manuscript.

**Acknowledgments:** The authors would like to thank the sensory panelists and volunteers who took part in this study as well as Nina Eggers for helping to conduct the study.

## References

1. Johnson, R.K.; Appel, L.J.; Brands, M.; Howard, B.V.; Lefevre, M.; Lustig, R.H.; Sacks, F.; Steffen, L.M.; Wylie-Rosett, J. Dietary sugars intake and cardiovascular health a scientific statement from the American heart association. *Circulation* **2009**, *120*, 1011–1020. [CrossRef] [PubMed]
2. Hu, F.B. Resolved: There is sufficient scientific evidence that decreasing sugar-sweetened beverage consumption will reduce the prevalence of obesity and obesity-related diseases. *Obes. Rev.* **2013**, *14*, 606–619. [CrossRef] [PubMed]

3. DuBois, G.E.; Prakash, I. Non-Caloric Sweeteners, Sweetness Modulators, and Sweetener Enhancers. *Annu. Rev. Food Sci. Technol.* **2012**, *3*, 353–380. [CrossRef] [PubMed]
4. Markey, O.; Lovegrove, J.A.; Methven, L. Sensory profiles and consumer acceptability of a range of sugar-reduced products on the UK market. *Food Res. Int.* **2015**, *72*, 133–139. [CrossRef]
5. Labbe, D.; Damevin, L.; Vaccher, C.; Morgenegg, C.; Martin, N. Modulation of perceived taste by olfaction in familiar and unfamiliar beverages. *Food Qual. Prefer.* **2006**, 582–589. [CrossRef]
6. Stieger, M.; Velde, F. van de Microstructure, texture and oral processing: New ways to reduce sugar and salt in foods. *Curr. Opin. Colloid Interface Sci.* **2013**, *18*, 334–348. [CrossRef]
7. Alcaire, F.; Antúnez, L.; Vidal, L.; Giménez, A.; Ares, G. Aroma-related cross-modal interactions for sugar reduction in milk desserts: Influence on consumer perception. *Food Res. Int.* **2017**, *97*, 45–50. [CrossRef]
8. Hutchings, S.C.; Low, J.Y.Q.; Keast, R.S.J. Sugar reduction without compromising sensory perception. An impossible dream? *Crit. Rev. Food Sci. Nutr.* **2019**, *59*, 2287–2307. [CrossRef]
9. Labbe, D.; Rytz, A.; Morgenegg, C.; Ali, S.; Martin, N. Subthreshold olfactory stimulation can enhance sweetness. *Chem. Senses* **2007**, *32*, 205–214. [CrossRef]
10. Labbe, D.; Martin, N. Impact of novel olfactory stimuli at supra and subthreshold concentrations on the perceived sweetness of sucrose after associative learning. *Chem. Senses* **2009**, *34*, 645–651. [CrossRef]
11. Tournier, C.; Sulmont-Rossé, C.; Sémon, E.; Vignon, A.; Issanchou, S.; Guichard, E. A study on texture-taste-aroma interactions: Physico-chemical and cognitive mechanisms. *Int. Dairy J.* **2009**, *19*, 450–458. [CrossRef]
12. Burseg, K.M.M.; Camacho, S.; Knoop, J.; Bult, J.H.F. Sweet taste intensity is enhanced by temporal fluctuation of aroma and taste, and depends on phase shift. *Physiol. Behav.* **2010**, *101*, 726–730. [CrossRef] [PubMed]
13. Boakes, R.A.; Hemberger, H. Odour-modulation of taste ratings by chefs. *Food Qual. Prefer.* **2012**, *25*, 81–86. [CrossRef]
14. Wang, G.; Hayes, J.; Ziegler, G.; Roberts, R.; Hopfer, H. Dose-Response Relationships for Vanilla Flavor and Sucrose in Skim Milk: Evidence of Synergy. *Beverages* **2018**, *4*, 73. [CrossRef]
15. Wang, G.; Bakke, A.J.; Hayes, J.E.; Hopfer, H. Demonstrating cross-modal enhancement in a real food with a modified ABX test. *Food Qual. Prefer.* **2019**, *77*, 206–213. [CrossRef]
16. Schifferstein, H.N.J.; Verlegh, P.W.J. The role of congruency and pleasantness in odor-induced taste enhancement. *Acta Psychol.* **1996**, *94*, 87–105. [CrossRef]
17. Djordjevic, J.; Zatorre, R.J.; Jones-Gotman, M. Odor-induced changes in taste perception. *Exp. Brain Res.* **2004**, *159*, 405–408. [CrossRef]
18. Oliveira, D.; Antúnez, L.; Giménez, A.; Castura, J.C.; Deliza, R.; Ares, G. Sugar reduction in probiotic chocolate-flavored milk: Impact on dynamic sensory profile and liking. *Food Res. Int.* **2015**, *75*, 148–156. [CrossRef]
19. Charles, M.; Endrizzi, I.; Aprea, E.; Zambanini, J.; Betta, E.; Gasperi, F. Dynamic and static sensory methods to study the role of aroma on taste and texture: A multisensory approach to apple perception. *Food Qual. Prefer.* **2017**, *62*, 17–30. [CrossRef]
20. Cliff, M.; Noble, A.C. Time-Intensity Evaluation of Sweetness and Fruitiness and Their Interaction in a Model Solution. *J. Food Sci.* **1990**, *55*, 450–454. [CrossRef]
21. Stevenson, R.J.; Prescott, J.; Boakes, R.A. Confusing tastes and smells: How odours can influence the perception of sweet and sour tastes. *Chem. Senses* **1999**, *24*, 627–635. [CrossRef] [PubMed]
22. Frank, R.A.; Byram, J. Taste-smell interactions are tastant and odorant dependent. *Chem. Senses* **1988**, *13*, 445–455. [CrossRef]
23. Barba, C.; Beno, N.; Guichard, E.; Thomas-Danguin, T. Selecting odorant compounds to enhance sweet flavor perception by gas chromatography/olfactometry-associated taste (GC/O-AT). *Food Chem.* **2018**, *257*, 172–181. [CrossRef] [PubMed]
24. Auvray, M.; Spence, C. The multisensory perception of flavour. *Conscious. Cogn.* **2008**, *17*, 1016–1031. [CrossRef]
25. Prescott, J. Chemosensory learning and flavour: Perception, preference and intake. *Physiol. Behav.* **2012**, *107*, 553–559. [CrossRef]
26. Doty, R.L.; Cameron, E.L. Sex differences and reproductive hormone influences on human odor perception. *Physiol. Behav.* **2009**, *97*, 213–228. [CrossRef]

27. Cooper, R.M.; Bilash, I.; Zubek, J.P. The effect of age on taste sensitivity. *J. Gerontol.* **1959**, *14*, 56–58. [CrossRef]
28. Moore, L.M.; Nielsen, C.R.; Mistretta, C.M. Sucrose taste thresholds: Age-related differences. *Journals Gerontol.* **1982**, *37*, 64–69. [CrossRef]
29. Mojet, J.; Christ-Hazelhof, E.; Heidema, J. Taste Perception with Age: Generic or Specific Losses in Threshold Sensitivity to the Five Basic Tastes? *Chem. Senses* **2001**, *26*, 845–860. [CrossRef]
30. Forde, C.G.; Delahunty, C.M. Understanding the role cross-modal sensory interactions play in food acceptability in younger and older consumers. *Food Qual. Prefer.* **2004**, *15*, 715–727. [CrossRef]
31. Yeomans, M.R.; Tepper, B.J.; Rietzschel, J.; Prescott, J. Human hedonic responses to sweetness: Role of taste genetics and anatomy. *Physiol. Behav.* **2007**, *91*, 264–273. [CrossRef] [PubMed]
32. Deglaire, A.; Méjean, C.; Castetbon, K.; Kesse-Guyot, E.; Hercberg, S.; Schlich, P. Associations between weight status and liking scores for sweet, salt and fat according to the gender in adults (The Nutrinet-Santé study). *Eur. J. Clin. Nutr.* **2015**, *69*, 40–46. [CrossRef] [PubMed]
33. Mahar, A.; Duizer, L.M. The effect of frequency of consumption of artificial sweeteners on sweetness liking by women. *J. Food Sci.* **2007**, *72*, 714–718. [CrossRef] [PubMed]
34. Garneau, N.L.; Nuessle, T.M.; Mendelsberg, B.J.; Shepard, S.; Tucker, R.M. Sweet liker status in children and adults: Consequences for beverage intake in adults. *Food Qual. Prefer.* **2018**, *65*, 175–180. [CrossRef] [PubMed]
35. Proserpio, C.; Laureati, M.; Invitti, C.; Cattaneo, C.; Pagliarini, E. BMI and gender related differences in cross-modal interaction and liking of sensory stimuli. *Food Qual. Prefer.* **2017**, *56*, 49–54. [CrossRef]
36. Philipsen, D.H.; Clydesdale, F.M.; Griffin, R.W.; Stern, P. Consumer Age Affects Response to Sensory Characteristics of a Cherry Flavored Beverage. *J. Food Sci.* **1995**, *60*, 364–368. [CrossRef]
37. Lavin, J.G.; Lawless, H.T. Effects of color and odor on judgments of sweetness among children and adults. *Food Qual. Prefer.* **1998**, *9*, 283–289. [CrossRef]
38. Storey, M.L.; Forshee, R.A.; Anderson, P.A. Beverage consumption in the US population. *J. Am. Diet. Assoc.* **2006**, *106*, 1992–2000. [CrossRef]
39. Ervin, R.B.; Ogden, C.L. Consumption of added sugars among U.S. adults, 2005–2010. *NCHS Data Brief* **2013**, *122*, 1–8.
40. Pedersen, A.N.; Christensen, T.; Matthiessen, J.; Knudsen, V.K.; Rosenlund-Sørensen, M.; Biltoft-Jensen, A.; Hinsch, H.-J.; Ygil, K.H.; Kørup, K.; Saxholt, E.; et al. *Dietary habits in Denmark 2011–2013. Main results*; National Food Institute, Technical University of Denmark: Copenhagen, Denmark, 2015.
41. Mendy, V.L.; Vargas, R.; Payton, M.; Cannon-Smith, G. Association between Consumption of Sugar-Sweetened Beverages and Sociodemographic Characteristics among Mississippi Adults. *Prev. Chronic Dis.* **2017**, *14*, 1–8. [CrossRef]
42. Poinot, P.; Arvisenet, G.; Ledauphin, J.; Gaillard, J.L.; Prost, C. How can aroma-related cross-modal interactions be analysed? A review of current methodologies. *Food Qual. Prefer.* **2013**, *28*, 304–316. [CrossRef]
43. Zheng, M.; Rangan, A.; Olsen, N.J.; Andersen, L.B.; Wedderkopp, N.; Kristensen, P.; Grøntved, A.; Ried-Larsen, M.; Lempert, S.M.; Allman-Farinelli, M.; et al. Substituting sugar-sweetened beverages with water or milk is inversely associated with body fatness development from childhood to adolescence. *Nutrition* **2015**, *31*, 38–44. [CrossRef]
44. Williams, E.J. Experimental Designs Balanced for the Estimation of Residual Effects of Treatments. *Aust. J. Chem.* **1949**, *2*, 149–168. [CrossRef]
45. Addinsoft. XLSTAT Statistical and Data Analysis Solution. Long Island, NY, USA. Available online: https://www.xlstat.com (accessed on 7 January 2020).
46. Chifala, W.M.; Polzella, D.J. Smell and taste classification of the same stimuli. *J. Gen. Psychol.* **1995**, *122*, 287–294. [CrossRef]
47. Stevenson, R.J.; Prescott, J.; Boakes, R.A. The acquisition of taste properties by odors. *Learn. Motiv.* **1995**, *26*, 433–455. [CrossRef]
48. Stevenson, R.J.; Boakes, R.A.; Prescott, J. Changes in Odor Sweetness Resulting from Implicit Learning of a Simultaneous Odor-Sweetness Association: An Example of Learned Synesthesia. *Learn. Motiv.* **1998**, *29*, 113–132. [CrossRef]
49. Keast, R.S.J.; Breslin, P.A.S. An overview of binary taste–taste interactions. *Food Qual. Prefer.* **2002**, *14*, 111–124. [CrossRef]

50. Clark, C.C.; Lawless, H.T. Limiting response alternatives in time-intensity scaling: An examination of the halo-dumping effect. *Chem. Senses* **1994**, *19*, 583–594. [CrossRef]
51. Sakai, N.; Kobayakawa, T.; Gotow, N.; Saito, S.; Imada, S. Enhancement of Sweetness Ratings of Aspartame by a Vanilla Odor Presented Either by Orthonasal or Retronasal Routes. *Percept. Mot. Skills* **2001**, *92*, 1002–1008. [CrossRef]
52. Uddannelses- og; Forskningsministeriet. Optag 2017—Alder (Enrolment 2017—Age). Available online: https://ufm.dk/uddannelse/statistik-og-analyser/sogning-og-optag-pa-videregaende-uddannelser/grundtal-om-sogning-og-optag/ansogere-og-optagne-fordelt-pa-kon-alder-og-adgangsgrundlag (accessed on 21 January 2020).
53. Hyde, R.J.; Feller, R.P. Age and sex effects on taste of sucrose, NaCl, citric acid and caffeine. *Neurobiol. Aging* **1981**, *2*, 315–318. [CrossRef]
54. Weiffenbach, J.M.; Baum, B.J.; Burghauser, R. Taste thresholds: Quality specific variation with human aging. *J. Gerontol.* **1982**, *37*, 372–377. [CrossRef] [PubMed]
55. Yamauchi, Y.; Endo, S.; Yoshimura, I. A new whole-mouth gustatory test procedure: II. Effects of aging, gender and smoking. *Acta Otolaryngol.* **2002**, *122*, 49–59. [CrossRef] [PubMed]
56. Mojet, J.; Christ-Hazelhof, E.; Heidema, J. Taste perception with age: Pleasantness and its relationships with threshold sensitivity and supra-threshold intensity of five taste qualities. *Food Qual. Prefer.* **2005**, *16*, 413–423. [CrossRef]
57. Dinnella, C.; Monteleone, E.; Piochi, M.; Spinelli, S.; Prescott, J.; Pierguidi, L.; Gasperi, F.; Laureati, M.; Pagliarini, E.; Predieri, S.; et al. Individual variation in PROP status, fungiform papillae density, and responsiveness to taste stimuli in a large population sample. *Chem. Senses* **2018**, *43*, 697–710. [CrossRef]
58. Siefarth, C.; Tyapkova, O.; Beauchamp, J.; Schweiggert, U.; Buettner, A.; Bader, S. Influence of polyols and bulking agents on flavour release from low-viscosity solutions. *Food Chem.* **2011**, *129*, 1462–1468. [CrossRef]
59. Iatridi, V.; Hayes, J.E.; Yeomans, M.R. Reconsidering the classification of sweet taste liker phenotypes: A methodological review. *Food Qual. Prefer.* **2019**, *72*, 56–76. [CrossRef]
60. Tuorila, H.; Keskitalo-Vuokko, K.; Perola, M.; Spector, T.; Kaprio, J. Affective responses to sweet products and sweet solution in British and Finnish adults. *Food Qual. Prefer.* **2017**, *62*, 128–136. [CrossRef]
61. Yang, Q.; Kraft, M.; Shen, Y.; MacFie, H.; Ford, R. Sweet Liking Status and PROP Taster Status impact emotional response to sweetened beverage. *Food Qual. Prefer.* **2019**, *75*, 133–144. [CrossRef]
62. Baker, A.H.; Wardle, J. Sex differences in fruit and vegetable intake in older adults. *Appetite* **2003**, *40*, 269–275. [CrossRef]
63. Rappoport, L.; Peters, G.R.; Downey, R.; McCann, T.; Huff-Corzine, L. Gender and Age Differences in Food Cognition. *Appetite* **1993**, *20*, 33–52. [CrossRef]
64. Roininen, K.; Tuorila, H.; Zandstra, E.H.; De Graaf, C.; Vehkalahti, K.; Stubenitsky, K.; Mela, D.J. Differences in health and taste attitudes and reported behaviour among finnish, Dutch and British consumers: A cross-national validation of the health and taste attitude scales (HTAS). *Appetite* **2001**, *37*, 33–45. [CrossRef]
65. Wardle, J.; Haase, A.M.; Steptoe, A.; Nillapun, M.; Jonwutiwes, K.; Bellisle, F. Gender Differences in Food Choice: The Contribution of Health Beliefs and Dieting. *Ann. Behav. Med.* **2004**, *27*, 107–116. [CrossRef]
66. Peterson, J.M.; Bartoshuk, L.M.; Duffy, V.B. Intensity and Preference for Sweetness is Influenced by Genetic Taste Variation. *J. Am. Diet. Assoc.* **1999**, *99*, A28.
67. Methven, L.; Xiao, C.; Cai, M.; Prescott, J. Rejection thresholds (RjT) of sweet likers and dislikers. *Food Qual. Prefer.* **2016**, *52*, 74–80. [CrossRef]
68. Looy, H.; Callaghan, S.; Weingarten, H.P. Hedonic response of sucrose likers and dislikers to other gustatory stimuli. *Physiol. Behav.* **1992**, *52*, 219–225. [CrossRef]
69. Kim, J.-Y.; Prescott, J.; Kim, K.-O. Patterns of sweet liking in sucrose solutions and beverages. *Food Qual. Prefer.* **2014**, *36*, 96–103. [CrossRef]
70. Amsellem, S.; Ohla, K. Perceived odor-taste congruence influences intensity and pleasantness differently. *Chem. Senses* **2016**, *41*, 677–684. [CrossRef]

# 7

# The Analysis of Marketing Factors Influencing Consumers' Preferences and Acceptance of Organic Food Products—Recommendations for the Optimization of the Offer in a Developing Market

**Boban Melovic [1], Dragana Cirovic [1], Branislav Dudic [2,3,*], Tamara Backovic Vulic [1] and Michal Gregus [2]**

1 Faculty of Economics, University of Montenegro, 81000 Podgorica, Montenegro; bobanm@ucg.ac.me (B.M.); dcirovic@ucg.ac.me (D.C.); tassabacc@ucg.ac.me (T.B.V.)
2 Faculty of Management, Comenius University in Bratislava, 82005 Bratislava, Slovakia; michal.gregusml@fm.uniba.sk
3 Faculty of Economics and Engineering Management, University Business Academy, 21000 Novi Sad, Serbia
* Correspondence: branislav.dudic@fm.uniba.sk.

**Abstract:** Considering the benefits of the organic production system, it is recognized as one of the main drivers of future economic development. However, the imbalance between demand and supply at the local market level represents one of the serious obstacles that prevents its future growth. Therefore, this article examines the key factors related to the main elements of the offer that have the strongest impact on consumer preferences and acceptance of organic food products. In that sense, organic product, price, distribution channel, and promotion are considered the main elements of the offer and are analyzed in this paper from the consumer preferences perspective. Further, this article provides insight into some of the sensory properties of the offer that are important to consumers. Finally, it gives recommendations for optimization of the offer on the organic food market based on the analysis of the influence of each of those elements (product, price, distribution, and promotion) on consumer acceptance of organic products and making purchasing decisions. The data were collected using a questionnaire, and analyzed using the structural equation model (SEM). The results revealed that price and promotion have the strongest impact on consumer acceptance and buying decisions. Further analysis revealed that attitudes towards organic food products, price/quality ratio, distribution barriers, and modern media as a promotion instrument are the factors that have the most significant impact on consumer perception and attitudes towards the available market offer. These findings can help producers and other decision makers to better understand what creates added value of the organic food products in consumers' mind and therefore make an offer that is in line with their expectations and preferences, which is recognized as one of the main prerequisites for the acceptance and purchase of organic food products.

**Keywords:** organic food; market; consumer preferences; product acceptance; sensory properties; optimization

---

## 1. Introduction

Organic production, as a specific system that sustains the health of soil, ecosystems, and people [1,2], appeared as a result of different types of concerns, amongst which the excessive pollution of the environment, imbalance of ecosystems, and human health protection are most important [3,4]. Compared to conventional production, it is an innovative production system that maximizes the performance of renewable energy sources; remarkably reduces the emission of $CO_2$, soils, metals, and other harmful substances into the environment; and optimizes nutrient and energy flow in agroecosystems [2]. Additionally, organic production implies the prohibition of the use of pesticides, genetically modified organisms, food additives, and other potentially harmful substances, which results in the production of healthy products of high nutritional value [1,2]. Therefore, fostering organic production has a significant contribution to solving some of the main problems of the global modern society, including obesity (especially amongst young people), environmental pollution, unsustainable production, and the necessity of giving assistance to local farm producers [1–5]. The organic production development has been fostered by income growth and a growing awareness of people about the importance of healthy lifestyles. It is based on several principles, whose implementation ensures the production of safe and healthy products while fully preserving the environment, its natural resources, and fostering equity, respect, and justice throughout society [2–4]. Therefore, it is recognized as one of the main drivers of future agricultural and overall economic development on sustainable grounds in many developing and developed countries.

Although the estimated value of the organic food market reached 97 billion US dollars in 2017, it is characterized by significant inequalities in growth rates, and reached a level of development in different countries [6]. About 90% of all organic products are sold in North America and Europe. Observed by countries, the biggest market of these products is in the USA, followed by Germany, France, and China. On the other hand, the vast majority of organic producers are located in less developed countries in Asia, Africa, and Latin America [6]. An additional obstacle to the intensive development of this sector is reflected in the fact that a small number of loyal consumers buy most of the total organic products produced, while other consumers buy them less frequently and in small quantities. That is why this market still retains some of the niche market features. These data clearly indicate that there is a serious imbalance between supply and demand in the individual geographic markets, caused by a misunderstanding of consumer preferences and key factors that influence the acceptance of organic food products in the market, which is a very important obstacle to the further development of organic food production. Therefore, harmonization of the supply with consumer preferences at the level of local markets is imposed as a necessary prerequisite for its further development. In order to achieve that, it is necessary to optimize the elements of the offer based on the results of more detailed surveys on consumer preferences, as well as on factors that influence the acceptance of organic food products and the consumer decision to purchase them.

However, in order to adjust the offer to the characteristics of demand in the developing organic food market, it is necessary for producers to achieve specific knowledge of marketing policies and marketing mix. Marketing mix optimization, among other things, includes the adjustment of product features, price level, distribution channel strategy, and promotion to the consumer preferences. In that sense, optimization of the offer that will foster demand on the developing organic market can be achieved through optimization of the marketing mix.

It is common in theory that product, price, distribution channels, and promotion are regarded as the main instruments of the marketing mix in any market, including organic. That is why the optimization of these elements can be regarded as the optimization of the total offer [7,8]. Therefore, key marketing mix elements are the focus of research in this paper. However, previous studies regarding these instruments on the organic food market mostly included the analysis of consumers' perception of one of them. However, there is a lack of research comparing the relative strength of the influence of each of these elements on consumer acceptance of organic food products and purchasing decisions, especially in developing markets. Additionally, although there are certain studies that investigate how different factors (such as health consciousness, price sensitivity, buying habits, etc.) influence consumer perception of the offer on the organic market [9–13], according to the authors' knowledge, there is a lack of research analyzing the relative significance of each of those factors.

Therefore, considering all the factors mentioned above, as well as the importance of fostering local organic markets' development, the aim of this paper was to fill these two gaps. First, this research identifies some of the factors that have the strongest influence on consumer perception of each of these main elements of marketing on the developing organic food markets in general. Secondly, it analyzes which of the mentioned elements (product, price, distribution, or promotion) has the strongest impact on the acceptance of organic food products (expressed through the consumer's decision to buy them). Additionally, this study also gives insight into the relative importance of some of the sensory properties of these products and their influence on consumer perception of the offer and the overall acceptance of the organic food products in developing markets.

The data analyzed in this paper were collected through a survey conducted in Montenegro. The motive for conducting research in this country is based on two reasons. Montenegro was one of the first countries in the world that was declared as an ecological state [14]. This, among other things, shows its potential for sustainable development and sustainable production, such as organic. However, despite its great potential, the organic food market in this country is still in its early stages of development. Therefore, it is necessary to understand what drives consumer behavior on the food market and what the main factors that influence the acceptance of organic food products are, in order to foster the development of this sector. Additionally, the obtained results can be useful for researchers and decision makers in other countries with similar production possibilities and an underdeveloped organic market.

Besides the theoretical contribution, the obtained results can help organic producers to better understand the main factors that shape consumer behavior on the developing organic food market and better allocate limited resources on each of the investigated elements of the offer, therefore creating more successful market strategy at the local and global level. Additionally, the obtained results can also help policy makers to better understand factors that drive organic market growth as well as those that prevent its further development, therefore allowing them to create and implement measures that will have significantly positive effects. However, it should be noted that fostering organic food production in developing markets can have significantly positive consequences that go beyond this economic sector. First of all, organic production is a system based on natural renewable resources that has no negative impact on the environment [1,2]. Therefore, it encourages economic development on sustainable grounds and a reduction of environmental pollution as one of the biggest global problems. Secondly, fostering organic agriculture development at the local level can help to reduce the problem of poverty and food waste, as it would increase employment rates and the production of food in the places of its consumption [2,15]. Finally, there are recent studies indicating that frequent consumption of organic food reduces the risk of obesity, which is also an important health issue that the global population faces nowadays [16].

## 2. Literature Review

### 2.1. Theoretical Background

Research conducted by Kriwy et al. as well as the research of Kashani-Nazari et al. showed that women are more prone to purchasing organic products compared to men [17,18]. On the other hand, the study of Kokmaz et. al. obtained the opposite results, indicating that men are more prone to purchase these products compared to women [19]. Although most of the studies showed that consumers who are younger, highly educated, and have higher level of income are more likely to buy organic products [17,20–23], there are some studies that revealed different results, showing that age, profession, and education do not have a significant impact on consumers' acceptance of organic food [19,24]. Considering the fact that that demographic analysis did not provide the information needed for successful targeting of organic food consumers, many authors applied segmentation analysis in order to identify the characteristics of different segments of organic food consumers. The results revealed that the same segments of organic food consumers cannot be found in all investigated markets, i.e., that organic food consumers have different demographic, psychographic, and behavioral characteristics, observed in different markets and different countries [9,20,25,26]. Therefore, different studies were conducted regarding consumer perception and barriers that prevent them from buying these products in order to better understand what drives the consumption of organic products. Currently, there is a growing research body on the motives and barriers that determine the behavior of organic consumers and influence their decision-making process. Most of those studies showed that consumers are faced with quite similar motives and barriers when purchasing organic food products, although the degree of their influence can significantly differ. The most important reasons for buying organic food products can be divided into three groups [27]: Avoiding the negative properties of conventional food products (such as the use of different harmful chemicals in production), unbiased features of organic food (such as safety), and positive features that consumers believe these products have (better taste, bigger nutritional value compared to conventional products, etc.). This is confirmed by most of the studies conducted in different markets that suggest that the most important motives for buying organic food are related to consumer perception of these products as healthy, tasty, safe, and of high nutritional value [9,10,20]. Additionally, consumers believe that organic production ensures protection of the environment, animal welfare, and preservation of natural resources [9–11,13,20,28]. Additionally, it should be pointed out that health and environmental concerns are far more important than other reasons, as well as the fact that the relative importance of each of these reasons did not significantly change over time [29]. A study conducted by Kereklas [30] showed that, unlike the purchase of conventional food products that is mostly based on personal considerations and habits, the acceptance of organic food products is under the significant influence of altruistic concerns, as well as under the influence of social pressure and social approval. However, the impact of these factors is rather indirect and is of less importance for price-conscious consumers [31,32]. Although most consumers have a positive attitude towards organic food products, there is only a small number of them who act in accordance with those attitudes, i.e., there is a far lower number of consumers who often buy these products. Therefore, the link between attitudes and buying intentions is much stronger than the link between intentions and the real behavior of consumers [33]. The reasons for such a condition can be found in the barriers that consumers face when buying organic food. Most authors state premium prices as the most important barrier, followed by the narrow range of products, poor availability in stores, consumer mistrust in certification labels, lack of time in the buying process, as well as negative attitudes towards organic products caused by a lack of information and satisfaction with conventional food [24,33–35].

### 2.2. *Key Marketing Determinants Affecting Organic Food Consumption*

A number of authors conducted research related to consumer perception of one specific marketing instrument as an important element of the offer (product, price, distribution, or promotion), in order to make suggestions for improvement of the offer on the organic food market [36–42]. Thanks to the specificities of organic production, organic food products have additional value that motivates consumers to buy them. However, that additional value cannot be experientially confirmed, which is why consumers search for ways in which organic food products can be distinguished from conventional ones [36–38]. This points out the importance of the visual cues of the product that can positively or negatively affect consumers' senses and perceptions of the product features. However, although certification is very important for consumer confidence, it is still uncertain which visual attributes organic food products should have and what is the role of packaging. The study of Enax revealed that it is of great importance to consumers, who find products with recognizable packaging healthier and tasty [39]. Similar results were obtained by Puelles et al. [40], emphasizing the difference between producers' and distributors' brands. The producers' brands are perceived as of better quality compared to distributors' brands, while distributors' marks are directed at the group of price-conscious consumers. A study of Aschemann-Witzel investigated the importance of nutrition and health claims visible on packaging and their effects on consumers' perception of the sensory properties of organic food products [41]. The results revealed that products with claims were neither significantly more preferred nor rejected when dealing with regular or loyal organic food consumers. However, highlighting these claims on packaging had significantly positive effects on occasional buyers. On the other hand, research conducted by van Herper et al. [42] obtained the opposite results. It revealed that packaging of any type can have negative effects on organic food sales (especially on organic fruits and vegetables), pointing out the fact that offering products without packaging increases the choice of consumers and that this effect is not limited to organic products only. Considering the results of previous studies, it still remains unclear what the visible features of organic food products that (besides of formed attitudes) have the strongest effect on consumer perception and senses are, and how much those effects are important for the acceptance of organic food products.

Price is considered to be one of the main obstacles that consumers face in the process of making buying decisions in the organic food market [24,33–35]. Bearing in mind that these products have premium prices due to higher production costs and limited supply [43], as well as the fact that its added value is intangible, Iyer et al. [31] suggest that the emphasis should be placed on the promotion of current and future benefits, especially when targeting price- and value-conscious consumers. Similarly, Bezawada et al. [44] point out that a reduction of regular prices and increasing assortments are very effective for non-core organic food consumers, and they also stimulate purchasing by core consumers. This is especially evident in the sales of organic food products with a high purchase frequency, as well as for those products that come directly from farms. However, promotional price cuts do not have significant effects on the purchase frequency of this type of product [44,45]. Marian et al. obtained similar results [46], showing that the least purchasing frequency is achieved for the most expensive organic brands. This implies that when consumers have some other quality cue (such as an organic certification label), they perceive higher prices only as a cost, not as an indicator of additional product value. Previous findings suggest that price reduction would be very effective for the stimulation of organic food purchasing. However, it still remains unclear whether consumers perceive the premium prices of these products as unjustified, or if these prices are a purchasing barrier due to insufficient income for most consumers. Answering this question is one of the main prerequisites for creating a successful business strategy on the organic food market.

An additional important barrier to organic food market growth is insufficient development of distribution channels [33–35]. However, different channel types attract different segments of consumers. Specialized organic stores are especially important for attracting core organic consumers [47,48], as these stores have a wide range of certified organic food products. On the other hand, although the internet can be an efficient channel for organic sales aimed at young people [48,49], the development of conventional stores is especially important for attracting non-core organic consumers, even though there is also a small percentage of loyal organic buyers who purchase in this type of store too [43,45]. Although consumers form an attitude towards groceries and supermarkets not only according to their personal habits but also according to the range of products that these shops sell [50,51], including organic products in the offer can have a positive influence on retail image, through improving its social dimension. Additionally, it can have positive effects even on the image of low-pricing outlets. According to the study of Diallo et al. [52], although organic products have premium prices, customers do not see a link between organic products and the store price level, probably because they believe that organic premium prices are related to production costs and high quality, and not related to the price strategy of the store. The results of previous studies suggest that consumers perceive an insufficient development of distributive channels as one of the main barriers to further development of the organic food market. However, the obtained results do not indicate whether consumers perceive the lack of these products in their favorite grocery stores as the main problem, or the consumers' inability to find organic food products at any other type of store despite the effort they are ready to make when purchasing these products is an even more important barrier.

Previous studies related to the promotion of organic food products were focused on creating messages line with the motivations of organic consumers. Kereklas at al. [30] revealed that the promotional messages featuring both altruistic and egoistic appeals are most effective when targeting organic consumers. Similar results were obtained by several other authors [53,54]. However, although some studies suggest that social networks and other Internet-based channels could be efficient instruments for the promotion of organic food products [49,55], it is still unclear whether consumers prefer traditional or modern media as a source of information about this type of product and what the relative impact of different promotion channels on consumer preferences and acceptance of this product type is. Considering all factors previously mentioned, this paper fills one of the literal gaps regarding marketing mix instruments that are addressed above, as basic elements of the offer on the developing organic food market.

*2.3. Conceptual Framework*

The results of previous studies revealed that the existing imbalance between demand and supply at the local market level represents a very important obstacle to the further development of the organic food market in general [6]. Additionally, consumers state that there are significant deficiencies regarding all main elements of the offer (addressed in this paper) that prevent them from buying organic food products, despite the positive attitudes they have towards them [23,32–34]. The authors consulted previous research in the field in order to consider all relevant factors that can influence consumers' perception towards four main marketing elements and their purchasing decision, in order to formulate the research questions and the conceptual model (some of the most important are presented in Table A1 given in Appendix A of this paper). The previous literature review indicated that there are many factors influencing consumer attitudes and perception of these elements, but which of them has the prevalent role in that process has not been investigated enough. Therefore, the first research question in this paper is formulated as follows:

**RQ1:** What are the factors that have the strongest influence on consumer perception of each segment of the offer in the developing organic food market?

Additionally, although previous studies investigated consumer perception towards some of these elements, which of them has the strongest impact on consumers' buying decision has not been investigated enough. Therefore, the second research question is formulated as follows:

**RQ2:** Which element of the offer, viewed from the marketing aspect (product, price, distribution channel, or promotion) has the strongest impact on consumer acceptance of organic food products, expressed through the decision to buy them?

The conceptual model was formed based on the research questions and it is given in Figure 1 presented below.

Our basic structural model contains three main constructs: (1) Factors that influence consumer perception of the stated elements of the offer (product, price, distribution channels, and promotion), (2) the consumer perception of these elements, and (3) the acceptance of products expressed through purchasing decisions.

Attitudes towards organic food products, producer, packaging, and frequency of purchase were identified as factors that influence consumers' senses and perception of the organic food products as the first element of the offer. The importance of price and the price/quality ratio were identified as factors that influence consumer perceptions of price. The distribution barriers (understood as the inability of consumers to find organic products in retail outlets in general) and consumer satisfaction with shops were identified as factors that influence consumer perceptions of distribution as an element of the offer. Finally, the modern media and the traditional media were identified as factors that influence consumer perceptions of promotion and its efficiency in fostering an acceptance of organic food products. In that sense, Internet-based media, such as social networks, email advertising, mobile marketing, pop-up advertising, etc., are considered as modern media, while TV, radio, newspapers, magazines, flyers, etc. are considered as traditional media [56,57]. While modern media are used for direct marketing communication and more precise targeting at relatively low prices, traditional media are still the best way of spreading information and gaining a large number of potential consumers, but the relative impact of these types of media varies depending on the industries of advertisers [58,59].

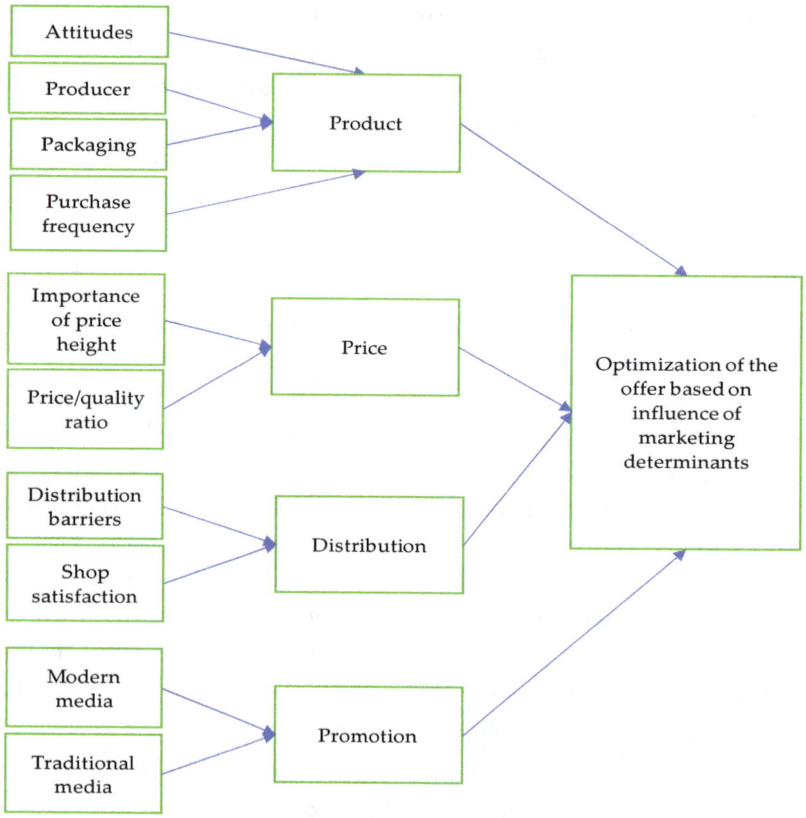


**Figure 1.** Conceptual model of research.

Each of these factors was chosen based on the previous research in the field related to their impact on the consumer perception of the offer on the food market. Therefore, each of them has a certain (positive or negative) influence on consumer perceptions and attitudes towards the named elements of the offer. On the other hand, consumer perceptions and attitudes towards these elements affect their acceptance of organic food products, i.e., purchasing decision. The causality between the identified factors and consumer perceptions and attitudes towards the four main elements of the offer, as well as the causality between consumer perceptions and attitudes towards these elements and the acceptance of organic food products expressed through purchasing decision, is what the authors explore in this paper. A better understanding of the causality and strength of the link that exists between these constructs will give insight into the actions that should be taken towards optimization of the offer, in order to foster the purchasing of organic food products and their acceptance.

## 3. Materials and Methods

Bearing in mind the research goals, formulated research questions, methods, and results of previous studies in this field, as well as their previous research experience in the organic food and marketing, the authors developed a questionnaire (which is given in Appendix B of this paper). A pilot survey was realized to examine the validity of the content of the questionnaire and was conducted on 40 respondents. Based on their suggestions, the final form of the questionnaire was prepared. It consisted of 22 questions and it was prepared in the Montenegrin and English language. The questionnaire included two types of questions: The multiple-choice question and a 5-point Likert scale. The questionnaire was formed considering the aims of research and the formulated research questions, in order to provide the data needed for measurement of the influence of factors addressed in the model. Most of the questions from the final questionnaire (questions 6, 7, 8, 9, 10, 11, 12, 20, 21, and 22) were related to the factors that influence consumer perceptions of organic products as a key element of the offer: The attitudes towards organic food products, the producer, the product packaging, and the purchase frequency. Considering the fact that price is one of the most important purchasing barriers, the questionnaire also contained questions related to factors affecting consumer perceptions of price: Price/quality ratio and the importance of price (questions 10, 11, 12, and 16). As the development of distribution channels is one of main prerequisites for organic food purchasing, questions 10, 12, 13, 14, 15, and 20 were related to factors affecting consumer perceptions of these channels (distribution barriers and shop satisfaction), while questions 10, 17, 18, and 19 were related to modern and traditional media in order to determine which of them has a stronger impact on the consumer perceptions of promotion.

In cooperation with the chamber of commerce in the mentioned country, the questionnaire was forwarded online through the bases of corporative group mails of retail companies to the 3530 potential respondents. The poll lasted for 30 days and it was conducted in the third quarter of 2019 in Montenegro. In total, 1634 respondents took part in the survey, which represents an answer rate of 46.3%. However, the total sample consisted of 1051 respondents. It is formed of respondents who buy organic food products at least a few times a year and their selection was made by using a filter question in the first part of the questionnaire, regarding the frequency of organic food product purchasing. So, if the questionnaire was completed by the respondent who never buys organic products, it is not considered valid. In that sense, all 1051 accepted questionnaires are valid and represent a validity rate of 64.3%. Based on the explanation of the sampling procedure, we can conclude that this is a stratified random sample. The motive for this selection is found in several reasons. First of all, it is expected that consumers who purchase organic food products at least sometimes are better informed about the principles of organic production and the characteristics of organic food products. Secondly, these consumers have more information on the real features of the available offer in the organic food market, and therefore can give more precise answers. Finally, compared to the non-consumers, it is easier to encourage the existing consumers to buy these products more often and in larger quantities.

Considering the fact that based on the results of previous studies it is not possible to determine the level of influence of education, occupation, income level, or other consumer demographic characteristics on the purchase frequency of organic food products, the stratified random sample method was applied. This choice of sample was also supported by the fact that there are only a few studies in Montenegro related to the organic food consumer, and the obtained results did not create a reasonable base for choosing another type of sample. Therefore, this method allows respondents of different demographic and psychographic characteristics to be represented in the sample. Additionally, considering the fact that there is a lack of research regarding this topic in Montenegro, there are no available data about existing specific consumer segments in the organic market that could be used as a basis for different sampling.

Hence, the application of a random sample method provides an acceptable degree of probability that the sample structure corresponds to the structure of the population. Additionally, we point out the fact that national statistics body, Statistical Office of Montenegro, conducts research of a whole population based on a sample of 1000 respondents [60], which also confirms the representativeness of the sample in this study. In terms of gender, 60.3% of the respondents were female while 39.6% of them were male. This sample structure regarding gender is representative, considering the fact that women do the purchasing of the food products more often compared to men. Additionally, this gender structure of the sample is very similar to the gender structure of the population in Montenegro [60], which also confirms its reliability. Organic food products are suitable for all ages, which is why the sample included all respondents who are over 15 years old (younger consumers rarely purchase themselves). A more detailed overview of the respondents' characteristics regarding their age, level of education, level of income, and purchase frequency is given in Table 1.

**Table 1.** An overview of the respondents' characteristics.

| Sample Characteristics | N | % | Sample Characteristics | N | % |
|---|---|---|---|---|---|
| **Gender** | | | **Purchase frequency** | | |
| Male | 417 | 39.7 | Every day | 180 | 11.0 |
| Female | 634 | 60.3 | At least once a week | 371 | 22.7 |
| **Age structure** | | | 1–3 times a month | 317 | 19.4 |
| 15–25 | 394 | 37.5 | Few times a year | 183 | 11.2 |
| 26–40 | 378 | 36.0 | Never | 583 | 35.7 |
| 41–60 | 182 | 17.3 | | | |
| More than 60 | 97 | 9.2 | | | |
| **Level of education** | | | **Average monthly income** | | |
| Primary school | 68 | 6.5 | Less than 500 euros | 705 | 67.1 |
| Secondary school | 437 | 41.6 | 500–800 euros | 242 | 23.0 |
| Bachelor | 470 | 44.7 | 800–1300 euros | 86 | 8.2 |
| Master's Degree | 59 | 5.6 | 1300–2000 euros | 18 | 1.7 |
| PhD | 17 | 1.6 | More than 2000 euros | 0 | 0 |

Further data analysis was preceded by determining the reliability of the survey results obtained on the basis of a questionnaire. The reliability coefficient of the whole questionnaire is 0.868 and represents the acceptable value of this coefficient in social science research. The data were analyzed using structural equation modeling (SEM). The validity of this model specification can be tested using a large number of tests. The most commonly cited tests refer to the model validity indexes: Goodness of fit index (GFI) and adjusted goodness of fit index (AGFI). The obtained values of these indexes for the model given in this paper are presented in Table 2 given below.

**Table 2.** Reliability indexes' values for the model specification.

| Index | Value |
|---|---|
| Goodness of Fit Index (GFI) | 0.951 |
| Adjusted Goodness of Fit Index (AGFI) | 0.9 |
| Root Mean Square Error of Approximation (RMSEA) | 0.08 |

In order to consider this model valid, these two indexes should have values equal or greater than 0.9. This condition is fulfilled for the presented model. Table 2 also gives the result of another widely accepted test, the root mean square error of approximation (RMSEA) index, whose recommended maximal value is 0.1. In our model it takes the value of 0.08, which is also acceptable. Thus, the results of the structural equation model (SEM) evaluation can be taken as valid. The obtained results are presented in the following section.

## 4. Results

The structural equation model (SEM) comprises a set of statistical methods designed to explain the complex relationship between one or more independent variables and one or more dependent variables. In this model, we used the rank-based strategy, proposed by Woods [61], to select anchor items (variable whose path relation will have value 1 as a constraint). Through the process of estimation of the structural equation model in this paper, 10 factors were identified that represent the elements of the offer. The factors were formed on the grounds of the joint influence of 45 independent variables, which were derived from the questionnaire and grouped in four new factors on the basis of their common characteristics. In evaluating the structural equation model, the factor that displays the characteristics of the respondents had to be excluded from the model because the parameters of this factor were not statistically significant. Thus, the structural equation model includes 10 factors. Factors that represent respondents' views, producer characteristics, packaging of organic food products, and purchase frequency were presented by their common factor named *Product*. Factors that indicate the importance of price height when buying organic food products and the price/quality ratio are presented by a common factor *Price.* The factor *Distribution* represents the impact of barriers in the distribution of organic food products and store satisfaction. The fourth common factor named *Promotion* indicates the impact of promotional activities (using traditional or modern media) when making a decision to buy organic food products. The regression coefficients obtained in the SEM model are presented in Figure 2, which is given below.

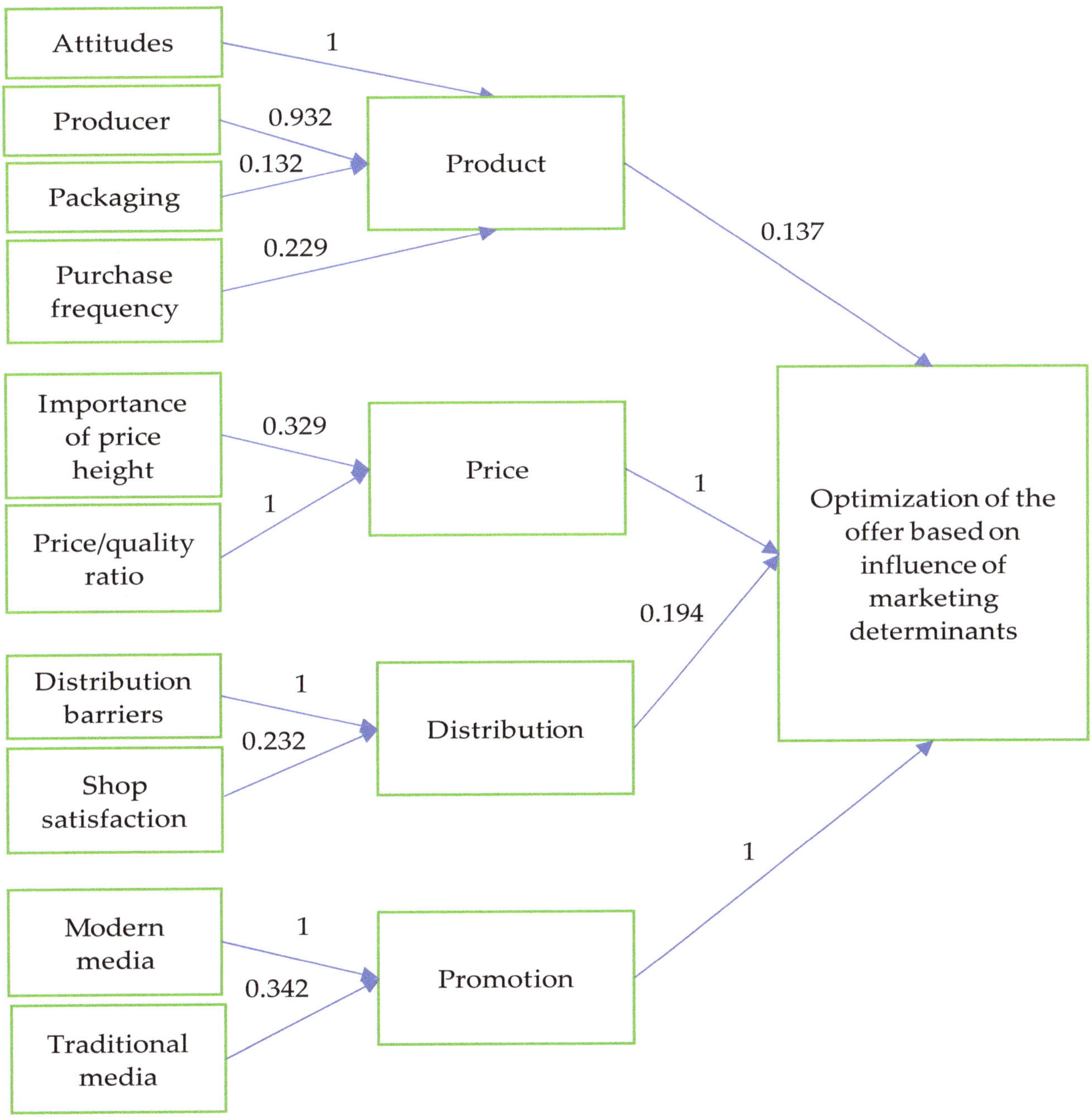


**Figure 2.** Structural equation modeling results.

The values of the standardized regression coefficients are given above the straight lines in the path diagram. The higher their value, the more the specific factor is considered a good indicator of the decision to buy organic food products and, therefore, an indicator of good product acceptance.

In order to justify the results of the SEM analysis, we examined whether the regression coefficients in SEM are statistically significant, as well as the validity of the specification of the defined model. The estimation results of this model are given in Table 3.

**Table 3.** Estimation results of the structural equation model (SEM).

| Exogenous Variable | Path Direction | Endogenous Variable | Estimate | Standard Error (SE) | Critical Ratio (CR) | Probability (P) |
|---|---|---|---|---|---|---|
| Attitudes | <— | Product | 1.000 | | | |
| Producer | <— | Product | 0.932 | 0.046 | 20,212 | *** |
| Packaging | <— | Product | 0.132 | 0.02 | 6,6 | *** |
| Purchase frequency | <— | Product | 0.229 | 0.017 | 13.47 | *** |
| Importance of price height | <— | Price | 0.329 | 0.046 | 7.191 | *** |
| Price/quality ratio | <— | Price | 1.000 | 0.051 | 4.821 | *** |
| Distribution barriers | <— | Distribution | 1.000 | | | |
| Shop store | <— | Distribution | 0.232 | 0.047 | 4.933 | *** |
| Modern media | <— | Promotion | 1.000 | | | |
| Traditional media | <— | Promotion | 0.342 | 0.065 | 5.261 | *** |
| Promotion | <— | Purchasing decision | 1.000 | | | |
| Product | <— | Purchasing decision | 0.137 | 0.029 | 4.724 | *** |
| Distribution | <— | Purchasing decision | 0.194 | 0.034 | 5.705 | *** |
| Price | <— | Purchasing decision | 1.000 | | | |

*** The regression coefficient, which is statistically significant with the level of significance of 1% (two-sided test).

All the regression parameters of the estimated structural equation model are statistically significant. This is confirmed by the corresponding probability values of the regression coefficients, which are marked with asterisks. Thus, with an error risk of 5%, all regression coefficients are statistically significant. For the estimation of SEM for each factor, a limit is defined according to which the value of the regression coefficient between the observed factors is 1. Since the value of a given regression coefficient is predefined, the statistical significance is not tested for it. In our model, value 1 is defined for the coefficients that relate the impact of attitudes and *product*, then the impact of the distribution barriers on the factor *distribution*, the influence of the factor *promotion* on the purchase decision, and the effect of *price* on the purchase decision.

Another advantage of the SEM model is the fact that it can be used to determine which factor has the strongest influence on respondents' decision to buy organic food products. Based on the estimated values of the regression coefficients for individual factors, it is concluded that the formation of the factor *product* was influenced the most by the variable that reflects customer attitudes towards organic food products. These findings are also supported by the results of the descriptive statistics in this research, which reveal that the main motivation for buying organic food products mostly relies on positive attitudes towards them and to a much lesser extent on the properties of these products that actually affect consumer senses (such as the visual appearance, taste, packaging, label, etc.). About two thirds of the respondents (58%) stated that organic food products are healthier compared to conventional ones. Additionally, 71% of them indicated the positive influence of these products on health as the main motive for their purchase. The additional reasons for purchasing are the nutritional value of the organic food products, environmental protection, animal welfare, and habit. On the other hand, consumers state that the taste of the organic food has the strongest effect on their perception, compared to other features that can be experientially verified. For 11% of them, this is, at the same time, the most important reason for purchasing. Other features of these products that can affect consumer senses (such as visual appearance, labeling, etc.) are of less significance for the purchase decision.

For the factor *price*, the most significant variable is the one that represents the price/quality ratio of organic food products. They suggest that if consumers perceive organic products as being high quality, they will be ready to pay premium prices for them. This points out the importance of educating and informing consumers about all the benefits of organic production for them as individuals and for the whole society. In that sense, the results of the descriptive statistics should be pointed out, which revealed that only 9.1% of consumers stated that they are not willing to pay premium prices for organic food products. Most of them were ready to pay 0% to 20% higher prices for organic products, compared to conventional (43.4% of consumers), while almost a third of respondents (28%) stated they are willing to pay 20% to 40% higher prices for this type of food product. These results reveal that premium prices are not an insuperable barrier to a wider acceptance of organic food products, and its significance is under the influence of information and the education of consumers towards an organic production system and all the benefits of this type of food.

In the case of the factor *distribution*, the most important was the variable that represents distribution barriers. Additionally, these results show that consumers do not expect that organic food products will always be available in their favorite stores (as the regression coefficient of this factor is smaller: 0.232), which means that they are ready to make an additional effort of going to another store, when purchasing these products. The descriptive statistics of the consumer preferences towards stores where organic products should be available revealed that most of them (57.7%) consider large supermarkets as the establishments where these products should be represented, while 20.4% of them consider that there should be more specialized organic stores in the market.

The factor *promotion* is mostly defined by a variable, which represents modern methods of promotion, suggesting that the promotion of organic food products in developing markets should be mostly conducted using modern media, such as social networks, Google adwords, banners, emails, and other Internet-based promotion channels.

The results of the evaluation of the structural equation model, i.e., the values of the estimated coefficients in the model, showed that *price* and *promotion* have a decisive influence on the consumer acceptance of organic food products and on the decision to buy them, while the influence of the factors *product* and *distribution* is less significant in developing organic food markets. These results are not so unexpected considering the purchasing power of the respondents who participated in the survey, the fact that organic products are currently mostly attractive to buyers with a higher income and higher education, as well as the fact that promotional activities disseminate information about the importance, quality, and availability of these products. Therefore, price and promotional activities must play a key role in the process of creating consumer perception of organic food products and fostering acceptance of these products in the society in the developing markets. On the other hand, for the buyers of the organic food products in such markets, the most important thing is the fact that these products meet the criteria of being truly "organic", which is why they pay very little attention to the other characteristics of these products, such as packaging or the manufacturer. Promotional activities in developing markets aim to inform customers where the organic food products are available, which is why the distribution channels also do not play such an important role in the process of making decisions to purchase these products.

In addition to the previous analysis, the authors also conducted the SEM model in order to test if there are differences in the significance of the investigated factors between men and women. The obtained results of the regression coefficients are given in Table 4.

**Table 4.** Estimated results of the structural equation model (SEM) for women and men.

| Regression Coefficients | | | Female | Male |
|---|---|---|---|---|
| Attitudes | <— | Product | 1.000 | 1.000 |
| Producer | <— | Product | 0.191 | 0.195 |
| Packaging | <— | Product | 0.173 | 0.237 |
| Purchase frequency | <— | Product | −0.121 | 1.046 |
| Importance of price height | <— | Price | −1.000 | −1.000 |
| Price/quality ratio | <— | Price | 1.532 | 3.793 |
| Distribution barriers | <— | Distribution | −1.379 | −1.500 |
| Shop store | <— | Distribution | 0.203 | 0.576 |
| Modern media | <— | Promotion | 1.000 | 1.000 |
| Traditional media | <— | Promotion | 9.660 | 0.618 |
| Promotion | <— | Purchasing decision | 1.000 | 1.000 |
| Product | <— | Purchasing decision | 0.700 | 0.630 |
| Distribution | <— | Purchasing decision | 2.445 | 0.152 |
| Price | <— | Purchasing decision | −4.620 | −1.644 |

The results of the SEM model obtained by using the gender as a control variable indicated that there are similarities but also certain differences in the way in which identified factors affect the purchasing decision of men and women in the organic food market. The results presented in the table explain that male consumers are highly influenced by the frequency of consumption, price/quality ratio, and distribution barrier when making a final decision to buy organic food products. Therefore, it is very important to diminish distributive barriers and explain all the benefits that consumers get for money (especially when it concerns the type of products with the biggest purchase frequency) in order to foster men to purchase these products more often and in larger quantities. Similarly, women are also mostly influenced by the price/quality ratio and distribution barriers, but traditional media has a significant impact on their purchasing decision too while it has no impact on men. Both men and women are price sensitive, which is why price is regarded as a very important obstacle to the organic food purchase. Additionally, they are positively influenced by promotion through modern media. This indicates that the results of the previous general SEM model can stand for both genders, confirming its reliability.

## 5. Discussion

Although consumers make purchasing decisions considering all elements of the offer as a whole, the previous analysis revealed that price and promotion have the strongest influence in that process on developing markets. The results of the structural equation model applied in this research showed that product does not have such a strong influence, compared to price and promotion. These results are somewhat surprising, considering the fact that earlier studies confirmed the specificities of the organic production system and objective features of organic food products (such as their nutritive value, health benefits, environmental benefits, etc.) as the main motives for buying these products [9–11,13,20,28,29]. However, the explanation could be found in the analysis of the significance of the factors that influence consumer perception of organic food products. The previous analysis showed that attitude is the factor that has the strongest impact, followed by producer. Such results are expected considering the fact that previous studies have shown that consumers evaluate the value of organic products mostly on the basis of their objective properties, resulting from the specificity of their production method (such as health safety, nutritional value, production process that is not harmful for the environment, etc.), so that the additional features of these products, such as the packaging and visual appearance, are not of great importance [7–10,12,19,27,28,62,63]. These additional features gain importance mostly when there is no reliable evidence that given food products are organic [38,64]. This reveals that consumers view all types of organic food products as of a similar level of quality, regardless of their packaging and other added features. Therefore, their objective characteristics that affect consumer senses and that can be experientially confirmed are of less significance. However, the most important amongst them is the taste of the product. The information about the producer is also important, as consumers search for cues that certain products are really organic, especially in developing organic markets where other cues are often missing. Therefore, trust in the producer reduces their doubts. However, considering the fact that these products have premium prices and that consumers in less developed markets do not have enough information about these products (regarding all their benefits, the available range, the price range, or the stores where they can be found), positive attitudes towards organic food products do not have such a significant role in the real behavior of consumers.

The fact that price has a strong effect in the process of accepting organic food products is confirmed by previous studies that revealed the price as one of the main barriers [24,33,34,65]. These results are not surprising, considering the fact that these products are more expensive than conventional ones and that they usually have premium prices, while on the other hand, the purchasing power of the consumers in Montenegro is relatively modest. According to the data retrieved from Statistical Office of Montenegro, Monstat, the minimal consumer basket for Montenegro in December 2019 amounted to 646.70 EUR, out of which 271.2 EUR refers to expenditure on food and non-alcoholic beverages, while 375.5 EUR refers to the expenditure on other non-food products and services [60]. On the other hand, the average earnings amount to 520 EUR [60], which implies that consumers are price sensitive first of all due to a low income. However, the analysis of factors that influence consumer perceptions of price revealed that their perception and behavior towards these products strongly depends on the product quality/price ratio and this factor is important for both men and women. In other words, if they consider that the product is of high quality, they will be ready to pay higher prices. Therefore, educating and informing consumers in Montenegro (and in other developing markets) about the organic production concept, as well as about all the benefits of the consumption of organic food products for individuals and for the whole society, is one of the main prerequisites for accepting these products and converting positive attitudes into real purchasing.

Although previous studies pointed out the significance of distribution channels as one of the main barriers in organic food purchasing [33–35,43,45,47,48] and therefore, as one of the most important factors that influence consumer perceptions and purchasing decisions, the data analysis in our study gave the opposite results. Although women are to a certain extent influenced by this factor, it is not of such great importance for consumers in general. This indicates that consumers in developing organic markets do not expect that organic food products should be available in every type of food shop and that those consumers who intend to buy these products are ready to make additional effort in searching for specialized stores where they can find them. However, it is important for them to know exactly where these products can be found. The previously mentioned factor was confirmed by the analysis of the factors that influence the perception of distribution channels. The results revealed that satisfaction with grocery stores does not have a significant impact, while the lack of these products in most of the stores is regarded a very significant factor that prevents their further acceptance and real purchasing in larger quantities. The results of the descriptive statistics showing that organic food products should be available more in the supermarkets and specialized stores are not surprising. Poor distribution channel development is a common problem in most developing organic food markets, such as Serbia or Poland [66,67]. Additionally, supermarkets and specialized stores are the most important distribution channel even in developed and mature organic food markets, such as Germany, France, and Canada, where supermarkets offer large varieties of product types and brands, while the specialized stores are usually better adapted to special consumer segments by offering a wide range of products customized and adapted from bulk or small packing containers [42,58]. Therefore, in the mature markets, producers can focus on in-depth development of their organic brand lines instead of fostering the development of distribution channels and generic promotion of organic food products as it should be the focus in emerging and developing markets [45,67,68].

The strong effect of promotion on the purchasing decision of consumers is also understandable. As Song et al. found in their research, the additional value of organic food products that makes them different from conventional ones cannot be experientially confirmed. This is why consumers search for ways in which organic food products can be differentiated from conventional ones [36–38]. Therefore, a very important role of promotion is to inform consumers of how they can distinguish them from non-organic products [65]. Additionally, in underdeveloped or developing markets of organic products, such as the market in Montenegro, it is expected that consumers are not well informed about the concept of organic production. Hence, promotion has a very important role of educating and informing consumers about the characteristics of organic products, the certification process, the available offer, and all other features that could be significant for them in the process of accepting these products and making purchasing decisions. The analysis in this paper also revealed that modern instruments of promotion in developing markets have a stronger effect compared to traditional media. More precisely, although traditional media has a significant impact on women, it does not influence the purchasing decisions of men. However, both genders are influenced by modern media, which is why this marketing element should be the focus of producers in developing organic food markets. These results are in line with the research of Pechrová and Jesus Medina-Viruel, [49,55], showing that the Internet and other modern promotion tools could be effective in promoting this type of product. However, it should be noted that most respondents in the sample were younger than 40 and had high school or university education, since younger and more educated people tend to use modern media much more than traditional ones.

Considering all previously mentioned factors, suggestions for marketing mix optimization in the developing organic food market can be made. Although all four marketing mix instruments as main elements of the offer should be significantly improved, the results in this paper reveal that most of the efforts should be made regarding price and promotion. Although consumers exhibit different levels of price sensitivity, they consider organic food prices related to the quality of products. Therefore, decision makers should point out all the benefits of organic farming and organic food consumption. If consumers understand the value of these products, they will be ready to pay premium prices. The fact that a significant number of respondents in the sample had an average income or income below the average confirms that the previous conclusion holds for price-sensitive consumers as well. However, in order to achieve the mentioned goal, promotion should be improved and conducted using predominantly modern media. Considering the fact that in developing markets, consumers mostly lack information about the organic production concept, promotional activities have a very important role in educating and informing consumers about the quality of organic food products and justification of their premium prices. Additionally, through the education of consumers, promotion can have a significant impact on their attitudes towards organic food products (which is in line with our research that promotion has a stronger impact on purchasing decisions, compared to product). Promotional activities also inform consumers about the available offer in the market, as well as about the stores where organic food products are being sold, which explains why distribution does not have such a strong influence on the purchasing decision of consumers. This way, the impact of the price and distribution as two main obstacles for the acceptance of organic food products in developing markets will be removed, or at least reduced. At the same time, by diminishing the impact of these barriers, the prerequisites for further development of the organic market will be fulfilled. Considering the specificities of organic production, the development of this sector will further have a series of positive consequences. First of all, it will have a significant contribution to better allocation of renewable resources, as organic production is a system completely adjusted to natural conditions [1,2]. Secondly, it enables food production in the places of its consumption, therefore contributing to solving problems of food waste and poverty in rural areas that developing countries mostly face [2,15]. Thirdly, fostering organic agriculture can contribute to the efforts in promoting healthy lifestyles and especially in solving the problem of obesity, which is a common problem of numerous countries [2,16]. Finally, fostering the organic food market will enable economic development without a negative impact on the environment [1,4], which is in line with the efforts of many developing countries to achieve economic development on healthy and sustainable grounds.

## 6. Conclusions and Future Research Recommendations

The aim of this paper was to generate recommendations for optimization of the offer on developing organic food markets, based on the analysis of the impact of four main elements of the offer (product, price, distribution, and promotion) on consumer acceptance of organic food products in developing markets. Additionally, in order to better understand the intensity and nature of those effects, this paper investigated which factors have the biggest impact on consumer perceptions of each of these elements.

The results revealed that price and promotion have the biggest impact on the acceptance of organic food products and consumer purchasing decisions in developing markets, while the impact of product and distribution channels is relatively modest. These results are not so unexpected, considering the fact that organic food products have premium prices, which makes them hardly affordable compared to conventional products. Additionally, promotion has the role of educating consumers about health, environmental, and other benefits of organic production, as well as the role of informing consumers about all the aspects regarding the available offer of these products in the market. This is especially important in developing markets due to a lack of information about the organic production concept and organic food benefits that most of the consumers face.

The analysis of factors that impact consumer perceptions of each of the addressed elements of the offer gave the additional explanation for such results. This analysis showed that consumers do not make a difference between different types of organic food products. Therefore, factors, such as packaging or frequency of purchase, do not have an impact on their perception of organic food in general. However, having a reliable cue that given products are really organic is very important for consumers, which is why they look for a recognizable organic producer. This points out the significance of giving reliable visual cues to the consumers that will help them to differentiate organic products from non-organic.

The consumer perception of price depends on the perceived quality of products and this is very important for both men and women. If they perceive that the product is of high quality, they will be ready to pay premium prices, which is why additional effort should be made in order to justify the premium prices of organic food products. Although consumers do not expect organic food products to be available in most food shops, the poor development of distributive channels is perceived as a significant barrier, which points out the fact that its future development is very important for ensuring the long-term growth of this sector. On the other hand, promotion is expected to inform and educate consumers about all important features of organic food products and therefore diminish the deficiencies of the other three elements of the offer. Although women are more strongly influenced by traditional media, for the best results, it should be mostly based on modern media due to the fact that both genders are influenced by this media type.

Apart from the theoretical factors, this paper has a significant practical contribution. It reveals the relative impact of the main marketing elements in developing organic food markets on consumer purchasing decisions, therefore helping decision makers to allocate limited resources to each of them. Additionally, it analyzed the factors that influence consumer perceptions of each of those instruments, therefore helping the decision makers to better understand what creates the added value of the offer in the consumers' mind and, hence, drives the acceptance of the organic food products.

Although the limits of this research are related to the fact that it was conducted only in Montenegro, it should be noted that, according to the authors' knowledge, it is the first research of this type conducted in the mentioned country. Additionally, the obtained results can be significant for research in other countries too, considering at least two reasons. First, the organic food market in Montenegro is still underdeveloped, which can stand for organic food markets in many countries. Secondly, most respondents in this survey have an average income and are price sensitive, while the segment of price-sensitive consumers exists in every country.

A further limitation of this paper arises from the fact that there is not any difference between loyal and occasional buyers of organic food products nor between different consumer segments that may exist regarding their demographic or psychographic characteristics. In order to overcome these limitations, further studies on this topic should include respondents from several countries and analyze if there are differences in the influence of each of the four elements of the offer on the purchasing decision, observed by different consumer segments. Additionally, whether the relative influence of each of those four elements changes over time can be explored. Considering the fact that optimization of the offer is an indispensable prerequisite for organic food acceptance and organic market development, similar research should be conducted not only in developing but in developed countries as well.

**Author Contributions:** Conceptualization, B.M., D.C. and B.D.; methodology, B.M., D.C. and T.B.V.; software, T.B.V.; validation, B.M., D.C. and B.D.; formal analysis, B.M., D.C. and T.B.V.; investigation, B.M., D.C. and M.G.; resources, B.M, D.C. and B.D.; data curation, B.M., D.C., B.D. and M.G.; writing—original draft preparation, B.M., D.C. and T.B.V.; writing—review and editing, B.M., D.C and B.D.; visualization, B.M., D.C., M.G.; supervision, B.M., D.C. All authors have read and agreed to the published version of the manuscript.

## Appendix A

**Table A1.** Previous research undergrounding the conceptual model of research.

| References | Sample | Methodology | Object of Research |
|---|---|---|---|
| Ham et al. 2016 [33] | In person survey of 411 households in Croatia | Techniques of univariate and multivariate analysis | Relationship between organic food purchase and purchasing barriers (time, price, knowledge, attitudes) |
| McEachern & McClean 2002 [10] | Random sample of 200 respondents; conducted in Scotland | Correlation and principle component analysis | Attitudes towards organic food products; purchasing motivation; |
| Bezawada & Pauwels 2013 [44] | The store-level date retrieved from 75 outlets in the USA | The persistence modeling approach | How changes in promotion, price and product assortment affect organic and conventional food purchase |
| Aschemann-Witzel et al. 2013 [41] | A survey of 210 organic consumers in Germany | Techniques of univariate and multivariate analysis | How nutrition, health or risk reduction claims visible on product packaging affect organic food purchase |
| Iyer et al. 2016 [31] | Online survey of 249 respondents in the USA | Structural Equation Modeling (SEM) | The effect of economic-driven motives (price and value consciousness) and altruistic motives on green purchase intentions |
| Henryks & Pearson 2011 [51] | A snowball sample of 21 respondents in Australia | A grounded theory approach | Identifies the variables affecting consumer choice of retail outlet and finds the ones that play a significant role in organic food purchase |
| Islam, S. 2014 [50] | A survey of 646 random shoppers visiting retail shops in Canada | Multiple regression analysis | Examines the choice of retail outlet and consumers' willingness to pay premium prices for organic food products |
| Hidalgo-Baz et al. 2017 [53] | Random sample consisting of 311 respondents in Spain | Linear regression model | Impact of different claims on organic food purchasing |

## Appendix B —Questionnaire

This survey is designed in order to investigate the specificities of the optimization of the offer on the organic food market. Please, take a few minutes to complete it. The survey is anonymous, and the data will be used for the purpose of writing a research paper.

1. Gender:

(a) male
(b) female

2. Age:

(a) 15–25
(b) 26–40
(c) 41–60
(d) more than 60

3. Your level of education:

(a) primary school

(b) secondary school
(c) Bachelor
(d) Master's Degree
(e) PhD

4. Your average monthly income:

(a) less than 500 €
(b) 500–800 €
(c) 800–1300 €
(d) 100–2000 €
(e) more than 2000 €

5. How often you buy organic food products:

(a) every day
(b) at least once a week
(c) 1–3 times a month
(d) few times a year
(e) never

If your answer to the question no. 5 is "never", this is the end of a questionnaire for you. If you chose any other given answer, please proceed with questionnaire.

6. What is the type of organic food products that you buy most often:

(a) organic fruit and vegetables
(b) organic dairy products
(c) organic cereals
(d) organic meat products
(e) organic bread and pasta
(f) organic juices and other soft drinks
(g) organic baby food
(h) other ________________________(Please, specify)

7. In your opinion, what type of organic food products should be more available on the market (you can choose more answers):

(a) organic fruit and vegetables
(b) organic dairy products
(c) organic cereals
(d) organic meat products
(e) organic bread and pasta
(f) organic juices and other soft drinks
(g) organic baby food
(h) other ________________________(Please, specify)

8. Rate the following characteristics of the organic food products on a scale 1–5 (1—*completely disagree*, 2—*disagree*, 3—*indifferent*, 4—*agree*, 5—*completely agree*):

(a) They are healthier than conventional products 1 2 3 4 5

(b) They are suitable for the prevention and treatment of certain diseases 1 2 3 4 5

(c) They are free from GMO, pesticides, antibiotics and other additives 1 2 3 4 5

(d) They have higher nutritional value than conventional products 1 2 3 4 5

(e) They have better taste than conventional products 1 2 3 4 5

(f) Their production is not harmful for the environment 1 2 3 4 5

(g) Organic production assures animal welfare 1 2 3 4 5

(h) By purchasing of organic food products you can give support to the local producers 1 2 3 4 5

9. Which reason for purchasing organic food is most important for you:

(a) its positive impact on health
(b) its nutritional value
(c) taste
(d) habit
(e) to contribute to the environmental protection
(f) animal welfare
(g) I don't know
(h) other ______________________(Please, specify)

10. Rate the following statements on a scale 1–5 (1—*completely disagree*, 2—*disagree*, 3—*indifferent*, 4—*agree*, 5—*completely agree*):

I do not buy organic food products in larger quantities because of:

(a) high prices compared to my level of income 1 2 3 4 5

(b) high prices compared to the quality of these products 1 2 3 4 5

(c) they taste worse than conventional 1 2 3 4 5

(d) the available offer is scarce (not diversified) 1 2 3 4 5

(e) they are not sufficiently represented in the stores, so their purchase requires too much time 1 2 3 4 5

(f) I don't know where to find them 1 2 3 4 5

(g) they have worse appearance than conventional products 1 2 3 4 5

(h) there is no real difference between organic and conventional products 1 2 3 4 5

(i) I don't have enough information about characteristics of these products 1 2 3 4 5

11. From the items offered below, select the three most important that you consider when buying organic food products and rank them by importance from 1 to 3 (1—*the most important*, 2—*of less importance*, 3—*of the least importance*)

| | | |
|---|---|---|
| (a) | price | 1 2 3 |
| (b) | nutritional value | 1 2 3 |
| (c) | impact on health | 1 2 3 |
| (d) | taste | 1 2 3 |
| (e) | diversity of the offer | 1 2 3 |
| (f) | packaging appearance | 1 2 3 |
| (g) | packaging size | 1 2 3 |
| (h) | information about producer | 1 2 3 |
| (i) | other ________________(Please, specify) | 1 2 3 |

12. What would particularly encourage you to buy more organic food products:

(a) price reduction
(b) better quality of the products
(c) greater diversity of the offer
(d) better representation of these products in supermarkets and other food stores
(e) opening new stores specialized in the organic food selling
(f) better quality and diversity of its packaging
(g) in no case I would buy them in larger quantities than I do now

13. Where do you usually buy organic food products:

(a) in supermarkets
(b) in small shops
(c) in stores specialized for selling organic food
(d) at the markets
(e) directly from the producer
(f) on the organic food fairs
(g) over the Internet
(h) other ________________(Please, specify)

14. In which types of stores would you prefer organic products to be available for purchase on a permanent basis:

(a) in the supermarkets
(b) in the small stores
(c) wish there were more stores specialized in organic food selling
(d) in the markets
(e) internet shops
(f) other ________________(Please, specify)

15. Rate the following statements on a scale 1–5 (1—*not important at all*, 2—*not important*, 3—*indifferent*, 4—*important*, 5—*very important*)

It is very important for organic food purchasing if these products are:

(a) available in the shops near the place where I live 1 2 3 4 5

(b) available in the shops where I usually purchase food products 1 2 3 4 5

(c) available in the shops with long working hours 1 2 3 4 5

(d) available in the shops that have friendly personnel 1 2 3 4 5

(e) available in the shops that offer high value for money 1 2 3 4 5

(f) available in the shops with high quality purchase services 1 2 3 4 5

16. How much higher price of organic food products (compared to conventional ones) are you ready to pay:

(a) I am not ready to pay higher prices for organic food products
(b) 0%–20%
(c) 20%–40%
(d) 40%–60%
(e) 60%–80%
(f) 80%–100%

17. From which sources do you usually collect information about organic food products:

(a) Friends and family members
(b) TV shows
(c) TV and radio commercials
(d) social networks
(e) internet portals or specialized websites
(f) magazines
(g) flyers
(h) other ____________________(Please, specify)

18. In your opinion, which media are more convenient for informing consumers about specificities and benefits of organic production system:

(a) traditional media (such as TV, radio, newspapers, magazines, flyer etc.)
(b) modern media (Internet based media, such as social networks, emails, websites, pop-up advertisement, mobile marketing etc.)
(c) other ____________________(Please, specify)

19. In your opinion, which media are more convenient for informing consumers about available offer of the organic food products in the market:

(a) traditional media (such as TV, radio, newspapers, magazines, flyers etc.)
(b) modern media (Internet based media, such as social networks, emails, websites or pop-up advertisement, mobile marketing, etc.)
(c) other ____________________(Please, specify)

20. How do you determine if food products labeled as "organic" are truly organic?

(a) based on official certification mark placed on the packaging

(b) based on information about producer (I buy it from well-known organic producer)
(c) based on the trust in the shop
(d) based on the product appearance
(e) based on label "organic"
(f) other ______________________ (Please, specify)

21. Rate the following items on a scale 1–5 according to their significance in purchasing organic food products (1—*not important at all*, 2—*not important*, 3—*indifferent*, 4—*important*, 5—*very important*):

(a) the way organic food products are differentiated from conventional ones in the stores 1 2 3 4 5

(b) appearance and the size of packaging 1 2 3 4 5

(c) label design 1 2 3 4 5

(d) information displayed on the product packaging 1 2 3 4 5

22. Rate how much important are the following items to you when choosing the organic producer whose products you will buy (1—*not important at all*, 2—*not important*, 3—*indifferent*, 4—*important*, 5—*very important*)

(a) the producer possesses the appropriate product quality certificate 1 2 3 4 5

(b) the producer has years of experience in organic food production 1 2 3 4 5

(c) the producer packs its products in recognizable packaging with a specific label 1 2 3 4 5

(d) the producer has a variety of organic food products in its offer 1 2 3 4 5

## References

1. A Website of IFOAM–Organics International. Available online: https://www.ifoam.bio/en/organic-landmarks/definition-organic-agriculture (accessed on 15 December 2019).
2. A Website of Food and Agriculture Organization of United States. Available online: http://www.fao.org/organicag/oa-faq/oa-faq6/en/ (accessed on 31 January 2020).
3. Cranfield, J.; Henson, S.; Holliday, J. The motives, benefits, and problems of conversion to organic production. *Agric. Hum. Values* **2010**, *27*, 291–306. [CrossRef]
4. Reganold, J.P.; Wachter, J.M. Organic agriculture in the twenty-first century. *Nat. Plants* **2016**, *2*, 15221. [CrossRef] [PubMed]
5. Gardasevic, J.; Vasiljevic, I.; Bjelica, D.; Popovic, S. Analysis of nutrition of boys and girls, adolescents from Montenegro. *J. Phys. Educ. Sport* **2015**, *4*, 702–705.
6. Willer, H.; Lernoud, J. *The World of Organic Agriculture. Statistics and Emerging Trends*; Research Institute of Organic Agriculture (FiBL), Frick, and IFOAM–Organics International: Bonn, Germany, 2019.
7. Goi, C.L. A review of marketing mix: 4Ps or more? *Int. J. Mark. Stud.* **2009**, *1*, 2. [CrossRef]
8. Tolušić, Z.; Zmaić, K.; Deže, J. Marketing-mix in the function of the organic food of Eastern Croatia. *Ekon. Pregl.* **2002**, *53*, 782–794.
9. Seegebarth, B.; Behrens, S.H.; Klarmann, C.; Hennigs, N.; Scribner, L.L. Customer value perception of organic food: Cultural differences and cross-national segments. *Br. Food J.* **2016**, *118*, 396–411. [CrossRef]
10. McEachern, M.G.; Mcclean, P. Organic purchasing motivations and attitudes: Are they ethical? *Int. J. Consum. Stud.* **2002**, *26*, 85–92. [CrossRef]
11. Persaud, A.; Schillo, S.R. Purchasing organic products: Role of social context and consumer innovativeness. *Mark. Intell. Plan.* **2017**, *35*, 130–146. [CrossRef]
12. Hoffmann, S.; Schlicht, J. The impact of different types of concernment on the consumption of organic food. *Int. J. Consum. Stud.* **2013**, *37*, 625–633. [CrossRef]
13. Schuldt, J.P.; Hannahan, M. When good deeds leave a bad taste. Negative inferences from ethical food claims. *Appetite* **2013**, *62*, 76–83. [CrossRef]

14. The Government of Montenegro–the National council for sustainable development. *The Analysis of Accomplishments and the Challenges of an Ecological State*; The Government of Montenegro: Podgorica, Montenegro, 2012; p. 3.
15. Crowder, D.W.; Reganold, J.P. Financial competitiveness of organic agriculture on a global scale. *Proc. Natl. Acad. Sci. USA* **2015**, *112*, 7611–7616. [CrossRef] [PubMed]
16. Kesse-Guyot, E.; Baudry, J.; Assmann, K.; Galan, P.; Hercberg, S.; Lairon, D. Prospective association between consumption frequency of organic food and body weight change, risk of overweight or obesity: Results from the NutriNet-Santé Study. *Br. J. Nutr.* **2017**, *117*, 325–334. [CrossRef] [PubMed]
17. Kriwy, P.; Mecking, R.A. Health and environmental consciousness, costs of behavior and the purchase of organic food. *Int. J. Consum. Stud.* **2012**, *36*, 30–37. [CrossRef]
18. Kashani-Nazari, M.; Rasli, A.; Fei Goh, C. Exploring consumer behavior towards ecological food consumption. *Lahore J. Econ.* **2016**, *28*, 1813–1825.
19. Korkmaz, M.; Alacahan, N.D.; Ataseven, Y. Testing of the impact of education level on the organic product buying preferences. *Eur. J. Exp. Biol.* **2014**, *4*, 680–690.
20. Pearson, D.; Henryks, J.; Sultan, P.; Anisimova, T. Organic food: Exploring purchase frequency to explain consumer behavior. *J. Org. Syst.* **2013**, *8*, 50–63.
21. Dettmann, R.L.; Dimitri, C. Who's buying organic vegetables? Demographic characteristics of US consumers. *J. Food Prod. Mark.* **2009**, *16*, 79–91. [CrossRef]
22. Paul, J.; Rana, J. Consumer behavior and purchase intention for organic food. *J. Consum. Mark.* **2012**, *29*, 412–422. [CrossRef]
23. Lockie, S.; Lyons, K.; Lawrence, G.; Grice, J. Choosing organics: A path analysis of factors underlying the selection of organic food among Australian consumers. *Appetite* **2004**, *43*, 135–146. [CrossRef]
24. Nasir, V.A.; Karakaya, F. Underlying motivations of organic food purchase intentions. *Agribusiness* **2014**, *30*, 290–308. [CrossRef]
25. Peštek, A.; Agic, E.; Cinjarevic, M. Segmentation of organic food buyers: An emergent market perspective. *Br. Food J.* **2018**, *120*, 269–289. [CrossRef]
26. Tleis, M.; Callieris, R.; Roma, R. Segmenting the organic food market in Lebanon: An application of k-means cluster analysis. *Br. Food J.* **2017**, *119*, 1423–1441. [CrossRef]
27. Iris Food Board. *Organic Consumer Research Study*; Ipsos MRBI: Dublin, Ireland, 2014.
28. von Meyer-Höfer, M.; Nitzko, S.; Spiller, A. Is there an expectation gap? Consumers' expectations towards organic: An exploratory survey in mature and emerging European organic food markets. *Br. Food J.* **2015**, *117*, 1527–1546. [CrossRef]
29. Hartman Group. Organic & Natural. Available online: http://store.hartman-group.com/content/organic-and-natural-2016-report-overview.pdf (accessed on 16 December 2019).
30. Kareklas, I.; Carlson, J.R.; Muehling, D.D. "I eat organic for my benefit and yours": Egoistic and altruistic considerations for purchasing organic food and their implications for advertising strategists. *J. Advert.* **2014**, *43*, 18–32. [CrossRef]
31. Iyer, P.; Davari, A.; Paswan, A. Green products: Altruism, economics, price fairness and purchase intention. *Soc. Bus.* **2016**, *6*, 39–64. [CrossRef]
32. Bhat, S.; Darzi, M.; Parrey, S.H. Green marketing: A driver for green brand equity and sustainable development. *Int. J. Humanit. Soc. Stud.* **2014**, *2*, 330–337.
33. Ham, M.; Pap, A.; Bilandžić, K. Percieved barriers for buying organic food products. In Proceedings of the 18th International Scientific Conference on Economic and Social Development–"Building Resilient Society", Zagreb, Croatia, 9–10 December 2016.
34. Bryła, P. Organic food consumption in Poland: Motives and barriers. *Appetite* **2016**, *105*, 737–746. [CrossRef] [PubMed]
35. Henryks, J.; Pearson, D. Attitude behavior gaps: Investigating switching amongst organic consumers. In Proceedings of the International Food Marketing Research Symposium Conference, Philadelphia, PA, USA, 19–20 June 2014.
36. Song, B.L.; Safari, M.; Mansori, S. The marketing stimuli factors influencing consumers' attitudes to purchase organic food. *Int. J. Bus. Manag.* **2016**, *11*. [CrossRef]
37. Anisimova, T.; Sultan, P. The role of brand communications in consumer purchases of organic foods: A research framework. *J. Food Prod. Mark.* **2014**, *20*, 511–532. [CrossRef]

38. Argyropoulos, C.; Tsiafouli, M.A.; Sgardelis, S.P.; Pantis, J.D. Organic farming without organic products. *Land Use Policy* **2013**, *32*, 324–328. [CrossRef]
39. Enax, L.; Weber, B. Marketing Placebo Effects–From Behavioral Effects to Behavior Change? *J. Agric. Food Ind. Organ.* **2015**, *13*, 15–31. [CrossRef]
40. Puelles Gallo, M.; Llorens Marín, M.; Talledo Flores, H. Perceptions of Control as a Determinant of the Intention to Purchase Organic Products. *Innovar* **2014**, *24*, 139–152. [CrossRef]
41. Aschemann-Witzel, J.; Maroscheck, N.; Hamm, U. Are organic consumers preferring or avoiding foods with nutrition and health claims? *Food Qual. Prefer.* **2013**, *30*, 68–76. [CrossRef]
42. van Herpen, E.; Immink, V.; van den Puttelaar, J. Organics unpacked: The influence of packaging on the choice for organic fruits and vegetables. *Food Qual. Prefer.* **2016**, *53*, 90–96. [CrossRef]
43. Cavaliere, A.; Peri, M.; Banterle, A. Vertical coordination in organic food chains: A survey based analysis in France, Italy and Spain. *Sustainability* **2016**, *8*, 569. [CrossRef]
44. Bezawada, R.; Pauwels, K. What is special about marketing organic products? How organic assortment, price, and promotions drive retailer performance. *J. Mark.* **2013**, *77*, 31–51. [CrossRef]
45. van Herpen, E.; van Nierop, E.; Sloot, L. The relationship between in-store marketing and observed sales for organic versus fair trade products. *Mark. Lett.* **2012**, *23*, 293–308. [CrossRef]
46. Marian, L.; Chrysochou, P.; Krystallis, A.; Thøgersen, J. The role of price as a product attribute in the organic food context: An exploration based on actual purchase data. *Food Qual. Prefer.* **2014**, *37*, 52–60. [CrossRef]
47. Hamzaoui-Essoussi, L.; Sirieix, L.; Zahaf, M. Trust orientations in the organic food distribution channels: A comparative study of the Canadian and French markets. *J. Retail. Consum. Serv.* **2013**, *20*, 292–301. [CrossRef]
48. Liébana-Cabanillas, F. *Electronic Payment Systems for Competitive Advantage in E-commerce*; Muñoz-Leiva, F., Sánchez-Fernández, J., Fiestas, M.M., Eds.; Business Science Reference: Hershey, PE, USA, 2014.
49. Medina-Viruel, M.J.; Bernal-Jurado, E.; Mozas-Moral, A.; Moral-Pajares, E.; Fernández-Uclés, D. Efficiency of organic farming companies that operate in an online environment. *Custos Gronegócio Online* **2015**, *11*, 265–289.
50. Islam, S. Marketing organic foods through conventional retail outlets. *J. Mark. Dev. Compet.* **2014**, *8*, 98.
51. Henryks, J.; Pearson, D. Retail outlets: Nurturing organic food consumers. *Org. Agric.* **2011**, *1*, 247–259. [CrossRef]
52. Diallo, M.F.; Coutelle-Brillet, P.; Riviere, A.; Zielke, S. How do price perceptions of different brand types affect shopping value and store loyalty? *Psychol. Mark.* **2015**, *32*, 1133–1147. [CrossRef]
53. Hidalgo-Baz, M.; Martos-Partal, M.; González-Benito, Ó. Is advertising helpful for organic businesses? Differential effects of packaging claims. *Int. J. Advert.* **2017**, *36*, 542–561. [CrossRef]
54. Salai, S.; Sudarević, T.; Đokić, N.; Pupovac, L. Marketing research for choosing the promotional message content for domestic organic products. *Econ. Agric.* **2014**, *61*, 501–515. [CrossRef]
55. Pechrová, M.; Lohr, V.; Havlicek, Z. Social media for organic products promotion. *Agris Line Pap. Econ. Inform.* **2015**, *7*, 40–50.
56. Chan, K.; Fang, W. Use of the internet and traditional media among young people. *Young Consum.* **2007**, *7*, 153–159. [CrossRef]
57. Salehi, M.; Mirzaei, H.; Aghaei, M.; Abyari, M. Dissimilarity of E-marketing VS traditional marketing. *Int. J. Acad. Res. Bus. Soc. Sci.* **2012**, *2*, 510–517.
58. Coulter, K.S.; Bruhn, M.; Schoenmueller, V.; Schäfer, D.B. Are social media replacing traditional media in terms of brand equity creation? *Manag. Res. Rev.* **2012**, *35*, 770–790.
59. Danaher, P.J.; Rossiter, J.R. Comparing perceptions of marketing communication channels. *Eur. J. Mark.* **2011**, *45*, 6–42. [CrossRef]
60. A Website of Statistical Office of Montenegro–Monstat. Available online: http://www.monstat.org/eng/index.php (accessed on 3 February 2020).
61. Woods, C.M. Empirical selection of anchors for tests of differential item functioning. *Appl. Psychol. Meas.* **2009**, *33*, 42–57. [CrossRef]
62. Slamet, A.; Nakayasu, A.; Bai, H. The determinants of organic vegetable purchasing in Jabodetabek region, Indonesia. *Foods* **2016**, *5*, 85. [CrossRef] [PubMed]
63. Petrescu, A.; Oncioiu, I.; Petrescu, M. Perception of organic food consumption in Romania. *Foods* **2017**, *6*, 42. [CrossRef] [PubMed]

64. Cirovic, D.; Vukcevic, M.; Melovic, B.; Mitrovic-Veljkovic, S. Organic production in Montenegro -analysis of the state and opportunities in the function of branding the products. In Proceedings of the VIII Jahorina Business Forum Conference, Jahorina, Bosnia and Herzegovina, 27–29 March 2019.
65. Bryła, P. Organic food online shopping in Poland. *Br. Food J.* **2018**, *120*, 1015–1027. [CrossRef]
66. Nikolić, M. Specifics of organic food markets' demand. *Contemp. Agric.* **2018**, *67*, 103–109. [CrossRef]
67. Renko, S.; Vignali, C.; Żakowska-Biemans, S. Polish consumer food choices and beliefs about organic food. *Br. Food J.* **2011**, *113*, 122–137.
68. Gottschalk, I.; Leistner, T. Consumer reactions to the availability of organic food in discount supermarkets. *Int. J. Consum. Stud.* **2013**, *37*, 136–142. [CrossRef]

# 8

# Factors Influencing Consumers' Perceptions of Food: A Study of Apple Juice using Sensory and Visual Attention Methods

**Katarzyna Włodarska [1], Katarzyna Pawlak-Lemańska [1], Tomasz Górecki [2] and Ewa Sikorska [1,*]**

1 Institute of Quality Science, Poznań University of Economics and Business, al. Niepodległości 10, 61-875 Poznań, Poland; katarzyna.włodarska@ue.poznan.pl (K.W.); katarzyna.pawlak-lemanska@ue.poznan.pl (K.P.-L.)

2 Faculty of Mathematics and Computer Science, Adam Mickiewicz University, Uniwersytetu Poznańskiego 4, 61-614 Poznań, Poland; tomasz.gorecki@amu.edu.pl

* Correspondence: ewa.sikorska@ue.poznan.pl

**Abstract:** The aim of this study was to evaluate the influence of intrinsic product characteristics and extrinsic packaging-related factors on the food quality perception. Sensory and visual attention methods were used to study how consumers perceive the quality of commercial apple juices from four product categories: clear juices from concentrate, cloudy juices from concentrate, pasteurized cloudy juices not from concentrate, and fresh juices. Laboratory tests included the assessment of sensory liking in blind and informed conditions and expected liking based on packages only. The results showed that brand and package information have a large impact on consumers' sensory perceptions and generate high sensory expectations. An innovative visual attention tracking technique was used in online experiments to identify packages and label areas on individual packages, which attracted consumer attention. During an online shelf test, consumers mostly focused on not from concentrate juices from local producers, which were perceived as more natural, healthy, and expensive than juices reconstituted from concentrate. When individual labels were analyzed, consumers predominantly focused on nutritional data, brand name, and information about the type of product. The present results confirm a large impact of information and visual stimuli related to packaging on product perception.

**Keywords:** apple juice; consumer perception; internal preference mapping; visual attention; packaging; label

---

## 1. Introduction

The sources that consumers use to form their impression of a product are typically classified as intrinsic or extrinsic cues [1,2]. The intrinsic cues are those characteristics that are part of the physical product. Some of these may be assessed before consumption (e.g., color, size, damage), while others may only be experienced through consumption, i.e., sensory properties. There is no single defined set of sensory attributes that are important across all of the food products. The importance of flavor, texture, and appearance attributes is product-dependent [3]. The extrinsic sources of information are those that are related to the product, but are not a part of it physically, such as brand name, label, packaging, price, the location where it is sold, and marketing communications [3,4]. These external cues generate consumer expectations about food products and affect their choices, sensory perception, and hedonic liking [5,6]. For these reasons, both sensory and non-sensory aspects should be included into the consumer research of food quality perception.

Sensory consumer research has confirmed that extrinsic product cues affect consumer perception of the food product sensory quality [4,7,8]. It has been reported that the brand name shown on a package influences consumer liking of taste of various food products [9,10]. The nutritional labeling also affects the estimation of taste [11]. In contrast, little is known about the relative effect of the extrinsic cues on the informed product evaluation when multiple cues are present [4].

The consumer perception of food packages and labels has been traditionally based on the self-reported estimates, although these estimates are subject to different biases and are reportedly only a poor indicator of the consumer behavior in real-life situations [6]. Some recent studies demonstrated that eye-tracking methodology is adequate for examining the effect of the package features on the consumer attention [12]. Food packages and nutrition labels were investigated with eye-tracking tools to monitor consumer visual attention to the respective product and nutritional information [13–15]. The eye-tracking technology precisely tracks the location and duration of gaze, providing detailed information on the consumer's visual attention and is more and more widely used in product marketing and consumer science, and recently also in consumer sensory studies [16,17].

Apple juice is one of the most popular products among juices all over the world, due to its pleasant sensory qualities and high nutritional value. Presently, the juice market offers a wide range of differently processed products. New trends in the juice market are determined by the consumer demand [18,19]. One of the fastest growing trends is that for minimally processed products, characterized by sensory properties similar to those of the fresh material and preserving valuable components of the fresh fruit. Apple juices available on the market include natural juices obtained directly from fruit by squeezing, not from concentrate (NFC) juices, both pasteurized and unpasteurized, and juices reconstituted from concentrate (FC), clear varieties and pulp-enriched cloudy varieties. Within each of the groups, one can distinguish ecological and conventional products. According to the European Fruit Juice Association (AIJN) Liquid Fruit Market Report [20], the consumption of non-concentrate fruit juices has increased significantly over the last five years in the European Union. The information on the production process is mandatory on the juice package. The results of previous studies showed varied consumer preferences for apple juices from different product categories [21,22].

Commercial apple juices are available in a variety of packages, including glass or plastic bottles and cartons. Food packaging is primarily used to protect food products from environmental contamination and other influences, and to ensure food quality and safety [23]. However, the packaging also has an important communicative role. The effective communication of product advantages through the packaging design determines consumer impressions of the product quality [24]. Thus, consumers mostly prefer products that attract their visual attention; therefore, food packaging must also be attractive or eye-catching. Both shape, material, color, and layout of the packaging elements have been shown to impact consumer perception [25]. Therefore, more research is needed in order to identify the packaging features that attract attention and enhance the buying potential.

The aim of the present work was to study the influence of the intrinsic product properties and the extrinsic cues related to packaging upon the food quality perception and liking. We studied commercially available apple juices from various product categories. We investigated the influence of the intrinsic cues and packages on liking and expectations using a consumer sensory study. Consumer visual attention to the apple juice packages and labels was examined with the use of an online visual attention tracking technique.

## 2. Materials and Methods

### 2.1. *Apple Juice Samples*

Eight apple juices from four product categories available on the Polish market were evaluated in this study. Two representative juices produced in Poland were selected in each of the four market categories, based on our previous studies of the physicochemical and sensory properties of a large number of juices and public recognition of their brands [26,27]. The selected products included juices reconstituted from concentrate (clear and cloudy varieties) and direct naturally cloudy juices (pasteurized and freshly squeezed varieties). Taking into account the type of production, the package, and the price, the juices studied may be classified into the standard and premium market segments. The detailed description of the samples is presented in Table 1.

**Table 1.** Description of the apple juices used in the study.

| Juice [1] | Product Category [2] | Package | Market Segment |
|---|---|---|---|
| A | FC, clear | 1.00 L box | Standard |
| B | FC, clear | 1.00 L light plastic bottle | Standard |
| C | FC, cloudy | 1.00 L box | Standard |
| D | FC, cloudy | 0.70 L light glass bottle | Standard |
| E | NFC, cloudy, pasteurized | 0.25 L dark glass bottle | Premium |
| F | NFC, cloudy, pasteurized | 0.25 L light glass bottle | Premium |
| G | NFC, freshly squeezed | 0.25 L light glass bottle | Premium |
| H | NFC, freshly squeezed | 0.25 L light glass bottle | Premium |

[1] The set of A–H juices was the same as studied previously [28]. [2] FC—from concentrate, NFC—not from concentrate.

*2.2. Consumer Studies*

Two experiments were performed to evaluate consumer perceptions of the commercial apple juices using different complementary methodologies. All of the participants declared to consume apple juices at least several times a month and to have apple juice as the first preference among fruit juices. No ethical approval was required for this study. The first experiment was performed in laboratory conditions and the second experiment online. The experiments were conducted as follows:

Experiment 1. A group of 96 consumers of apple juices participated in the laboratory sensory study (64.6% female and 35.4% male). The participants were recruited from staff and students of the Poznań University of Economics and Business based on their availability and interest in participation in the study. Of this number, 55.2% were under 25 years old, 36.5% were between 26 and 45 years old, and 8.3% were over 45 years old. The same group of consumers examined the quality of apple juice samples in different exposure conditions. To prevent communications between the members of the consumer panel, they were informed about the rules of the experiment, each person occupied a separate cubicle, and the sessions were supervised.

Two sessions were held over two consecutive weeks. Three types of data were gathered with the same group of consumers: (1) overall liking of the samples in blind-testing conditions, (2) expected liking of juices in packages, and (3) overall liking of the samples in an informed testing conditions; all samples were evaluated on 9-point hedonic scales (1 = dislike extremely, 5 = neither like nor dislike, and 9 = like extremely). The samples were presented to the consumers following a balanced rotation scheme. In blind-testing experiments, juices were coded by three-digit random numbers.

Fifty milliliters of each beverage were served to consumers at 20 °C in transparent plastic containers in blind and informed sensory evaluations, respectively, without and with packages. Mineral water was available during studies. The results of this experiment were already presented in our previous paper [28], where blind liking scores were discussed in detail and the relationships between the consumer liking and the physicochemical and sensory properties of juices were determined.

In an expected liking evaluation, consumers were asked to inspect full packages of eight apple juices and to score their expected sensory liking. To evaluate spontaneous perception of the labels, without making participants focus on specific aspects, they were asked to answer check-all-that-apply (CATA) questions with terms describing the product characteristics: natural, artificial, tasty, tasteless, not very healthy, very healthy, expensive, inexpensive, a familiar product, an innovative product, and a trendy product. Consumers were asked to check all the terms they considered appropriate to describe each of the products.

Experiment 2. A group of 171 consumers (68.8% female and 31.2% male) participated in the online study using Attensee software (https://www.attensee.com). Of these, 72.9% were under 25 years old, 23.0% were between 26 and 45 years old, and 4.1% were over 45 years old.

Attensee is an application used for the study of visual attention, that mimics an eye-tracker [29]; it provides an interface for computers and tablets and does not require any special hardware. This innovative approach to record human information lets one see what consumers look at on labels, packaging, or pictures. To get the required information, the reference areas (RA) on packaging and labels should be marked, and the percentage of consumers focusing their attention on these RA are recorded [30]. Based on these results, final calculations of the visual path tracking parameters may be made. The following RA were used in our experiments: brand names, logos, pictograms, photos, tables and lists of nutritional values, slogans regarding the content of pro-health ingredients, eco/bio labeling, and packaging size statement.

This experiment was conducted online. Consumers were asked to take part in the study via social media, and after agreeing, they received an individual link to the online experiment and were informed that the study involved the simulation of visual perception on a computer screen by means of cursor movements. The participants used the mouse pointer to bring certain parts of the screen into clarity (focused circle 2 cm in diameter). The software updated the information in real time as participants changed their area of focus and the way of exploration. After receiving instructions and doing some trial runs, participants performed the actual test. First, an image of all of the analyzed products on the shop shelf was presented on a computer screen for 45 s, and then the images of individual juice labels were presented, each for 30 s. The sequence of showing the individual labels was proportional and randomized across participants to eliminate the order effects. Each of the participants examined an image with all of the products (shelf picture), and randomly selected four of the eight individual product images. The photographic images of juice packages (with 4592 × 3056 pixels resolution) were captured using a digital camera (Sony A3056, Tokyo, Japan).

The Attensee software can also produce different types of questionnaires. Thus, the participants were asked to answer a check-all-that-apply questionnaire with the terms describing the product, the same as in Experiment 1.

### 2.3. *Data Analysis*

Consumer liking data were analyzed by multivariate analysis of variance (MANOVA). When a significant difference ($p < 0.05$) was detected, a $t$-test with Holm correction was applied to evaluate the difference between the samples. Pearson coefficients were calculated to evaluate the correlations between the consumer liking scores gained in different evaluation conditions. Principal component analysis (PCA) was performed on the individual consumer scores in three different conditions using the correlation matrix. PCA models were validated using the full cross-validation procedure. The correspondence analysis (CA) was performed on the frequency table containing responses to the check-all-that-apply questions. Statistical analyses were performed using the R-language and XLSTAT software (Addinsoft, Paris, France) packages.

## 3. Results

### 3.1. *Consumer Liking of Apple Juices in Different Evaluation Conditions*

The same consumer group evaluated the quality of eight apple juices in three testing conditions in our laboratory study. The results of consumer evaluations expressed on the 9-point hedonic scale are presented in Table 2

**Table 2.** Mean consumer liking scores[1] of apple juices evaluated in blind, expected, and informed conditions.

| Juice | Evaluation Conditions [1] | | |
|---|---|---|---|
| | Blind [2] | Expected | Informed |
| A | 4.5 $^{c,d,C}$ | 5.8 $^{c,A}$ | 5.1 $^{e,B}$ |
| B | 4.9 $^{b,c,C}$ | 6.1 $^{b,c,A}$ | 5.4 $^{d,eB}$ |
| C | 5.8 $^{a,A}$ | 6.4 $^{a,b,c,A}$ | 6.3 $^{b,c,A}$ |
| D | 5.6 $^{a,bB}$ | 6.6 $^{a,b,A}$ | 6.1 $^{b,c,d,A,B}$ |
| E | 4.1 $^{d,C}$ | 5.8 $^{c,A}$ | 5.3 $^{d,e,B}$ |
| F | 4.2 $^{c,d,C}$ | 6.4 $^{b,c,A}$ | 5.7 $^{c,d,e,B}$ |
| G | 6.2 $^{a,B}$ | 6.7 $^{a,b,A,B}$ | 6.9 $^{a,b,A}$ |
| H | 5.7 $^{a,B}$ | 7.1 $^{a,A}$ | 7.3 $^{a,A}$ |

[1] Evaluated on a nine-point hedonic scale. [2] The consumer ratings in blind conditions were available from our previous study [28]. Different lowercase superscripts (a–e) within a column indicate significant differences according to the *t*-test ($p < 0.05$). Different capital superscripts (A–C) within a row indicate significant differences according to the *t*-test ($p < 0.05$).

The results of the consumer study in blind condition were described in detail and discussed in a previous paper, as regards physicochemical and sensory characteristics of juices [28]. Presently, we used these results to compare them with the consumer ratings in expected and informed conditions. The consumers reacted differently to the sensory characteristics of the juices. The overall liking scores in blind conditions ranged from 4.1 corresponding to dislike slightly to 6.2 corresponding to like slightly. Significant differences were found in consumer liking of the samples assessed in a blind test ($p < 0.05$). Freshly squeezed juices (samples G and H) and reconstituted cloudy juices (samples C and D) were the most liked juices, while cloudy NFC juices (samples E and F) and reconstituted clear juices were the most disliked. The results are in line with those reported earlier [31], where average consumer ratings of the studied apple juice were at 5.5–6.0 on a 9-point scale, and these rather low ratings were explained by consumer habits for certain products.

The expected liking scores were in the range of 5.8–7.1 and higher than blind scores. These results indicate that the package with its product information, including brand, type of production, nutritional value, etc., generates high consumer expectations. Significant differences were observed in mean consumer scores; fresh juices (G, H) and cloudy FC juice (D) gained the highest ratings, while clear juice (A) and naturally cloudy (E) the lowest. The NFC juice (E) which is an organic product (information on the package) received the lowest rating among the studied juices, as it had a large amount of natural sediment. Note that sediment and turbidity of fruit juice may be perceived negatively by consumers as a product defect [32]. On the other hand, consumers were reported to have higher expectations for local apple juices as opposed to the mainstream juices available on the market, and also expected the higher quality of fresh apple juices as opposed to juices subject to thermal, high hydrostatic pressure, or pulsed-electric field treatment [21,22].

Our present study explored consumer perceptions of apple juices based on the comprehensive product information available on the label. The consumers, knowing the product brand and product information available on the package, rated the sensory quality of fresh juices (G, H) the highest. Significantly lower scores were obtained by clear juices (A, B) and naturally cloudy NFC (E, F) juices. Similar results were reported by Lee, Lusk, Mirosa, and Oey [22], indicating that consumers mostly valued the untreated apple juices.

When comparing consumer liking of juices in various testing conditions, the highest influence of product information on liking was observed for NFC juices (E, F). These samples correspond to the premium brands available in glass bottles. Clear FC juices (A, B) and fresh ones (G, H) were also assessed higher in informed as compared to blind tests. The blind and informed liking did not differ significantly only for cloudy FC juices (C, D). These latter juices were available in cartons and received high ratings for sensory quality at the first stage of the study. On the other hand, the expected quality scores were higher than the sensory scores for coded samples for six of the eight juices tested (except samples C and G, which received high sensory evaluation). The expected liking scores were significantly higher than the informed liking scores for clear (A, B) and naturally cloudy NFC (E, F) juices. The discrepancy between these assessments indicates that the sensory profiles of these juices do not meet consumer expectations generated by the product information contained on the packaging.

The mean consumer liking scores are useful for determining the overall trends, but do not provide information on the formation of the consumer groups (segments) with similar preferences [33]. Consumer liking data were analyzed using principal component analysis (PCA), performed on the individual consumer scores obtained in three different conditions. Internal preference maps were generated, illustrating the main preference directions [34], presented as biplots in Figure 1.

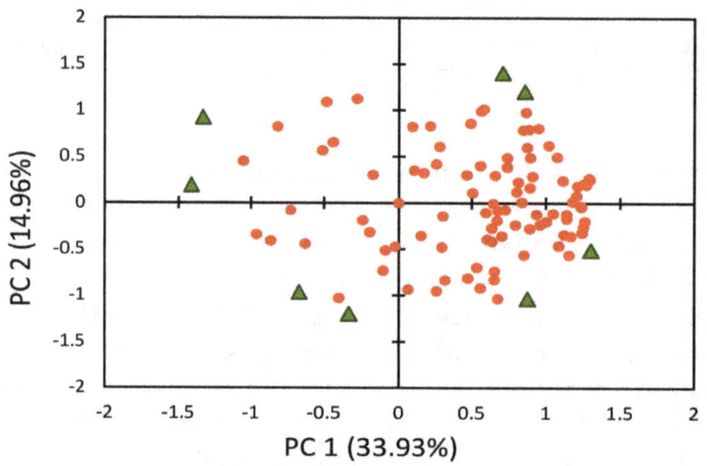


**(a)**

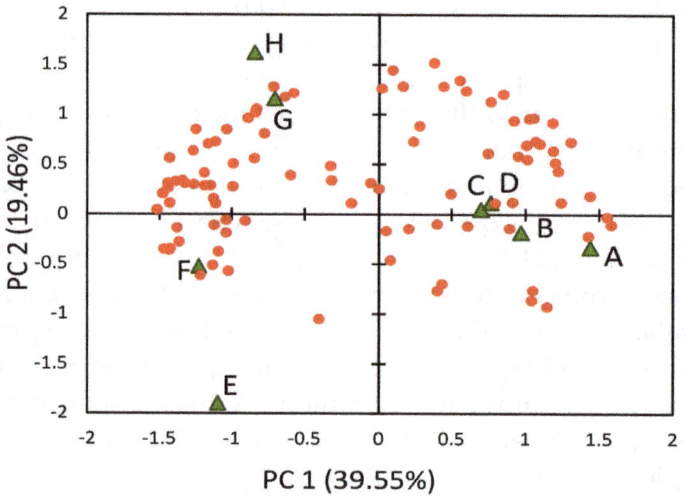


**(b)**

**Figure 1.** *Cont.*

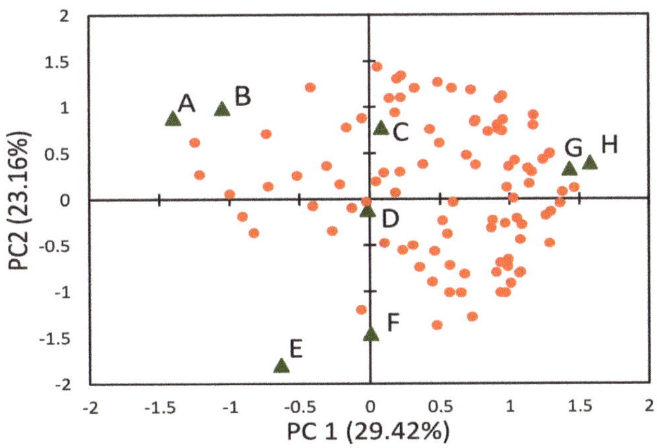


(c)

**Figure 1.** Internal preference maps for different testing conditions: (**a**) blind, (**b**) expected, (**c**) informed. A,B—clear FC juices; C,D—cloudy FC juices; E,F—cloudy NFC juices; G,H—fresh juices. Red dots correspond to the individual consumers. PC1 and PC2 correspond to the first and second principal component, respectively.

The distribution of individual consumer scores varied depending on the testing conditions. Regardless of the testing conditions, the two juices in each of the pairs belonging to a certain product category were rated similarly by consumers; thus, the points corresponding to these juices are located close together in the two-dimensional space.

Most of the consumers preferred cloudy FC and freshly squeezed juices in the blind test (Figure 1a). The first PC1 component distinctly differentiated the juices according to sensory preferences (direct vs. reconstituted juices). A very different distribution of consumers and juices was obtained for expected liking (Figure 1b), where two main groups of the consumers may be distinguished. The first group assigned higher ratings to NFC juices, available in glass bottles. The second group, more scattered, rated FC juices, clear and cloudy varieties, available in 1 L boxes higher. As shown in Figure 1c, combined internal and external attributes provided poorer differentiation of the juices studied. Note that PC1 clearly differentiated clear FC juices from fresh juices, but not pasteurized cloudy FC and NFC juices. Most consumers, basing on sensory quality and external attributes, assessed the quality of fresh juices (G, H) as higher. The preference map for informed testing differed significantly from that for blind testing. These results confirmed the strong influence of the external attributes on the sensory evaluation of apple juice. Similar results were obtained by Lee, Lusk, Mirosa, and Oey [22].

There was a strong positive correlation ($p < 0.05$) between the expected and informed liking scores ($r = 0.944$) and between the blind and informed consumer scores ($r = 0.800$). No correlation was observed between blind and expected liking, indicating that the two approaches measure different aspects of product quality. These results are in agreement with the findings reported by Varela, Ares, Giménez, and Gámbaro [35].

### *3.2. Perception of Apple Juices Based on External Attributes*

In order to gather supplementary data on the consumer perception of the juices studied, a check-all-that-apply (CATA) questionnaire with terms related to characteristics of the product and the brand image was used. CATA questions are increasingly more often used in consumer research to investigate the perceptions of a variety of attributes perceived by consumers and to obtain a rapid product profile from consumers [36–38].

Consumers used between one and seven terms to describe the apple juices in both laboratory and online studies. The most frequently used terms were natural and tasty; the least used term was tasteless. Most consumers indicated both cloudy NFC (E, F) and fresh (G, H) juices as natural. These juices were rated also as very healthy and expensive. The differences between FC and NFC juices in their naturalness were clearer in the laboratory as compared to the online study. This may be due to increased attention of the respondents and the lack of a time limit in the laboratory testing. Furthermore, consumers perceived the juices studied as more innovative and trendier in the classic laboratory study as compared to the online experiment. This may be due to the fact that the photos did not fully reflect the differences between the products. Based on the results obtained, we concluded that apple juices are generally perceived by the consumers as natural and healthy products.

The correspondence analysis was used to visualize and interpret the results of the check-all-that-apply questions [39]. The plots of the first two factors for the consumer laboratory study and the online study data are shown in Figure 2.

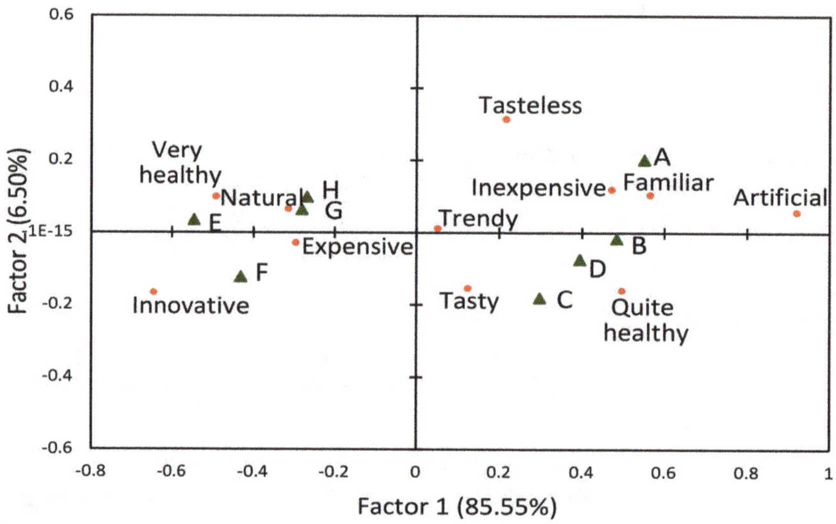


(a)

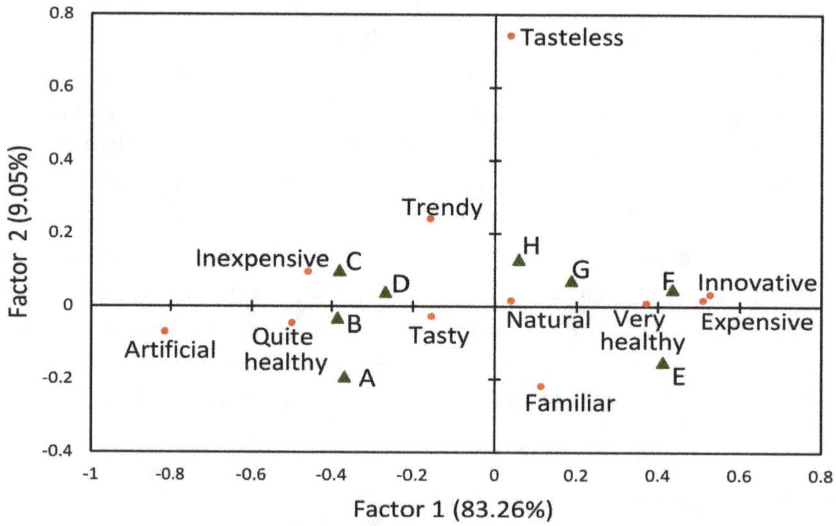


(b)

**Figure 2.** Correspondence analysis of check-all-that-apply responses: (**a**) laboratory consumer study, (**b**) online consumer study. A,B—clear FC juices; C,D—cloudy FC juices; E,F—cloudy NFC juices; G,H—fresh juices.

A two-dimensional representation of the data was obtained, describing respectively 92.05% and 92.31% of the variability of laboratory and online consumer data. The results of the correspondence analysis showed that the external attributes differentiated the studied juices. The first direction differentiating the products was related to the perceived naturalness. The second dimension was related to the expected taste of the juice and was positively correlated to tasteless and negatively to tasty. As shown in Figure 2, the first and the second dimensions sorted the samples into the two main groups, according to consumer expectations: the first group contained FC juices (A, B, C, D) and the second group direct juices (E, F, G, H). The sample distribution was distinctly different from that obtained for blind and informed liking scores, while similar to the distribution of expected liking scores (Figure 2). As a result of both experiments, FC juices were perceived as more artificial, quite healthy, and inexpensive, whereas NFC juices were perceived as more healthy, natural, innovative, and expensive. These characteristics may be explained by the type of packaging and the production information declared by the manufacturer on the package. Note that NFC juices are available in glass bottles, which makes the natural sediment more visible.

CATA results obtained in the laboratory and online studies were convergent. This suggests that an online experiment could be an effective alternative to the classic consumer laboratory study, being a faster and cheaper way of obtaining consumer perception data.

### *3.3. Visual Attention to Product Labels*

To examine the labels' zones that attract consumer attention, an online study was conducted using the innovative Attensee.com application. Based on the defined reference areas (RA), the application generated three types of data which are similar to the data obtained by the eye-tracking tool: (1) path of engagement specifying locations on packaging which attracted the attention of the respondents; (2) heatmap for time, showing the average time that respondents focused on an area; and (3) heatmap for attention, specifying the number of people who focused on a particular area of an image.

Firstly, the image of all the analyzed products was presented to consumers. The purpose of this presentation was to investigate which products attract the most consumer attention during the decision process on the store shelf.

During the on-shelf product recognition survey, consumers mostly focused on NFC juices (samples E and F) from local producers and focused the least on FC juices from well-known brands. According to the CATA questionnaire results discussed above (Section 3.2), it can be presumed that customers were focusing longer on products that they considered more attractive (healthy, natural, innovative) and from the premium segment (expensive). Consumer interest could also be due to the fact that NFC juices produced by local producers are less recognized than FC juices from well-known brands available nationwide. This feature may be seen by consumers as positive. Hempel and Hamm [40] examined consumer preferences for different food products of varying places of origin and observed that consumers prefer locally produced food.

Based on our results, we may distinguish the packaging elements that attract consumer attention. First of all, consumers paid attention to brand names of products, but it should be emphasized that they focused more attention on local juices as opposed to well-known brands. Subsequently, consumers noticed the information about product type and quality. More specifically, over 99% of consumers noticed the green BIO-label on NFC juice E. Studies show that consumers have a high preference for ethical or 'green' products and favor environmentally labeled packaging as the most important criteria in food choice [41,42]. Furthermore, consumers focused for over 2 s on average on the area of NFC juice F informing about type of product ("pressed apple juice").

One more experimental survey concerned the individual label pictures of juices from different product categories. Table 3 presents the aggregated results of this study.

**Table 3.** Average percentage of consumers who focused their attention on the label's specified reference area defined individually for each type of label, and the average time of attention for apple juices assessed using a visual attention tracking tool.

| Juice | Average Percentage of Consumers (%) | | | Average Time of Attention (s) | | |
|---|---|---|---|---|---|---|
| | Brand | Nutritional Label | Type of Product | Brand | Nutritional Label | Type of Product |
| A | 87 | 94 | 74 | 0.3 | 2.2 | 1.2 |
| B | 87 | 100 | 100 | 0.6 | 0.8 | 0.8 |
| C | 85 | 97 | 70 | 0.6 | 2.1 | 0.7 |
| D | 76 | 84 | 84 | 0.5 | 0.9 | 1.0 |
| E | 93 | 83 | 91 | 0.7 | 0.4 | 0.7 |
| F | 95 | 97 | 86 | 0.2 | 1.2 | 0.7 |
| G | 99 | 100 | 100 | 0.4 | 0.3 | 1.0 |
| H | 95 | 95 | 87 | 0.4 | 1.4 | 0.7 |
| Mean | 89.6 | 93.8 | 86.5 | 0.5 | 1.2 | 0.8 |

Generally, the longest average attention time (1.2 s) was devoted to the part of labels with nutritional information and involved almost 94% of consumers. Based on a detailed analysis of the consumer results and reference areas for the individual products, we noticed that the more clearly designed and sharply outlined nutritional information caused shorter attention time to this part of the label.

The brand names of juices were explored by almost 90% of the consumers who focused on this part of the label for an average of 0.5 s. Analyzing the percentage of consumers who focused their attention on the label's specified reference area, we found that the products from the better-known brands (clear and cloudy FC juices) attracted less attention than little-known craft brands (freshly squeezed and NFC juices).

The third main area of the consumer interest was the part of the label describing the type of product. About 87% of consumers focused their attention on that area for 0.8 s on average. In this case, the differences in the time of gaze fixation for all of the studied products was not significant and fluctuated between 0.7 and 1.2 s.

The next two areas, also important from the consumer point of view, were the information about the manufacturer/producer and places with special pictograms and signs including certificates and pictograms describing social action. The interest in manufacturer information was visible especially in the case of the NFC juices (samples F and E)—88% and 78% of consumers visited these areas and spent about 0.5 and 0.9 s, respectively, focusing on it. Special signs about sports and social action available on the package of juice D attracted the attention of 59% consumers, who spent about 0.7 s focusing on it. Furthermore, the culinary certificate on the label of juice F was noticed by 59% consumers who focused their gaze on it for 0.7 s on average.

When analyzing the heatmaps of a single juice label, we noted that consumers did not study some parts of the product information. This suggests that consumers did not assess the presented information comprehensively. These observations are in agreement with findings published by Oliveira et al. [15]. Generally, consumers mainly turn their attention to the information given in larger letters, information presented in an intense color (especially red), and special pictograms like certificates. It should be emphasized that we presented the labels to the consumers on a computer screen one by one. The consumers had possibility to also investigate the information presented on the sides and back of the packages. The consumer attention to product information in real-life conditions is expected to be even lower than presently reported.

It was previously reported that regardless of the type of product and the label design, consumers direct their attention to selected label zones such as brand, ingredients, and nutritional information [15,43]. The results obtained in the present study for real market products are convergent with those reports. The proposed approach would be useful for assessing the effects of a variety of commercial juice packaging designs on consumer perception.

## 4. Conclusions

The present study combined sensory evaluation and visual attention experiments to investigate how intrinsic and extrinsic factors affected consumers' perceptions of apple juice quality. Eight commercial apple juices were studied, representing the four product categories currently available on the Polish market, which were evaluated by consumers. In a blind test, consumers rated the experienced quality of fresh and cloudy FC juices as the highest and the quality of cloudy NFC juices as the lowest. The packages with the product description not only generated high consumer expectations, but also impacted the sensory liking in a markedly positive way. For most of the tested juices, consumers rated the expected quality higher than the experienced quality of the coded samples. The product information influenced the rating of the quality of NFC juices, fresh juices, and clear FC juices in a significant and positive manner. These results confirmed the importance of the information presented on the packaging. On the other hand, they indicated the need to improve the sensory quality of juices to meet consumer expectations and encourage them to buy the product again.

An innovative online visual attention tracking technique was used for a more detailed analysis of the perception of packaging and labels. The online shelf test revealed that consumers focused the most on NFC juices from local producers. Additional information about the perceived juice quality in both the laboratory and online studies was provided by CATA questionnaires. The results obtained in all experiments were convergent and revealed that consumers perceived NFC juices as more natural, healthy, and expensive than juices reconstituted from concentrate.

The study of individual packaging enabled to identify zones that attracted consumers' visual attention. The consumers predominantly focused on nutritional data, brand name, and information about the type of product. This knowledge may be of practical utility for packaging and label design, to improve the attractiveness and competitiveness of the product. However, the observed effects might be modified by many factors including participants' previous knowledge of the product, text information presented with the image, and the level of congruency of the images. These aspects will be explored more fully in future research.

We conclude that the experimental approach proposed in this study provided a comprehensive view of the influence of the product information presented on packaging on the consumer expectations and sensory perception; moreover, it enabled to identify the important features of the packaging and labels. The results confirmed the effect of intrinsic and extrinsic factors on the product quality perception.

**Author Contributions:** Conceptualization, K.W., K.P.-L., and E.S.; Data curation, T.G.; Formal analysis, K.W., T.G., and E.S.; Investigation, K.W., and K.P.-L.; Methodology, K.W., K.P.-L., T.G., and E.S.; Supervision, E.S.; Validation, K.W., K.P.-L., and E.S.; Visualization, K.W.; Writing–original draft, K.W., and K.P.-L.; Writing–review & editing, K.W., K.P.-L., T.G., and E.S.

**Acknowledgments:** The authors thank Dawid Wiener (Cogision Ltd.) for the opportunity to use the Attensee.com software. The authors are very grateful to all consumers participating in the study.

## References

1. Grunert, K.G. Current issues in the understanding of consumer food choice. *Trends Food Sci. Technol.* **2002**, *13*, 275–285. [CrossRef]
2. Piqueras-Fiszman, B.; Spence, C. Sensory expectations based on product-extrinsic food cues: An interdisciplinary review of the empirical evidence and theoretical accounts. *Food Qual. Prefer.* **2015**, *40*, 165–179. [CrossRef]
3. Li, X.E.; Jervis, S.M.; Drake, M.A. Examining extrinsic factors that influence product acceptance: A review. *J. Food Sci.* **2015**, *80*, R901–R909. [CrossRef] [PubMed]

4. Mueller, S.; Szolnoki, G. The relative influence of packaging, labelling, branding and sensory attributes on liking and purchase intent: Consumers differ in their responsiveness. *Food Qual. Prefer.* **2010**, *21*, 774–783. [CrossRef]
5. Deliza, R.; MacFie, H.J.H. The generation of sensory expectation by extermal cues and its effect on sensory perception and hedonic ratings: A review. *J. Sens. Stud.* **1996**, *11*, 103–128. [CrossRef]
6. Bartels, M.; Tillack, K.; Jordan Lin, C.-T. Communicating nutrition information at the point of purchase: An eye-tracking study of shoppers at two grocery stores in the United States. *Int. J. Consum. Stud.* **2018**, *42*, 557–565. [CrossRef]
7. Menichelli, E.; Olsen, N.V.; Meyer, C.; Næs, T. Combining extrinsic and intrinsic information in consumer acceptance studies. *Food Qual. Prefer.* **2012**, *23*, 148–159. [CrossRef]
8. Combris, P.; Bazoche, P.; Giraud-Héraud, E.; Issanchou, S. Food choices: What do we learn from combining sensory and economic experiments? *Food Qual. Prefer.* **2009**, *20*, 550–557. [CrossRef]
9. Olsen, N.V.; Menichelli, E.; Meyer, C.; Næs, T. Consumers liking of private labels. An evaluation of intrinsic and extrinsic orange juice cues. *Appetite* **2011**, *56*, 770–777. [CrossRef]
10. Wansink, B.; Payne, C.R.; North, J. Fine as North Dakota wine: Sensory expectations and the intake of companion foods. *Physiol. Behav.* **2007**, *90*, 712–716. [CrossRef]
11. Sörqvist, P.; Marsh, J.E.; Holmgren, M.; Hulme, R.; Haga, A.; Seager, P.B. Effects of labeling a product eco-friendly and genetically modified: A cross-cultural comparison for estimates of taste, willingness to pay and health consequences. *Food Qual. Prefer.* **2016**, *50*, 65–70. [CrossRef]
12. Orquin, J.L.; Bagger, M.P.; Lahm, E.S.; Grunert, K.G.; Scholderer, J. The visual ecology of product packaging and its effects on consumer attention. *J. Bus. Res.* **2019**. [CrossRef]
13. Graham, D.J.; Orquin, J.L.; Visschers, V.H.M. Eye tracking and nutrition label use: A review of the literature and recommendations for label enhancement. *Food Policy* **2012**, *37*, 378–382. [CrossRef]
14. Fenko, A.; Nicolaas, I.; Galetzka, M. Does attention to health labels predict a healthy food choice? An eye-tracking study. *Food Qual. Prefer.* **2018**, *69*, 57–65. [CrossRef]
15. Oliveira, D.; Machín, L.; Deliza, R.; Rosenthal, A.; Walter, E.H.; Giménez, A.; Ares, G. Consumers' attention to functional food labels: Insights from eye-tracking and change detection in a case study with probiotic milk. *LWT-Food Sci. Technol.* **2016**, *68*, 160–167. [CrossRef]
16. Gere, A.; Danner, L.; de Antoni, N.; Kovács, S.; Dürrschmid, K.; Sipos, L. Visual attention accompanying food decision process: An alternative approach to choose the best models. *Food Qual. Prefer.* **2016**, *51*, 1–7. [CrossRef]
17. Rebollar, R.; Lidón, I.; Martín, J.; Puebla, M. The identification of viewing patterns of chocolate snack packages using eye-tracking techniques. *Food Qual. Prefer.* **2015**, *39*, 251–258. [CrossRef]
18. Sonne, A.M.; Grunert, K.G.; Veflen, O.N.; Granli, B.S.; Szabó, E.; Banati, D. Consumers' perceptions of HPP and PEF food products. *Br. Food J.* **2012**, *114*, 85–107. [CrossRef]
19. Lee, P.Y.; Kebede, B.T.; Lusk, K.; Mirosa, M.; Oey, I. Investigating consumers' perception of apple juice as affected by novel and conventional processing technologies. *Int. J. Food Sci. Technol.* **2017**, *52*, 2564–2571. [CrossRef]
20. AIJN, E.F.J.A. Liquid Fruit Market Report. 2017. Available online: http://www.aijn.org/publications/facts-and-figures/aijn-market-reports/ (accessed on 15 April 2019).
21. Stolzenbach, S.; Bredie, W.L.P.; Christensen, R.H.B.; Byrne, D.V. Impact of product information and repeated exposure on consumer liking, sensory perception and concept associations of local apple juice. *Food Res. Int.* **2013**, *52*, 91–98. [CrossRef]
22. Lee, P.Y.; Lusk, K.; Mirosa, M.; Oey, I. Effect of information on Chinese consumers' acceptance of thermal and non-thermal treated apple juices: A study of young Chinese immigrants in New Zealand. *Food Qual. Prefer.* **2016**, *48*, 118–129. [CrossRef]
23. Han, J.-W.; Ruiz-Garcia, L.; Qian, J.-P.; Yang, X.-T. Food packaging: A comprehensive review and future trends. *Compr. Rev. Food Sci. Food Saf.* **2018**, *17*, 860–877. [CrossRef]
24. Venter, K.; van der Merwe, D.; de Beer, H.; Kempen, E.; Bosman, M. Consumers' perceptions of food packaging: An exploratory investigation in Potchefstroom, South Africa. *Int. J. Consum. Stud.* **2011**, *35*, 273–281. [CrossRef]

25. Van Rompay, T.J.L.; Deterink, F.; Fenko, A. Healthy package, healthy product? Effects of packaging design as a function of purchase setting. *Food Qual. Prefer.* **2016**, *53*, 84–89. [CrossRef]
26. Włodarska, K.; Pawlak-Lemańska, K.; Górecki, T.; Sikorska, E. Classification of commercial apple juices based on multivariate analysis of their chemical profiles. *Int. J. Food Prop.* **2017**, *20*, 1773–1785. [CrossRef]
27. Włodarska, K.; Pawlak-Lemańska, K.; Sikorska, E. Prediction of key sensory attributes of apple juices by multivariate analysis of their physicochemical profiles. *Br. Food J.* **2019**, *121*, 2429–2441. [CrossRef]
28. Włodarska, K.; Pawlak-Lemańska, K.; Górecki, T.; Sikorska, E. Perception of apple juice: A comparison of physicochemical measurements, descriptive analysis and consumer responses. *J. Food Qual.* **2016**, *39*, 351–361. [CrossRef]
29. Berger, S.; Wagner, U.; Schwand, C. Assessing advertising effectiveness: The potential of goal-directed behavior. *Psychol. Market.* **2012**, *29*, 411–421. [CrossRef]
30. Holmqvist, K.; Nyström, N.; Andersson, R.; Dewhurst, R.; Jarodzka, H.; Van de Weijer, J. *Eye Tracking: A Comprehensive Guide to Methods and Measures*; Oxford University Press: Oxford, UK, 2011.
31. Aguiar, I.B.; Miranda, N.G.M.; Gomes, F.S.; Santos, M.C.S.; Freitas, D.d.G.C.; Tonon, R.V.; Cabral, L.M.C. Physicochemical and sensory properties of apple juice concentrated by reverse osmosis and osmotic evaporation. *Innov. Food Sci. Emerg. Technol.* **2012**, *16*, 137–142. [CrossRef]
32. Beveridge, T. Opalescent and cloudy fruit juices: Formation and particle stability. *Crit. Rev. Food Sci.* **2002**, *42*, 317–337. [CrossRef]
33. Carbonell, L.; Izquierdo, L.; Carbonell, I.; Costell, E. Segmentation of food consumers according to their correlations with sensory attributes projected on preference spaces. *Food Qual. Prefer.* **2008**, *19*, 71–78. [CrossRef]
34. Van Kleef, E.; van Trijp, H.C.M.; Luning, P. Internal versus external preference analysis: An exploratory study on end-user evaluation. *Food Qual. Prefer.* **2006**, *17*, 387–399. [CrossRef]
35. Varela, P.; Ares, G.; Giménez, A.; Gámbaro, A. Influence of brand information on consumers' expectations and liking of powdered drinks in central location tests. *Food Qual. Prefer.* **2010**, *21*, 873–880. [CrossRef]
36. Ares, G.; Jaeger, S.R. Check-all-that-apply questions: Influence of attribute order on sensory product characterization. *Food Qual. Prefer.* **2013**, *28*, 141–153. [CrossRef]
37. Bruzzone, F.; Vidal, L.; Antúnez, L.; Giménez, A.; Deliza, R.; Ares, G. Comparison of intensity scales and CATA questions in new product development: Sensory characterisation and directions for product reformulation of milk desserts. *Food Qual. Prefer.* **2015**, *44*, 183–193. [CrossRef]
38. Dooley, L.; Lee, Y.-S.; Meullenet, J.-F. The application of check-all-that-apply (CATA) consumer profiling to preference mapping of vanilla ice cream and its comparison to classical external preference mapping. *Food Qual. Prefer.* **2010**, *21*, 394–401. [CrossRef]
39. Meyners, M.; Castura, J.C.; Carr, B.T. Existing and new approaches for the analysis of CATA data. *Food Qual. Prefer.* **2013**, *30*, 309–319. [CrossRef]
40. Hempel, C.; Hamm, U. Local and/or organic: A study on consumer preferences for organic food and food from different origins. *Int. J. Consum. Stud.* **2016**, *40*, 732–741. [CrossRef]
41. Rokka, J.; Uusitalo, L. Preference for green packaging in consumer product choices – Do consumers care? *Int. J. Consum. Stud.* **2008**, *32*, 516–525. [CrossRef]
42. Zepeda, L.; Sirieix, L.; Pizarro, A.; Corderre, F.; Rodier, F. A conceptual framework for analyzing consumers' food label preferences: An exploratory study of sustainability labels in France, Quebec, Spain and the US. *Int. J. Consum. Stud.* **2013**, *37*, 605–616. [CrossRef]
43. Ares, G.; Giménez, A.; Bruzzone, F.; Vidal, L.; Antúnez, L.; Maiche, A. Consumer visual processing of food labels: Results from an eye-tracking study. *J. Sens. Stud.* **2013**, *28*, 138–153. [CrossRef]

# 9

# The Role of Trust in Explaining Food Choice: Combining Choice Experiment and Attribute Best–Worst Scaling

**Ching-Hua Yeh [1,*], Monika Hartmann [1] and Nina Langen [2]**

[1] Department of Agricultural and Food Market Research, Institute for Food and Resource Economics, University of Bonn, 53115 Bonn, Germany; monika.hartmann@ilr.uni-bonn.de

[2] Department of Education for Sustainable Nutrition and Food Science, Institute of Vocational Education and Work Studies, Technical University of Berlin, 10587 Berlin, Germany; nina.langen@tu-berlin.de

* Correspondence: chinghua.yeh@ilr.uni-bonn.de.

† The paper was presented in the conference of ICAE 2018.

**Abstract:** This paper presents empirical findings from a combination of two elicitation techniques—discrete choice experiment (DCE) and best–worst scaling (BWS)—to provide information about the role of consumers' trust in food choice decisions in the case of credence attributes. The analysis was based on a sample of 459 Taiwanese consumers and focuses on red sweet peppers. DCE data were examined using latent class analysis to investigate the importance and the utility different consumer segments attach to the production method, country of origin, and chemical residue testing. The relevance of attitudinal and trust-based items was identified by BWS using a hierarchical Bayesian mixed logit model and was aggregated to five latent components by means of principal component analysis. Applying a multinomial logit model, participants' latent class membership (obtained from DCE data) was regressed on the identified attitudinal and trust components, as well as demographic information. Results of the DCE latent class analysis for the product attributes show that four segments may be distinguished. Linking the DCE with the attitudinal dimensions reveals that consumers' attitude and trust significantly explain class membership and therefore, consumers' preferences for different credence attributes. Based on our results, we derive recommendations for industry and policy.

**Keywords:** preference; trust; choice experiment; best-worst scaling; latent class analysis; hierarchical Bayesian mixed logit model

---

## 1. Introduction

Over the last decades, food safety and quality have been highly debated and investigated topics in policy, industry, and research. This holds for industrialized as well as emerging countries, such as Taiwan. In Taiwan, food scares and scandals, such as food adulteration [1], food-borne contamination [2], counterfeiting [3], and mislabeling [4], have induced consumer distrust and concerns regarding the quality and safety of food. Additionally, high—and in parts, improper—use of chemical inputs in Taiwanese agriculture [5] has led to illegal levels of chemical residues in food products, with considerable danger for the immediate and long-term health of consumers [6]. For example, in December 2011, the Greenpeace organization (http://www.greenpeace.org/taiwan/zh/publications/reports/food-agriculture/) sampled 58 fresh fruits and vegetables in eight supermarket chains across Taiwan, and detected 36 different pesticide residues above the maximum allowable levels in 43 types of fruits and vegetables. In the same year the Taiwan Food and Drug Administration discovered a major threat to public health caused by phthalate-contaminated foodstuffs sold on the Taiwanese market.

This outbreak event is known as the "2011 Taiwan Food Scandal" [7,8]. Most Taiwanese retailers reacted to the numerous food safety incidents by starting in 2011 to display chemical residue test information, particularly for fresh agricultural and food products. Parallel, consumers' interest in food labels associated with a higher level of product quality and safety has been gaining in relevance [9–11].

### 1.1. Thematic Background

There has been considerable interest in studying how consumers, across countries, evaluate and use food quality and safety information [12,13]. Research shows that consumption of organic food has increased [14–16], motivated by consumers' values and health concerns [10,17]. Consumers also associate the origin of foods with product and process quality [9,18]. Previous studies showed that domestically grown food is perceived as fresher and/or of higher quality [19,20]. More generally, research reveals that country (or region) of origin conveys the production country's (or area's) reputation for value and quality [19,21]. Food labels can be a source of information for consumers with respect to a product's quality and safety characteristics. However, the usage of labels for product choice crucially depends on consumers' perception and trust in the signals [22,23]. Thus, for decision makers in policy and businesses, it is central to understand how consumers perceive and trust food labels, and food product and process characteristics.

Trust is a complex notion, and a multifaceted concept. In the past 30 years, a growing body of literature has emerged across various scientific fields. In the empirical research on food, a number of studies have examined the role of consumer trust in different food institutions [24–30] for consumers' perceptions of food risk [26,31–33], such as the usage of pesticides [34,35], irradiated foods [36–39], nanotechnology foods [40–43], and genetically modified foods [31,44]. Beyond the wide-ranging investigations on consumer trust in risk management settings, there is also an increasing number of works dealing with consumer trust with respect to different typologies of food (see, e.g., [45–48]) and consumers' trust in food suppliers and retailers [49,50]. The term "trust" has often been linked to broader categories, such as confidence [26], preference [51–53], loyalty [54], risk taking [55], satisfaction [56], cooperation [57], and commitment [58].

According to Mayer et al. [59], trust can be defined as:

> ... the willingness of a party to be vulnerable to the actions of another party based on the expectation that the other will perform a particular action important to the trustor, irrespective of the ability to monitor or control that other party ([59], p. 712).

However, it should be noted that despite the extensive research on trust, there is neither consensus among scholars on the definition or conceptualization of trust [60] nor on its dimensions [61]. According to Siegrist et al. [62] and Ding et al. [63], trust can be measured in two dimensions, a specific and a more general manner. Specific trust refers to trust specifically related to the given referent (i.e., trust to the institution or the company) while general trust is presented as trust towards an object or a group entity. Based on the previous literature, consumer trust can also be divided into four conceptual dimensions: (1) Trust belief [64–68]; (2) trust intention [59,64,65,68–71]; (3) institutional-based trust [59,64,65,68,69,72]; and (4) general trust [28,59,64,65,68,70,73–75]. According to Moorman et al. [76], psychological and sociological aspects are the two key components characterizing trust in the marketing area. Taking the purchase of a food product as an example, the former refers to the confidence and belief in the trustworthiness of the related food actors, such as producers, retailers, certification bodies, and labels, and covers the dimension "trust beliefs" [64,65,67,68,70]. "Trust beliefs" stands for the perception of the trustworthiness of a person or object [59,64,65,68]. The latter implies the cognitive, affectional, or behavioral willingness to rely on such actors and relates to the dimension of "trust intention" [64,65,71]. Trust intention is recognized as the intention to engage in trust-related behavior with a specific willingness of the trustor to rely, or intend to rely, on other trustees based on the expectation about the behavior of others though those cannot be controlled [64,65,68,69]. Likewise, "institutional-based trust", so-called systems trust [70], is considered as an antecedent to trusting intentions and trusting beliefs [59,64,65]. It refers to an individual's perception that an action is constitutively embedded in an institutional environment that is conducive to a favorable outcome [59,65,68,71]. Thus, an individual intrinsically feels or believes that the macro-level organization or the social environment, in which (s)he performs a transaction, takes on a regulatory role and provides appropriate formal protection [59,64,65,68]. "General trust", so-called dispositional trust [65,70], is described as the attitude of the general trusting stance and natural tendency of an individual to trust other people or an object; thus, the trustor inherently possesses faith in optimism [64,68,70,71]. It is like a personal conscious choice or strategy to trust others until they prove to be untrustworthy [59,68,70,71]. Consumer trust is a subjective concept and is influenced by an individual's past experience and perceived reputational value of the object [77].

### 1.2. Methodological Background

Many food-related studies investigate via choice experimental settings the role of labels in consumers' purchase decisions (see the review by [78–80]), or the interaction between food labeling and consumer trust (see the review by [81]). The respective literature so far primarily focuses on western countries with only three studies referring to non-western countries (two peer-reviewed papers, one focusing on Japan ([82] and one on China [83]; as well as a dissertation with data collected in Kenya [84]). Parallel, hybrid choice models (HCMs) have been developed. Those models extend the standard discrete choice experiment (DCE) by explicitly incorporating consumers' psychological or sociological factors into a random utility framework, thereby better capturing the complexity of the choice process [85–87]. HCM covers a number of approaches encompassing latent class [88] or mixed logit models [89] that also include, besides the information obtained from the DCE, e.g., attitudinal factors. Integrated choice and latent variable models [90] also belong to this group. Those models improve the modeling of the decision-making behavior. Our approach goes beyond the HCMs as we combine the findings of two elicitation techniques—DCE and best–worst scaling (BWS)—to provide information about the role of consumers' trust in food choice decisions in the case of credence attributes in a discriminant manner. Although, Song et al. [91] also combined DCE and BWS, the authors used the identical seven attributes in the DCE and in the BWS settings and analyzed the choice data via HCM, treating each BWS importance score as a single variable. In our study, the attributes in the DCE (four attributes) and the BWS (25 statements) designs are different, with the former referring to characteristics of the product and the latter to 25 attitudinal and trust-based items, which are for the further analysis componentized to five attitudinal and trust dimensions. Thus, in contrast to Song et al. [91] we are able to link consumers' choice to the attitudinal and trust statements. Furthermore, our approach differs in that we apply DCE using latent class analysis and BWS employing Bayesian estimation. Thus, we use state of the art methods to provide in-depth insights in explaining consumers' choice for different consumer segments by attitudinal and trust dimensions.

Besides this methodological innovation, our study adds to the literature by revealing in the example of Taiwan—a newly industrialized country—the relevance selected process and food safety standards have for consumers in their purchase decisions, thereby in contrast to previous studies differentiating between consumer segments [92,93] and identifying the role of trust for consumers when buying food products. In summary, our analysis aimed at providing a better understanding of consumer choices, and thus allowing for more meaningful recommendations for marketers and policy makers.

This paper is structured as follows: First, the methods of the discrete choice experiment and best–worst scaling and their application are introduced, followed by a presentation of the empirical results. We conclude with a discussion of our empirical findings, derive practical implications, and suggest directions for future research.

## 2. Materials and Methods

The study combined two elicitation techniques: Discrete choice experiment (DCE) and best–worst scaling (BWS). Both methods are based on random utility theory [94,95].

### 2.1. Data Collection

The questionnaire was formulated in Mandarin and started with two screening questions. To qualify for taking part in the survey, respondents had to be red sweet pepper consumers who were (partly) responsible for the food purchases in their family. Subsequently, participants were asked to complete two stated preference experiments (DCE and BWS as discussed in the next sections). In the last section of the questionnaire, participants were requested to provide information with respect to their food purchase behavior, shopping frequency, and socio-demographic characteristics, such as gender, age, and income. The questionnaire was tested to ensure the comprehension of the questionnaire. The survey was conducted by five trained interviewers in front of supermarkets (the two hyper-supermarkets Taisuco (http://www.tsctaisuco.com.tw/) and Carrefour (http://www.carrefour.com.tw/) were chosen based on the convenience of the location, the customer flow, and the amount of fresh agri-food product categories) in the three largest cities (New Taipei, Kaohsiung, and Taichung) of Taiwan in the form of computer-assisted web interviews. The majority of the respondents completed the survey within approximately 15 min. To reduce possible self-selection bias, we trained our interviewers to actively encounter randomly every second person leaving the checkout counter of the supermarkets. This was done to ensure that, e.g., not only young or female consumers were approached. Combining computer-assisted personal interviews with traditional web survey techniques allowed us to overcome the problem of coverage error linked to the latter [96] and non-response error linked to the former [97].

### 2.2. Discrete Choice Experiment (DCE)

Discrete choice experiments are based on Lancaster's new demand theory [95,98], which assumes that consumers' derive utility from a variety of product characteristics. Participants are presented with multiple choice sets and asked to choose the product among a given choice set of alternatives that holds the combination of attributes that maximizes his/her utility [99,100].

Our DCE was conducted to investigate consumers' food preference and heterogeneity regarding different food quality and food safety information. In this study, fresh unpackaged red sweet peppers were selected as the study object. This product seemed especially suitable for our analysis because first, it is part of Taiwanese people's daily diet [101,102], and second, it is one of the few fresh agri-food products permitted to be imported into Taiwan from mainland China. In fact, red sweet peppers are available on the Taiwanese domestic market in conventional and organic quality from the three countries considered in the study: Taiwan, China, and Japan. Country of origin (COO) and production methods (organic and conventional) are both important attributes influencing perceived quality and trust in a product's overall quality [103,104]. Besides these two attributes, we considered price and chemical residue testing information (see Table 1) as characteristics in the DCE. Those four attributes were identified as the most relevant selection criteria for consumers in their purchase of red sweet peppers based on two focus group discussions held in March 2014 by the first author of this paper via video meetings with Taiwanese consumers primarily responsible for their household food purchase ($n = 17$). Those food market experts were recruited via social media networks. Out of 12 potential attributes extracted from previous literature (COO, easiness in preparation; chemical residue testing; visual appearance of the product; production methods, e.g., organic certification; product's shelf life; taste of the product; health claim; word of mouth information; seasonal product; and sense of touch) participants were asked to select those five attributes most important to them when buying fresh fruits and vegetables. In addition, participants were encouraged to express their opinions and reasons about why they choose the respective attributes.

**Table 1.** Attributes and levels used in the Discrete Choice Experiment (DCE).

| Attributes | Levels |
|---|---|
| Country of origin | • Taiwan<br>• Japan<br>• China |
| Production method | • Organic<br>• Conventional |
| Chemical residue testing information | • Chemical residue test approved in the production country<br>• Chemical residue test approved in Taiwan<br>• No chemical residue test information provided |
| Price | • NT 65<br>• NT 85<br>• NT 105<br>• NT 125 |

In July 2014, 1 US Dollar = 29.98 New Taiwanese (NT) Dollars. 1 Taiwanese catty = 600 g.

Using NGENE version 1.1 [105], we identified an efficient unlabeled choice design with a D-error value of 0.237, which is smaller than the D-errors of other design alternatives, such as the sequential orthogonal design (0.420) and the least efficient simultaneous orthogonal design (2.018) [105]. The D-error is the most common criterion for evaluating the efficiency of experimental choice designs [106].

A design with the lowest D-error measure is a D-efficient design. Efficient designs can be generated using two different approaches. The first approach assumes that prior parameters are known with certainty by the researchers (e.g., [107,108]), whereas the second one uses prior parameter distributions (Bayesian efficient design) [109]. We used the latter approach. For that, a pilot study was carried out in April 2014 with Taiwanese consumers ($n$ = 290) from a convenience sample using an internet-based choice experiment questionnaire in Taiwan.

The pilot study's parameter estimates were used as the prior information to derive the asymptotic variance-covariance matrix and subsequently the D-efficient design. Based on an iterative process, the most efficient design with the smallest D-error was derived. The final design consisted of 36 choice sets. However, as 36 choice tasks would lead to respondents' fatigue, the choices were allocated in six blocks by NGENE software [105] via the design generation process, each consisting of six choice situations. Respondents were randomly assigned to the six blocks.

In each choice task, consumers were asked to make a choice between three red sweet peppers that varied in the levels of the four product attributes presented in Table 1. Though, all attributes depicted on the products in the choice tasks exist in the market, this does not hold for all combinations of attribute levels (e.g., the link between the chemical residue test and the country where it is approved). We also provided participants with an "opt-out" option, which ensured that participants were not forced to choose a product they normally would not purchase. In order to make the choice experiment as tangible as possible, the attribute levels were visualized using pictures and text as shown in Figure 1 (translated version).

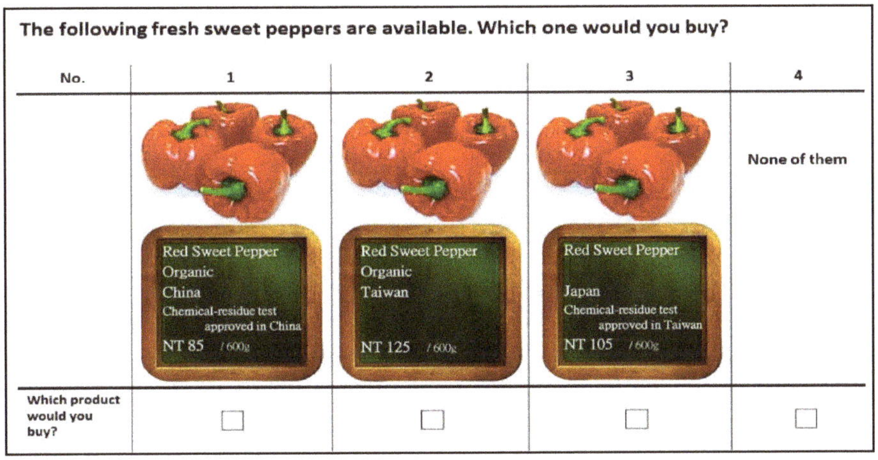


**Figure 1.** A choice task example in DCE (English translation).

We applied standard latent class analysis (LCA) [110,111] to the choice experimental data to identify different consumer segments. Latent class choice models assume that respondents can be categorized into two or more classes sharing unobserved characteristics that affect choice, in our case the choice of red sweet peppers, differentiated by different attribute levels. LCA allowed simultaneous determination of both the consumer's product choice and group membership, thereby segmenting the sample into internally homogenous subgroups regarding their preferences [110–115].

Following the random utility theory [95], the utility of individual $n$ ($n = 1, \ldots, N$) from choosing alternative $j$ ($j = 1, \ldots, J$) is the sum of a systematic observed component ($\beta_n X'_{nj}$) and a random error term ($\varepsilon_{nj}$) [110]:

$$U_{nj} = \beta_n X'_{nj} + \varepsilon_{nj},\ \beta_n \sim N(\alpha,\ D). \tag{1}$$

In the LCA model, the utility of alternative $j \in J$ to individual $n$, who belongs to a specific class, $s$, can be expressed as [112,113]:

$$U_{nj|s} = \beta_s X'_{nj} + \epsilon_{nj}, \tag{2}$$

where the $X'_{nj}$ is a vector of explanatory variables associated with alternative $j$ and individual $n$, $\beta_s$ is a class-specific parameter vector associated with the vector of explanatory attribute variables and $\epsilon_{nj}$ is the error term. The probability of individual $n$ choosing alternative $j$ from a particular choice set of alternative $J$ is conditional to the fitting of the individual to class $s$, which can be estimated using Equation (3) [110]:

$$P_{nj|s} = \frac{\exp\left(\beta_s X'_{nj}\right)}{\sum_{j=1}^{J} \exp\left(\beta_s X'_{nj}\right)}. \tag{3}$$

In addition, the belonging of individuals to the $S$ classes is determined by a classification model as a function of individual specific invariant characteristics. For each respondent, the class member probability ($C_{ns}$) of individual $n$ belonging to class $s$ can be computed as Equation (4):

$$C_{ns} = \frac{\exp(\gamma_s Z_n)}{\sum_{s=1}^{S} \exp(\gamma_s Z_n)}, \tag{4}$$

where $\gamma_s$ is the class membership vector estimates and $Z_n$ is a vector of individual specific invariant variables that enter the model for class membership. According to Boxall and Adamowicz [112], a vector-labelled $Z_n$ as covariates can be used as a proxy for individual motivating factors influencing the choice. This vector, Z, consists of both the observable indicators of the latent attitudes expressed by the respondent and of the observable socioeconomic characteristics, such as gender. The $N$ individuals can be divided into $S$ latent classes. Preferences are assumed to differ between latent classes but to be homogeneous within classes. The joint probability that individual $n$ belongs to a specific class, $s$, and selects alternative $j$ can be shown by Equation (5):

$$P_{nj} = \sum_{s}^{S} P_{nj|c} C_{ns} = \sum_{s=1}^{S} \frac{\exp\left(\beta_s X'_{nj}\right)}{\sum \exp\left(\beta_s X'_{nj}\right)} \times \frac{\exp(\gamma_s Z_n)}{\sum \exp(\gamma_s Z_n)}, \tag{5}$$

where $\beta_s$ is the parameter vector of individuals in the class $s$, and $\frac{\exp(\gamma_s Z_n)}{\sum \exp(\gamma_s Z_n)}$ is the probability of the individual $n$ falling into latent class $s$. The number of latent classes is determined by the researcher based on the statistical measures of fit, such as the Bayesian information criterion (BIC) and Akaike information criterion (AIC).

In addition, we calculated the segment-specific willingness-to-pay (WTP) values for each attribute level for the detected consumer segments by dividing the attribute level coefficient by the price coefficient. Due to the use of effect coding in the choice model, we calculated the mean WTP according to: WTP $= -\frac{2\beta_{\text{attribute level}}}{\beta_{\text{Price}}}$ [116–118]. We used the delta method introduced by [119] to generate the 95% confidence intervals for the WTP estimates.

### 2.3. Best–Worst Scaling (BWS)

Although the DCE method allows the combination of a product's attributes and levels to examine consumer preference with respect to a specific food product, it does not provide information on an individual's attitudinal and trust perceptions driving those choices. Therefore, many studies include rating scales (e.g., Likert measure points) to obtain information on consumers' attitudinal and trust perception. While rating tasks are easy for respondents to answer, they may ineffectually discriminate between rating statements [120], as respondents are not forced to make a choice between items, allowing them to rate multiple items as being of equally high importance. In addition, it is difficult to interpret what the rating scale values actually mean [121]. To overcome these weaknesses, we employed the best–worst scaling method [122] to uncover the attitudinal and trust factors underlying consumers' food choices on a reliable basis. Best–worst scaling (BWS), also known as maximum difference scaling, is an annotation scheme that exploits this comparative approach [123,124]. In BWS experiments, respondents are asked to choose the best and worst option among a number of statements. In this study, we used the so called 'object case' of BWS [122]. Thereby, respondents are forced to consider trade-offs and discriminate between options as in real life [125]. Choice frequencies are the metric that allow reveal information on the order and strength of the importance of all objects to be revealed. The method was introduced by [126] and first applied in the study of [122].

The BWS experiment covered nine attitudinal dimensions with a total of 25 statements related to attitudinal and trust factors. These were derived from the literature as well as our own consideration and were adapted to the context of the study (see Table 2). The balanced incomplete block design (BIBD) [127] has been frequently used in the BWS setting; however, as a BIBD is subject to the symmetry condition, the number of possible BIBDs is limited. In the present study, an orthogonal frequency balanced design using MaxDiff Designer v.6 [128] was generated to maximize the BWS design efficiency [129,130]. Orthogonality ensures that differences among items varied independently over choice sets while balance confirms that all items appeared with (nearly) equal frequency in the BWS questionnaire. Given a 25-statement BWS questionnaire, 300 BWS choice tasks were generated to control for context effects that may alter respondent choice processes [128]. To avert respondent fatigue, the choice tasks were divided into 30 blocks, where each version had 10 choice sets displaying 5 statements at a time. The generated BWS choice tasks satisfy the optimal design characteristics in terms of perfectly balanced frequency (on average, each BWS statement is displayed across all 30 blocking versions of the BWS questionnaire 60 times), orthogonality, positional balance (on average each statement appears 12 times at the same position (S.D. = 0.522)), and connectivity among tasks (all statements are linked directly) (see Appendix A) [128]. After ensuring a balanced and nearly orthogonal BWS design, the BWS situations were randomized and each respondent was randomly assigned to a BWS block while orthogonal properties hold after the randomization blocking process. Each BWS blocking version has the same sample size to maintain its statistical properties.

In addition, each statement appears equally often on each of the five positions within the BWS sets to prevent any position bias [131]. In each BWS task, respondents were asked to choose the statements that most and least represent their attitude when purchasing food (mix of questions regarding food purchases in general and red sweet peppers in particular) (see Figure 2).

**Table 2.** Nine dimensions of trust constructs with 25 Best-Worst Scaling (BWS) items.

| Trust Constructs | Items Used in BWS Experiment | No. of Items | References |
|---|---|---|---|
| 1. **Trust belief** in organic products from different COO | • Taiwanese/Chinese/Japanese organic sweet peppers are trustworthy. | 3 | Adapted from the studies of [65,68,132]. |
| 2. **(Dis)trust belief** in the superior nutritional value of organic food | • I feel sure that organic sweet peppers contain higher vitamin C and anti-cancer substances than conventional ones.<br>• I feel sure that organic sweet peppers contain the same vitamin C and anti-cancer substances as conventional ones. | 2 | Adapted from the studies of [65,132]. |
| 3. **(Dis)Trust belief** in the environmental benefit of organic food | • With purchasing organic sweet peppers, I help preserving the environment and natural resources.<br>• There are no differences between buying organic sweet peppers or conventional ones with respect to preserving the environment and natural resources. | 2 | Adapted from the studies of [65,132]. |
| 4. **Trust belief** in the monetary value of organic food from different COO | • Taiwanese/Chinese/Japanese organic sweet peppers have good value for money. | 3 | Adapted from the studies of [65,68]. |
| 5. **Trust intention** in purchasing products produced in different COO depending on chemical residue test information | • It is more likely that I buy Taiwanese/Chinese/Japanese sweet peppers if information on chemical residue testing is provided. | 3 | Adapted from the studies of [65,68]. |
| 6. **Trust intention** in purchasing products produced in different COO depending on a price discount | • It is more likely that I buy Taiwanese/Chinese/Japanese sweet peppers if it is on special offer. | 3 | Adapted from the studies of [65,68]. |

**Table 2.** *Cont.*

| Trust Constructs | Items Used in BWS Experiment | No. of Items | References |
|---|---|---|---|
| 7. **Institutional-based trust** in governments of different countries | • I feel assured that the Taiwanese/Chinese/Japanese institutions do a good job in adequately protecting consumers. | 3 | Adapted from the studies of [65]. |
| 8. **General trusting stance** regarding products produced from different COO | • I generally like to consume conventional sweet peppers produced in Taiwan/China/Japan. | 3 | Adapted from the studies of [133,134]. |
| 9. **General trusting stance** in organic products from different COO | • I generally like to consume organic sweet peppers produced in Taiwan/China/Japan. | 3 | |

Please consider only these five statements and imagine the situation of purchasing food. Please indicate the statement would you most agree to and the one you would least agree to.

| Agree least | Statements | Agree most |
|---|---|---|
| O | **Japanese organic sweet peppers** have good value for money. | O |
| O | It is more likely that I buy **Chinese sweet peppers** if **information on chemical residue** testing is provided. | O |
| O | It is more likely that I buy **Chinese sweet peppers** if it is on special offer. | O |
| O | I feel assured that the **Taiwanese institutions** do a good job in adequately protecting consumers. | O |
| O | **Chinese organic sweet peppers** are trustworthy. | O |

**Figure 2.** A choice task example of BWS (English translation).

First, hierarchical Bayesian estimation of the mixed logit model was applied to analyze the BWS choice data. We started with the general formula of random utility theory [95]:

$$U_{nj} = \beta_n X'_{nj} + \varepsilon_{nj}, \quad (6)$$

where $U_{nj}$ is the utility obtained by an individual $n$ choosing item $j$ ($j = 1, 2, \ldots, J$). $\beta_n$ is the individual-specific preference parameter vector, $X'_{nj}$ is the vector of observable explanatory variables including the chosen alternative $j$, and $\varepsilon_{nj}$ is the stochastic error term. In the BWS dataset, the most important item is coded as 1, where the least important item is coded as −1, and the non-chosen items are coded as 0.

It is assumed that an individual, $n$, chooses item $j$ and $j'$ as the most important and the least important item in a choice set, respectively, out of a choice set of $J$ items. Thus, the utility difference between $U_{njt}$ and $U_{nj't}$ is then greater than all other $J(J-1)$ possible differences among the other items in the choice set. Namely, in our case, each choice scenario of 5 items would have $5(5-1) = 20$ possible best–worst combinations a person could choose. Following Louviere et al. [124], the choice probability of the individual $n$ of choosing item $j$ as the best and $j'$ as the worst can be written as:

$$P = \frac{\exp\left(\beta_n X'_{nj} - \beta_n X'_{nj'}\right)}{\sum_{\substack{j,\, j' \in J \\ j \neq j'}} \exp\left(\beta_n X'_{nj} - \beta_n X'_{nj'}\right)}. \quad (7)$$

In hierarchical Bayesian estimation, based on Equation (6), we estimated the parameters at the individual level and the coefficient vector, $\beta_n$, for the individual, $n$, can be expressed by dividing it into an individual characteristic vector ($Z_n$) and the parameter matrix, $\Gamma$, shown in Equation (8):

$$\beta_n = \Gamma Z_n + \delta_n, \ \delta_n \sim N(0, \mathrm{D}), \tag{8}$$

where $\delta_n$ is the error term assuming that a normal distribution with a mean of 0, and D represents the covariance between the partworth estimated values [135]. During the analytical process, a parameter's posterior distribution is computed by combining its prior distribution for each partworth estimate with the likelihood determined based on the choice data. Thus, Equation (9) plays a role of identifying the prior distribution for the prior distributions of $\Gamma$ (follows a normal distribution) and D (follows an inverse-Wishart distribution) that were established to complete the hierarchical Bayesian modeling procedure [135]:

$$\begin{aligned} \Gamma &\sim N(a,\ A) \\ \mathrm{D} &\sim IW(w,\ W). \end{aligned} \tag{9}$$

Along with the above procedures and assumptions, choice data were drawn from a conditional distribution by generating the hierarchical structure shown in Equation (10):

$$\begin{aligned} &\Gamma \big| \ \mathrm{D}, \beta_n \\ &\mathrm{D} \big| \beta_n,\ \Gamma \\ &\beta_n \big| \Gamma, \mathrm{D}. \end{aligned} \tag{10}$$

When the covariance matrix (D) and the individual level value of an attribute ($\beta_n$) are given, the coefficient estimates ($\Gamma$) for the variable of an individual characteristics ($Z_n$) can be extracted. Therefore, we were also able to extract the D when $\beta_n$ and $\Gamma$ were obtained, as well as $\beta_n$ when $\Gamma$ and D were gained. This process repeats iteratively until a parameter value converges to draw a distribution of the individual parameters [135]. As a result, we obtained individual attitudinal importance scores for each of the BWS statements. Compared to the multinomial logit model or the mixed logit model with classical maximum likelihood estimation, the hierarchical Bayesian estimation allows for more precise estimates of individual-level partworth values [136,137] by combining information on the distribution of utility values across the entire sample with the specific choices of the individual [110,138].

Second, following the analytic methodology of [139], a principal component analysis (PCA) with oblimin rotation was used to identify latent constructs as the drivers of food purchasing decisions behind the 25 attitudinal items [140,141]. The number of components in the PCA was determined by the scree test, and the parallel analysis method [142–145] using STATA version 15. The former identifies the optimal number of components to retain by examining the scree plot of the eigenvalues and looking for the "elbow" point in the data where the curve flattens out. The latter compares the components' eigenvalues derived from the own data set with those of a large number of data matrices obtained from random values of data with the same dimensionality (same number of observations and same number of variables). As long as the estimated eigenvalues from the own data exceeds the mean of the eigenvalues of the random data the component is retained. The parallel analysis is considered to be among the most recommended factor/component retention methods [143–146] for assessment of the dimensionality of a variable set. Afterward, for each component the individual-level important scores were obtained from the respective values of the underlying BWS items.

### 2.4. Combining DCE and BWS

In order to better understand class membership as detected in the latent class analysis, we followed the approach of [112]. Thus, by estimating a multinomial logit model with each participant's latent class membership as the dependent variable and attitudinal as well as trust factors (from the BWS and PCA analysis) and sociodemographic characteristics as independent variables, we attempted to explain class membership. Figure 3 provides an illustration of the analysis flow.

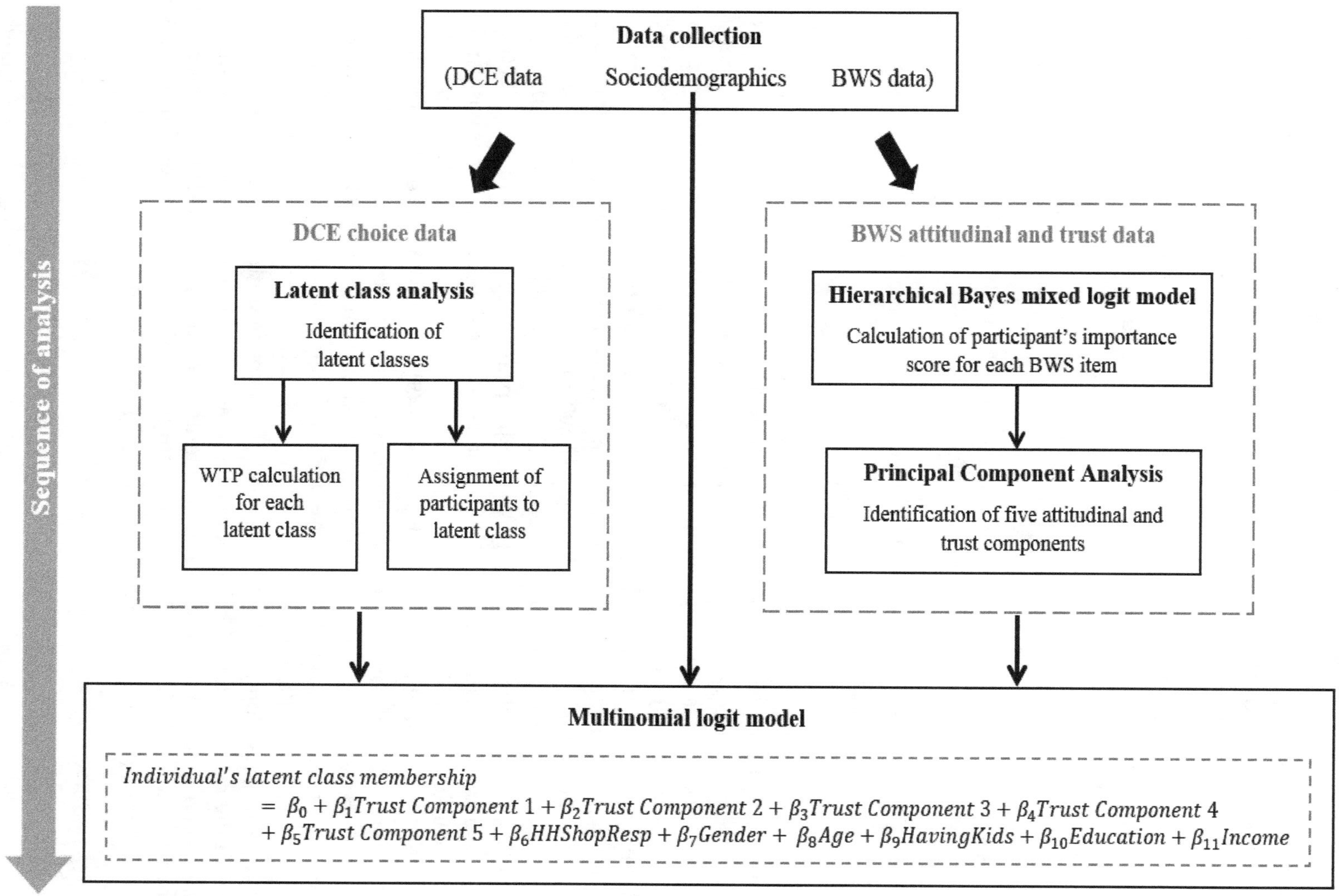


**Figure 3.** Research flow diagram (stepwise analysis (in the (long-format) dataset, as each individual contributed to 24 rows in the DCE data and 100 rows in the BWS data, the DCE and BWS analysis were not estimated simultaneously but sequentially. This is due to the reason that none of the exiting software, at current, is capable of estimating two different choice datasets concurrently in a model)).

## 3. Results

### 3.1. Sample Demographics

In total, 790 people joined the survey. Excluding those not responsible for their household's food shopping and/or not consuming red sweet pepper resulted in 459 (58% of all participants) valid responses. Thus, we far exceeded the target sample size ($N = 237$). The latter was determined using a power analysis assuming three alternatives for each choice task, a 5% margin of error, and a desired 95% confidence interval. Our target sample size is consistent with guidelines in the conjoint analysis methodology [147]. Of the 459 respondents, 72.1% were female, 61.9% were married, 33.3% had children under the age of 18 living in their household, and approximately 50.8% had completed university or higher education, and 39.0% stated that their household average monthly net income was below NT 60,000 (approximately US$2001) (and 45.3% above NT 60,000). The respondents' average age was 39.2 years. Compared to the Taiwanese population, the sample is biased towards younger and higher-educated segments. The high proportion of females in the sample is desirable because they are the primary food shoppers in most Taiwanese households [148]. Summary statistics for the demographic variables are presented in Table 3.

**Table 3.** Demographical statistics of the sample.

| | Respondents | | Taiwanese Population [a] |
|---|---|---|---|
| **Number of Respondents** | **459** | | |
| | **Freq.** | **(%)** | **(%)** |
| ***Gender*** | | | |
| Male | 128 | 27.9 | 49.9 |
| Female | 331 | 72.1 | 50.1 |
| ***Responsibility for household food shopping*** | | | |
| Fully | 220 | 47.9 | |
| Partly | 239 | 52.1 | |
| ***Age*** | | | |
| Up to 29 | 68 | 14.8 | 34.2 |
| 30–49 | 311 | 67.7 | 32.5 |
| 50 and over | 78 | 17.0 | 33.4 |
| Missing ¶ | 2 | 0.4 | |
| ***Marital status*** | | | |
| Single | 147 | 32.0 | 34.67 |
| Married | 284 | 61.9 | 51.12 |
| Other (widowed/divorced) | 23 | 5.0 | 14.21 |
| Missing | 5 | 1.1 | |
| ***Having children (<18 years old) in a household*** *(dummy coded: 1 = Yes; 0 = No)* | 208 | 45.3 | |
| ***Education*** | | | |
| Up to senior high school (12 years) | 95 | 20.7 | 58.2 |
| College (14 years) | 119 | 25.9 | 11.4 |
| University | 233 | 50.8 | 30.4 |
| Missing | 12 | 2.6 | |
| ***Avg. monthly net income of the household*** | | | |
| Up to NT 60,000 | 179 | 39.0 | |
| NT 60,001–120,000 | 152 | 33.1 | |
| NT 120,001 and over | 56 | 12.2 | |
| Missing | 72 | 15.7 | |

[a] Source: Ministry of the Interior, R.O.C. https://www.moi.gov.tw/; National Statistics, R. O. C. https://www.stat.gov.tw/. ¶ Two female participants did not give information about their age. As both already obtained a university degree, it seems reasonable to assume that they were above 18 years old, and thus eligible to participate in the survey. We therefore included their data in the analysis.

### *3.2. Identifying Consumer Segments Based on DCE Data*

Determining the optimal numbers of classes required a balanced assessment of the five major criteria reported in Table 4 [149]: Log-likelihood, percent certainty (Pct. Cert.) [150,151], consistent Akaike information criterion (CAIC) [152], Chi-square, and Bayesian information criterion (BIC) [153]. For this study, all indicators improved as more segments were added, supporting the presence of multiple segments in the sample. The four-segment solution (Table 4) provided the best fit to the data. Although indicators further improved as more segments were added, the changes were much smaller from a four- to a five-segment model compared to the move from a three- to a four-segment model. Furthermore, the model interpretability is as important as the statistical tests [113] and was best for the four-class model.

**Table 4.** Criteria for selecting the optimal number of classes.

| **Participants** | | **459** | | | |
|---|---|---|---|---|---|
| **Null Log-Likelihood** | | **−3817.85** | | | |
| **Groups** | **Log-Likelihood** | **Pct. Cert.** | **CAIC** | **Chi-Square** | **BIC** |
| 2 | −2691.81 | 29.49 | 5517.43 | 2252.09 | 5502.43 |
| 3 | −2557.13 | 33.02 | 5319.43 | 2521.46 | 5296.43 |
| 4 | −2454.62 | 35.71 | 5185.87 | 2726.48 | 5154.78 |
| 5 | −2397.36 | 37.21 | 5142.64 | 2840.99 | 5103.64 |
| 6 | −2358.04 | 38.24 | 5135.36 | 2919.63 | 5088.36 |

Table 5 summarizes the results of the four-class model and provides information on the attribute importance scores as well as on the partworth utilities (with respect to the attribute price information provided on the coefficient) and the corresponding standard errors for each attribute level for the four different consumer segments identified in the latent class analysis. To allow for comparability between classes, attribute importance scores were standardized to sum up to 100% across all attributes of each segment. For the same reason, partworth utilities were re-scaled and zero-centered. Positive (negative) partworth values indicate an increase (decrease) in utility relative to the average level of the respective attribute.

As shown in Table 5 attribute importance considerably differs between the four classes. For segments 1 (*Japan Lovers*) and 2 (*Domestic Supporters*), COO is by far the most important attribute (attribute importance: 59.8% and 71.7%, respectively). Consumers of both segments strongly dislike China as a COO. Price is the second most important attribute in both segments (attribute importance: 24.5% and 16.2%, respectively) followed by chemical residue testing (CRT) information (attribute importance: 8.9% and 6.9%), with production methods (attribute importance: 6.8% and 5.2%) being least important. As expected, *Japan Lovers* and *Domestic Supporters*, like the other groups, reveal a negative price elasticity and prefer organic to conventional red sweet peppers. Regarding the attribute levels for CRT information preferences, both groups prefer CRT information to no information. However, while *Japan Lovers* obtain an above-average utility from CRT approved in either the production country or Taiwan—though with a higher preference for the former—*Domestic Supporters* highly value CRT approved in Taiwan.

**Table 5.** Latent class analysis of DCE data.

| | | | | | | | | |
|---|---|---|---|---|---|---|---|---|
| Null log-likelihood | −3817.85 | | | | | | | |
| Restricted log-likelihood | −2454.62 | | | | | | | |
| Pct. Cert. | 35.71 | | | | | | | |
| Consistent Akaike Info Criterion | 5185.87 | | | | | | | |
| Chi-Square | 2726.38 | | | | | | | |
| Bayesian Information Criterion | 5154.87 | | | | | | | |
| Segmentation | 1. *Japan Lovers* | | 2. *Domestic Supporters* | | 3. *Price Conscious Consumers* | | 4. *Process Quality Supporters* | |
| Segment size ($N$ = 459) | 31.3% | | 26.1% | | 21.8% | | 20.8% | |
| | **Att. Imprt.** | **Rescaled Util. (S.E.)** | **Att. Imprt.** | **Rescaled Util. (S.E.)** | **Att. Imprt.** | **Rescaled Util. (S.E.)** | **Att. Imprt.** | **Rescaled Util. (S.E.)** |
| **COO** | 59.83 | | 71.66 | | 27.36 | | 32.26 | |
| Taiwan | | 64.38 (0.22) *** | | 141.18 (0.21) *** | | 28.98 (0.18) *** | | 54.18 (0.18) *** |
| Japan | | 87.46 (0.11) *** | | 4.25 (0.13) | | 40.22 (0.10) *** | | 20.67 (0.10) *** |
| China | | −151.85 (0.30) *** | | −145.44 (0.29) *** | | −69.20 (0.23) *** | | −74.85 (0.23) *** |
| **Production methods** | 6.83 | | 5.23 | | 3.36 | | 32.45 | |
| Organic | | 13.66 (0.06) *** | | 10.46 (0.07) *** | | 6.73 (0.07) * | | 64.90 (0.07) *** |
| Conventional | | −13.66 (0.06) *** | | −10.46 (0.07) *** | | −6.73 (0.07) * | | −64.90 (0.07) *** |
| **CRT** | 8.85 | | 6.90 | | 15.33 | | 27.98 | |
| CRT appr. in prod. country | | 14.47 (0.08) *** | | −1.39 (0.10) | | −23.50 (0.09) *** | | 5.58 (0.09) |
| CRT appr.in TW | | 6.46 (0.08) ** | | 14.50 (0.10) *** | | 37.84 (0.09) *** | | 53.16 (0.09) *** |
| No CRT | | −20.94 (0.10) *** | | −13.11 (0.12) ** | | −14.34 (0.12) ** | | −58.74 (0.12) *** |
| **Price** | 24.49 | −32.65 (0.13) *** | 16.21 | −21.61 (0.13) *** | 53.95 | −71.93 (0.10) *** | 7.32 | −9.75 (0.10) * |
| **No Choice** | | −55.30 (0.19) *** | | −26.87 (0.17) *** | | −320.63 (0.37) *** | | −156.05 (0.37) *** |
| **Share of No-Choice option** | 8.69% | | 13.87% | | 0.17% | | 1.24% | |

***, **, * Statistical significant level at 1%, 5%, and 10%. Att. Imprt. = Attribute importance. Rescaled part-worth utilities are zero-centered and normalized measures.

In contrast to segments 1 and 2, for respondents in segment 4 (*Process Quality Supporters*), the production method, COO, and CRT information is of similar importance (attribute importance: 32.5%, 32.3%, and 28.0%, respectively) while price plays a minor role in consumers' purchase decisions of red sweet peppers (attribute importance: 7.3%). *Process Quality Supporters* prefer Taiwanese over Japanese foods and dislike products from China.

They prefer organic products and those which are CRT approved in Taiwan. Finally, for consumers of class 3 (*Price Conscious Consumers*), price is by far the most important attribute (attribute importance: 54.0%). COO takes second place, but with an attribute importance score of 27.4%, it is of much lower relevance. *Price Conscious Consumers* especially prefer Japanese products but also like those from Taiwan while products originating from China are disliked, though to a lesser extent compared to the other three segments. CRT information is the third important attribute (attribute importance: 15.3%), with CRT approved in Taiwan being preferred.

No CRT is less disliked than CRT approved in the production country. The production method is of little relevance (attribute importance: 3.4%) for consumers of this group. For members of the *Price Conscious Consumers*, the no-choice option is linked to a high negative value (−320.63), implying that not purchasing any red sweet pepper is associated with a high utility loss. Accordingly, the opt-out option is hardly chosen (0.17% for the share of the opt-out decision). In comparison, the share of deciding for the no-choice option is 1.24%, 8.69%, and 13.87% for the *Process Quality Supporters, Japan Lovers*, and *Domestic Supporters*, respectively, which is 7 to 82 times higher.

To ease the comparison of the attribute-level importance between attributes of one consumer segment as well as between segments, WTP measures were calculated (see Table 6). Table 6 reveals that *Domestic Supporters* have a high WTP of NT 13.07 (approximately US$0.44) per 600 g for fresh red sweet peppers originating from Taiwan while they would buy red sweet peppers originating from China only at a high discount (NT −13.46) (approximately US$−0.45).

All other attribute levels just marginally influence *Domestic Supporters'* WTP. The latter also holds for consumers belonging to the *Japan Lovers* class. This group is, however, willing to pay extra for products originating from Taiwan (NT 3.94) (approximately US$0.13), and even more for those from Japan (NT 5.36) (approximately US$0.18). In line with the *Domestic Supporters*, they would only buy Chinese red sweet peppers at a high discount (NT −9.30) (approximately. US$−0.31).

Furthermore, *Process Quality Supporters* exhibit a high WTP for several attribute levels, e.g., organic products (NT 13.31) (approximately US$0.44) and Taiwan-authorized chemical residue testing information (NT 10.90) (approximately US$0.36) while they only would buy products from China (NT −15.35) (approximately US$−0.51), conventional products (NT −13.31) (approximately US $−0.44), or those with no CRT information (NT −12.05)(approx. US$−0.40) if they obtain a high discount. Finally, the *Price Conscious Consumers* segment is characterized by very low WTP values for every attribute level.

**Table 6.** Willingness to pay of different consumer segments.

| Segmentation | 1. *Japan Lovers* | | 2. *Domestic Supporters* | | 3. *Price Conscious Consumers* | | 4. *Process Quality Supporters* | |
|---|---|---|---|---|---|---|---|---|
| **Segment Sizes (*N* = 459)** | 31.3% | | 26.1% | | 21.8% | | 20.8% | |
| **(NT/600 g)** | **WTP** | **[95% C.I. Lower, Upper §]** | **WTP** | **[95% C.I. Lower, Upper]** | **WTP** | **[95% C.I. Lower, Upper]** | **WTP** | **[95% C.I. Lower, Upper]** |
| **COO** | | | | | | | | |
| Taiwan | 3.94 | [3.16, 4.73] | 13.07 | [9.45, 16.68] | 0.81 | [0.44, 1.17] | 11.11 | [4.11, 18.11] |
| Japan | 5.36 | [4.46, 6.15] | 0.39 | [−0.28, 1.06] | 1.12 | [0.96, 1.27] | 4.24 | [2.35, 6.12] |
| China | −9.30 | [−10.83, −7.77] | −13.46 | [−17.05, −9.86] | −1.92 | [−2.42, −1.43] | −15.35 | [−24.09, −6.61] |
| **Production methods** | | | | | | | | |
| Organic | 0.84 | [0.64, 1.04] | 0.97 | [0.51, 1.42] | 0.19 | [0.05, 0.32] | 13.31 | [4.27, 22.35] |
| Conventional | −0.84 | [1.04, −0.64] | −0.97 | [−1.42, −0.51] | −0.19 | [−0.32, −0.05] | −13.31 | [−22.35, −4.27] |
| **CRT** | | | | | | | | |
| CRT appr. in prod. country | 0.89 | [0.65, 1.12] | −0.13 | [−0.66, 0.41] | −0.65 | [−0.73, −0.58] | 1.14 | [−0.28, 2.57] |
| CRT appr. in TW | 0.40 | [0.21, 0.58] | 1.34 | [0.69, 1.99] | 1.05 | [0.85, 1.25] | 10.90 | [3.92, 17.88] |
| No CRT | −1.28 | [−1.59, −0.97] | −1.21 | [−1.95, −0.48] | −0.40 | [−0.60, −0.19] | −12.05 | [−20.13, −3.96] |

§ Upper and lower limits from the confidence intervals of WTP values, calculated with the delta method (Hole, 2007) at the 95% confidence level.

*3.3. Identifying Consumers' Attitude and Trust Based on BWS Data*

Consumers' attitude and trust as obtained from the BWS data were estimated through hierarchical Bayesian analysis. Information on the importance consumers attach to each of the 25 attitudinal and trust statements is provided in Appendix B. Appendix B reveals that respondents' trust with respect to products originating from China as well as trust in Chinese institutions is very low. All attitudinal and trust statements with reference to China (statements 2, 5, 8, 11, 18, 21, and 24) have an average importance score below one (between 0.05 and 0.75). In contrast, the average importance scores for all other statements range between 1.98 (I generally like to consume conventional red sweet peppers produced in Japan) to 8.42 (It is more likely that I buy Taiwanese red sweet peppers if they have information on chemical residue testing).

Following the analytic method presented in the paper of [139], we performed principal component analysis (PCA) based on the individual-level importance scores obtained from the hierarchical Bayesian estimation to check the dimensionality of the BWS data. Examining the eigenvalues (Table 7) and scree plot (Figure 4) with a permutation test approach and applying oblimin rotation (see Table 8), we obtained a factor solution containing five components. Factor loading of all attitudinal statements exceeded the cut-off value of 0.4. A cut-off for the statistical significance of the factor loadings of 0.40 was used, according to the guidelines presented by Hair et al. [154] for sample sizes greater than 200. A more conservative cutoff value would have been 0.50. The reasons for deciding on this lower value were first, that this was an exploratory factor analysis, and the factors detected were reasonable. Second, with a sample size of 459, it seems unlikely that the weaker factor we found ("I feel assured that the Chinese institutions do a good job in adequately protecting consumers" with a factor loading of 0.454) was random noise. Together, the five components explained 78.03% of the total variance. Table 9 shows the initial eigenvalues, the variance explained by each component and the cumulative variance.

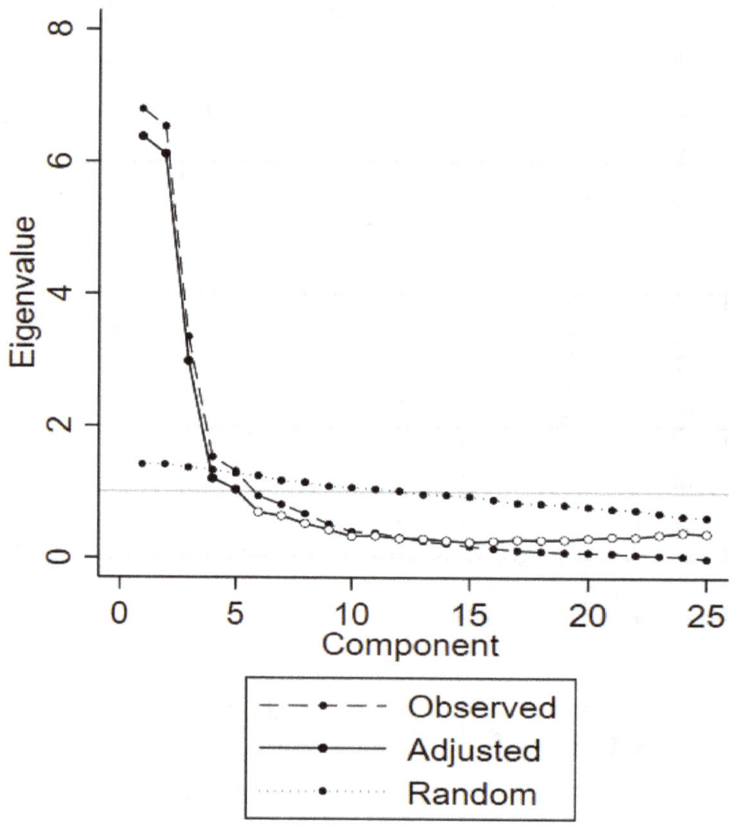


**Figure 4.** Diagnostic scree plotting graph (performed by parallel analysis) for principal component analysis (PCA) on estimated BWS importance scores data.

**Table 7.** Parallel analysis performance for component retention decisions across 25 BWS statements.

| Component | Adjusted Eigenvalue | Unadjusted Eigenvalue | *Estimated Bias* |
|---|---|---|---|
| 1 | 6.33 | 6.79 | *0.46* |
| 2 | 6.17 | 6.53 | *0.36* |
| 3 | 3.03 | 3.34 | *0.31* |
| 4 | 1.25 | 1.53 | *0.28* |
| 5 | 1.06 | 1.31 | *0.25* |

**Table 8.** Principal component analysis with oblimin rotation of 25 BWS items.

| BWS Statement | Comp. 1 Trust in Japan | Comp. 2 Trust in Taiwan and Organics | Comp. 3 Trust in Chinese Products | Comp. 4 No Trust in Organics | Comp. 5 Trust in Chinese Organic Products |
|---|---|---|---|---|---|
| 9. I generally like to consume organic sweet peppers produced in Japan. | 0.890 | | | | |
| 12. Japanese organic sweet peppers are trustworthy. | 0.863 | | | | |
| 4. I generally like to consume conventional sweet peppers produced in Taiwan. | −0.817 | | | | |
| 20. It is more likely that I buy Taiwanese sweet peppers if it is on special offer. | −0.794 | | | | |
| 25. Japanese organic sweet peppers have good value for money. | 0.700 | | | | |
| 10. Taiwanese organic sweet peppers are trustworthy. | | 0.866 | | | |
| 22. It is more likely that I buy Japanese sweet peppers if it is on special offer. | | −0.837 | | | |
| 7. I generally like to consume organic sweet peppers produced in Taiwan. | | 0.818 | | | |
| 15. With purchasing organic sweet peppers, I help preserving the environment and natural resources. | | 0.785 | | | |
| 6. I generally like to consume conventional sweet peppers produced in Japan. | | −0.758 | | | |
| 1. I feel assured that the Taiwanese institutions do a good job in adequately protecting consumers. | | 0.748 | | | |
| 19. It is more likely that I buy Japanese sweet peppers if information on chemical residue testing is provided. | | −0.704 | | | |
| 13. I feel sure that organic sweet peppers contain higher vitamin C and anti-cancer substances than conventional ones. | | 0.679 | | | |
| 23. Taiwanese organic sweet peppers have good value for money. | | 0.631 | | | |
| 3 I feel assured that the Japanese institutions do a good job in adequately protecting consumers. | | −0.517 | | | |
| 18. It is more likely that I buy Chinese sweet peppers if information on chemical residue testing is provided. | | | 0.906 | | |

**Table 8.** *Cont.*

| BWS Statement | Comp. 1 Trust in Japan | Comp. 2 Trust in Taiwan and Organics | Comp. 3 Trust in Chinese Products | Comp. 4 No Trust in Organics | Comp. 5 Trust in Chinese Organic Products |
|---|---|---|---|---|---|
| 21. It is more likely that I buy Chinese sweet peppers if it is on special offer. | | | 0.749 | | |
| 5. I generally like to consume conventional sweet peppers produced in China. | | | 0.615 | | |
| 14. I feel sure that organic sweet peppers contain the same vitamin C and anti-cancer substances as conventional ones. | | | | −0.816 | |
| 16. There are no differences between buying organic sweet peppers or conventional ones with respect to preserving the environment and natural resources. | | | | −0.777 | |
| 8. I generally like to consume organic sweet peppers produced in China. | | | | | 0.930 |
| 24. Chinese organic sweet peppers have good value for money. | | | | | 0.928 |
| 11. Chinese organic sweet peppers are trustworthy. | | | | | 0.903 |
| 17. It is more likely that I buy Taiwanese sweet peppers if information on chemical residue testing is provided. | | | | | −0.828 |
| 2. I feel assured that the Chinese institutions do a good job in adequately protecting consumers. | | | | | 0.454 |

The first component includes five items related to the trust and liking of organic products from Japan and skepticism toward Taiwanese conventional products (trust in Japan), while the 10 items under the second component refer to the trust and liking of organic and of Taiwanese products and institutions with a skepticism towards Japan (trust in Taiwan and organic). The third component covers three items referring to trust in Chinese products (trust in Chinese products). The two items of the fourth component include variables relating to a lack of trust in the superiority of organic products (no trust in organic). Finally, four of the five items of the last component refer directly to the trust and liking of organic products from China as well as the trust in Chinese institutions (trust in Chinese organic products). For the fifth item ('It is more likely that I buy Taiwanese red sweet peppers if they have information on chemical residue testing'), this is not the case.

**Table 9.** Total variance explained §.

| Component | Initial Eigenvalues | | |
|---|---|---|---|
| | Total | Percentage of Variance | Cumulative Percentage |
| 1 | 6.79 | 27.17 | 27.17 |
| 2 | 6.53 | 26.12 | 53.29 |
| 3 | 3.34 | 13.38 | 66.67 |
| 4 | 1.53 | 6.12 | 72.79 |
| 5 | 1.31 | 5.24 | 78.03 |
| 6 | 0.94 | 3.75 | 81.78 |
| 7 | 0.81 | 3.24 | 85.02 |
| 8 | 0.67 | 2.68 | 87.70 |
| 9 | 0.52 | 2.06 | 89.76 |
| 10 | 0.40 | 1.60 | 91.36 |
| 11 | 0.39 | 1.54 | 92.90 |
| 12 | 0.32 | 1.27 | 94.17 |
| 13 | 0.26 | 1.04 | 95.21 |
| 14 | 0.23 | 0.90 | 96.11 |
| 15 | 0.18 | 0.73 | 96.84 |
| 16 | 0.15 | 0.61 | 97.45 |
| 17 | 0.12 | 0.49 | 97.94 |
| 18 | 0.11 | 0.43 | 98.37 |
| 19 | 0.09 | 0.38 | 98.75 |
| 20 | 0.09 | 0.36 | 99.10 |
| 21 | 0.08 | 0.32 | 99.42 |
| 22 | 0.06 | 0.24 | 99.66 |
| 23 | 0.05 | 0.19 | 99.85 |
| 24 | 0.04 | 0.15 | 100.00 |
| 25 | 0.00 | 0.00 | 100.00 |

Extraction method: PCA. § The first five rows present the eigenvalues for the BWS individual level scores and percentage of variance for the five components.

### *3.4. Characterizing Consumer Segments with Respect to Attitudes and Trust*

To identify the relevance of attitudes and trust in determining differences between segments in the choice model (see Section 3.2), a latent segmentation model was estimated. For each individual in the sample, factor scores were calculated for all five attitudinal and trust components identified in Section 3.3. Subsequently, we estimated two multinomial logit models taking each participant's latent class membership as the dependent variable and the individual attitudinal and trust factor scores alone (Model 1) and in combination with sociodemographic information (Model 2) as independent variables. Both models fit the data well according to the likelihood ratio (LR) Chi-square test (see Table 10). Pseudo R-square measures indicate that the models explain 23% and 25% of the variance, respectively. Hence, the trust-related components alone explain 23% of the variance while sociodemographics only add two percentage points to the model fit.

The segment membership parameters are summarized in Table 10. The parameters of the class *Japan Lovers* were normalized to zero. This is necessary to allow for identification of class membership of the three other segments. It, however, implies that the results of the segments *Domestic Supporters, Price Conscious Consumers*, and *Process Quality Supporters* have to be interpreted relative to the segment of *Japan Lovers*.

**Table 10.** Multinomial logit models: DCE latent class membership regressed on five trust components (Model 1) and on five trust components plus sociodemographics (Model 2).

| (*N* = 459) | Model 1 | | Model 2 | |
|---|---|---|---|---|
| Log-likelihood of null model | −629.93 | | −629.93 | |
| Log-likelihood of restricted model | −484.92 | | −469.61 | |
| LR test Chi-square (33) | 196.05 | | 223.32 | |
| Prob > Chi-square | 0.00 | | 0.00 | |
| Pseudo R-squares | 0.23 | | 0.25 | |
| DCE four segments | **Coef.** | **Robust Std. Err.** | **Coef.** | **Robust Std. Err.** |
| ***Japan Lovers*** | *Reference group* | | | |
| ***Domestics Supporters*** | | | | |
| Trust in Japan | −0.84 *** | 0.17 | −0.81 *** | 0.17 |
| Trust in Taiwan & organic | 1.29 *** | 0.18 | 1.33 *** | 0.19 |
| Trust in Chinese products | −0.28 | 0.42 | −0.10 | 0.45 |
| No trust in organic | 0.09 | 0.18 | 0.04 | 0.18 |
| Trust in Chinese organic prod. | −0.10 | 0.29 | −0.14 | 0.29 |
| Full_HHShopResp | | | 0.63 * | 0.33 |
| Female | | | −0.16 | 0.34 |
| Age_below40 | | | 0.13 | 0.32 |
| Have_Kids | | | −0.48 | 0.30 |
| Edu_aboveCollege | | | −0.67 * | 0.36 |
| HHincome_above90k | | | −0.04 | 0.34 |
| Constant | −0.42 * | 0.22 | 0.12 | 0.52 |
| ***Price Conscious Consumers*** | | | | |
| Trust in Japan | −0.89 *** | 0.18 | −0.86 *** | 0.19 |
| Trust in Taiwan & organic | −0.03 | 0.17 | 0.09 | 0.17 |
| Trust in Chinese products | 1.19 *** | 0.41 | 1.26 *** | 0.49 |
| No trust in organic | 0.07 | 0.19 | −0.01 | 0.20 |
| Trust in Chinese organic prod. | 0.52 * | 0.29 | 0.49 * | 0.30 |
| Full_HHShopResp | | | −0.01 | 0.34 |
| Female | | | −0.63 * | 0.34 |
| Age_below40 | | | −0.48 | 0.35 |
| Have_Kids | | | −0.79 ** | 0.34 |
| Edu_aboveCollege | | | −0.90 ** | 0.44 |
| HHincome_above90k | | | −0.16 | 0.35 |
| Constant | −0.26 | 0.29 | 1.47 *** | 0.59 |
| ***Process Quality Supporters*** | | | | |
| Trust in Japan | −0.36 ** | 0.17 | −0.34 ** | 0.17 |
| Trust in Taiwan & organic | 0.80 *** | 0.17 | 0.86 *** | 0.18 |
| Trust in Chinese products | 0.67 * | 0.40 | 0.75 | 0.47 |
| No trust in organic | 0.23 | 0.20 | 0.18 | 0.20 |
| Trust in Chinese organic prod. | 0.93 *** | 0.23 | 0.90 *** | 0.23 |
| Full_HHShopResp | | | 0.25 | 0.33 |
| Female | | | −0.13 | 0.36 |
| Age_below40 | | | −0.48 | 0.31 |
| Have_Kids | | | −0.15 | 0.31 |
| Edu_aboveCollege | | | −0.63 | 0.42 |
| HHincome_above90k | | | −0.15 | 0.37 |
| Constant | −0.26 | 0.21 | 0.55 | 0.58 |

***, **, * Statistical significant level at 1%, 5%, and 10%.

As expected, attitude towards and trust in products, labels, origins, and institutions have a considerable influence on segment membership. Table 10 indicates that all components but 'no trust in organic' significantly influence class membership with respect to the purchase of red sweet pepper. Not surprisingly, high 'trust in Japanese organic products' reduces the likelihood of a consumer to belong into any other segment but *Japan Lovers*. Also, in line with expectations, high 'trust in Taiwan and organic' increases the likelihood to be part of the segments *Domestic Supporters* and *Process Quality Supporters* relative to the segment *Japan Lovers*, respectively. Those consumers with high 'trust in Chinese products' or 'high trust in Chinese organic products' have a greater likelihood to be a member of the *Price Conscious Consumers* or the *Process Quality Supporters* than of the *Japan Lovers*. The influence of the trust components on class membership holds for Model 1 as well as for Model 2. The latter model also investigates the influence of sociodemographic factors on classification. The findings indicate that compared to *Japan Lovers*, consumers belonging to the segment of *Domestic Supporters* are more likely to have full shopping responsibility for their household and to have a lower level of education. *Price Conscious Consumers* have, compared to *Japan Lovers*, a higher probability to be female and to have no kids. Also, this group has a lower probability to have an above-college educational level when compared to the reference segment. Comparing *Process Quality Supporters* and *Japan Lovers*, it becomes obvious that sociodemographics have no influence on class membership.

## 4. Discussion

The present study is the first that combines a latent class model for discrete choices with BWS to uncover attitudinal and trust dimensions. This approach allows us (1) to distinguish consumer segments that differ in their preferences for selected process and food safety standards, and (2) to reveal the importance of consumers' attitudes and trust in explaining consumers' group membership. Our study illustrates that combining DCE and BWS allows for a better understanding of the drivers of consumers' food choices and thus provides additional important information for decision makers in the economic and policy arena.

Based on our analysis, we could distinguish four segments of Taiwanese consumers based on differences they attach to various process attributes and price. The largest group can be described as *Japan Lovers*. According to Huang [155], the trend of Taiwanese consumers to buy Japanese products can be explained by the positive cultural image of "Japanese-ness". In our analysis, we showed that respondents belonging to the segment *Japan Lovers* reveal a higher level of trust in Japan relative to the other three consumer segments identified. Furthermore, Ma et al. [156] showed that Japanese products have established a reputation of being of high quality. Quality seems an important driver for this consumer segment's purchase decisions as this group is not interested in special offers and perceives the generally more expensive Japanese products [157] as better value for the money (see Tables 9 and 10).

The second largest group (26.1%) identified in our study, *Domestic Supporters*, has been detected in previous DCE-based research investigating the relevance of COO in consumers' food purchase decisions, however, with slightly smaller segments compared to our findings [103,158]. Our segment of *Domestic Supporters* has a high WTP for products originating from Taiwan. According to our results, this originates from a high level of institutional-based trust in the Taiwanese government and its regulatory and controlling power as well as a high level of trust belief in the quality of Taiwanese products. *Domestic Supporters* have a tendency towards ethnocentrism [156]. As revealed in previous research, consumers with a higher level of ethnocentrism favor domestic to foreign goods, perceive the quality of domestic food products higher than those of foreign goods, and have a higher purchasing preference for the former [19,159]. In line with [158], less-educated consumers were found to be more likely in the segment of *Domestic Supporters*.

For the *Price Conscious Consumers* identified in our study, process attributes are of relatively low relevance when buying red sweet peppers. The production method does not impact, and COO and CRT have only a small impact on consumers' WTP in this group. Price is the determinant that primarily drives consumers' purchase decisions in this segment (attribute importance 53.95%). *Price Conscious Consumer* segments have also been found in previous DCE-based studies investigating the relevance of process characteristics in consumers' food purchase decisions, with some studies identifying a segment of a similar size (around 20%, e.g., [160–162]) while others reveal a considerably larger group of *Price Conscious Consumers* [17,118,163,164]. The class determinant estimates revealed that male respondents were more likely to be in the class of *Price Conscious Consumers*, a result in line with the findings of [160,164]. In addition, our results indicate that respondents in this segment are less educated and more likely to not have children.

As indicated above, members of the price-conscious segment attach a higher importance to chemical residue test information compared to the production method. Considering that organic certification implies the logic of zero chemicals, this either reflects that respondents are not aware of the core standards behind the organic label and/or do not trust organic labels. According to Tung et al. [14] and Chen [165], Taiwanese consumers indeed have some doubt that products labeled as organic are always grown without using synthetic fertilizers, pesticides, and chemicals. Based on a literature review, Yiridoe et al. [166] came to the conclusion that consumers have difficulties in understanding the complexity of organic standards. Chemical residue test information is easier to understand when compared to organic standards, which might be of special relevance for the price-conscious segment as it consists of a higher share of less-educated consumers.

The analysis provides additional interesting insights regarding this group of consumers. *Price Conscious Consumers* are characterized by a relatively low trust in Japan and a relatively high trust in Chinese products in general, and in Chinese organic products more specifically (see Table 10). This holds despite the fact that this consumer segment also has the lowest WTP for Chinese products (see Table 6). Festinger's theory of cognitive dissonance might provide an explanation for this finding [167]. With Chinese products typically the cheaper alternative, and given the dominant relevance of price for members of this segment, *Price Conscious Consumers* will often end up with Chinese products in their shopping cart. A lack of trust in and liking of Chinese products and a high trust and liking of Japanese and Taiwanese products would lead to dissonance with the choice behavior. The stronger the latter, the higher the attractiveness of the unchosen alternative compared to the actual choice. One way for consumers to relax or correct this disturbing and unpleasant psychological state is a dissonance-related attitude change [168], implying a correction towards a more positive perception of Chinese products and a higher trust in China.

For the segment of *Process Quality Supporters* with a share of 20.8%, only slightly smaller than the group of *Price Conscious Consumers*, but in contrast to the latter, price is of little importance for the former. This group attaches a much higher importance to the attributes, production method and CRT information, than any other segment. Members of this group are willing to pay high price premia for organically produced red sweet peppers as well as for those with a CRT approved in Taiwan, indicating that for members of this segment, health and food safety are of high importance. This assumption is supported by [165], who found that health concerns are the most important determinant for Taiwanese consumers forming a positive attitude towards organic. Along the same line, Hasimu et al. [169] and Xie et al. [170] showed that health and food safety are the core motives for consumers in China to buy organic foods. Our analysis reveals that a high level of trust (component 1, 2, 3, and 4 in Table 10) increases the probability of a consumer belonging to the segment of *Process Quality Supporters*. This result is in line with [171], who showed that trust in labeling is essential for Taiwanese consumers' intent to purchase organic products.

The four segments identified reveal the difference in importance Taiwanese consumer groups attach to different attributes when buying red sweet peppers. From a marketing perspective, this implies that a 'one size fits all' marketing strategy is inappropriate. Thus, marketers need to develop segment-specific offerings in order to better target the needs of their customers. For the two largest segments, the *Japan Lovers* and *Domestic Supporters*, by far the most important attribute is COO, though with a clearly distinct preference for the attribute levels. In addition, both groups reveal very high opt-out ratios (*Japan Lovers*, 8.7%; *Domestic Supporters*, 13.9%). Thus, to secure the loyalty of customers and to extract a price premium in both groups, retailers need to have red sweet peppers from Japan as well as Taiwan in their assortment. The *Price Conscious Consumers* attach little value to any of the attributes considered in our purchase experiment but price. This group will switch products, in general choosing the cheapest alterative. According to Jing and Wen [172] and Koçaş and Bohlmann [173], retailers can win this group of consumers by providing low-price alternatives (e.g., conventional sweet red pepper) or by providing price discounts (e.g., potentially for larger size packages). Finally, for the *Process Quality Supporters*, all process attributes are not only of relatively high but also of about equal importance in consumers' purchase decisions while price seems to be of minor relevance. The low opt-out ratio of 1.7%, in addition, signals that this group can be retained if retailers offer red sweet peppers that fulfill at least one of the desired attribute levels. This, in fact, does not require a further differentiation beyond the one already suggested for the *Japan Lovers* and *Domestic Supporters*. However, large retailers that have the possibility of further differentiating their assortment can extract high price premia by offering organic red sweet peppers and/or red sweet peppers with a CRT approved in Taiwan. Along the same lines, smaller retailers may be able to run a successful niche strategy by focusing on the needs of this smaller segment.

The recommendation so far takes class membership as a given. Our findings, however, show that while sociodemographic factors provide little power to explain class membership (compare Models 1 and 2 in Table 10), a result in line with previous research [17,163,174], attitudinal and trust perceptions significantly influence class membership, explaining 23% of the models' variance. Thus, assuming labeling to be sufficient for consumers to consider process attributes in their purchase decisions will likely prove wrong. For labeling to receive credence, it seems essential to implement effective control authorities and secure a high level of transparency to (re)gain and maintain consumers' trust. Furthermore, our findings suggest a lack of knowledge regarding what organic implies, at least for the segment of *Price Conscious Consumers* (see discussion above). Accordingly, action is needed by private and public agencies to raise consumers' knowledge regarding the standards behind organic products.

Last, some potential limitations of this study must be acknowledged. First, in this study, we focused on one product, fresh red sweet pepper, in one country. Whether our findings are transferrable to other products part of Taiwanese people's daily diet should be addressed in further research. Also, a comparison of results across countries would be desirable. In particular, it would be interesting to carry out the same kind of study in China and/or Japan, and thus in competing markets, and investigate whether segments characterized by similar preferences and characteristics across locations exist. Respective insights could be valuable for producers in their decision to export or concentrate on local markets. Second, it has to be noted that the analysis is based on a hypothetical choice experiment. Thus, the results might suffer from social desirability bias. Third, to avoid consumer fatigue, we decided for four product attributes of red sweet pepper based on a pre-study. However, it must be considered that the selection of attributes and attribute levels may impact DCE outcomes. Along the same line, we limited the number of BWS statements to 25 as to cover all different dimensions of trust, with respect to all relevant characteristics of the products, would have resulted in more than 70 BWS statements, which would have been unmanageable. Nevertheless, there could be arguments for a different choice of statements than those selected that might have had an impact on the results. More specifically, we suggest for future studies to give more attention to capturing consumer animosity as a means to deepen the understanding about consumers' attitude and perception, especially in the case of a negative COO image. Furthermore, in the questionnaire, participants were always requested to first answer the DCE questions and then the once of the BWS. We decided on this sequence to prevent consumers from knowing, while conducting the DCE, that the focus of the study was on trust. However, due to this order, we cannot rule out that the DCE had some influence on respondents' answers in the BWS task.

Moreover, as we combined in our multinomial logit model the findings from a DCE latent class analysis (class membership entered as the dependent variable) and the outcome of the BWS Bayesian estimation (trust components entered as independent variables), we cannot rule out the problem of endogeneity. As we, however, lack an adequate instrumental variable, we cannot control for the endogenous issue. For future research, we suggest controlling for endogeneity and considering this issue by inclusion of an adequate covariate in the design of the survey. Finally, a simultaneous estimation of the DCE and BWS data would have been desirable. However, currently available software is incapable of estimating two large choice datasets (per individual 24 DCE rows and 100 BWS rows) concurrently in a single model. Therefore, such an analysis will only be possible in the future with the development of more sophisticated software.

## 5. Conclusions

Our study revealed that an analysis of what Taiwanese consumers value in their purchase of red sweet peppers falls short if it does not account for market heterogeneity. This holds as consumers (consumer segments) differ in the weights they assign to different product and process attributes. The findings of this paper, in addition, strongly support the argument that consumers' attitudinal and trust perceptions are important determinants of class membership. These insights help to understand what actually drives class membership and to derive targeted marketing and policy strategies. From a methodological point of view, our paper is the first to combine a latent class model for discrete choices with BWS. The latter allowed us to capture consumers' attitudinal and trust perceptions in a discriminant manner. Our results show that combining those methods enriches insights gained from latent class choice analysis.

**Author Contributions:** Conceptualization, C.-H.Y., N.L. and M.H.; data recruitment, C.-H.Y.; data analysis, C.-H.Y.; modelling strategy, C.-H.Y. and M.H.; writing—original draft preparation, C.-H.Y.; writing—editing, M.H., C.-H.Y. and N.L.; revising, C.-H.Y., M.H. and N.L.; supervision, M.H. All authors have read and agreed to the published version of the manuscript.

## Appendix A BWS Design Matrices

Number of statements = 25
Number of statements per choice set = 5
Number of sets per respondent: 10
Number of blocks: 30
**One Way Frequencies:**

| Statement | Frequencies Used |
|---|---|
| 1 | 60 |
| 2 | 60 |
| 3 | 60 |
| 4 | 60 |
| 5 | 60 |
| 6 | 60 |
| 7 | 60 |
| 8 | 60 |
| 9 | 60 |
| 10 | 60 |
| 11 | 60 |
| 12 | 60 |
| 13 | 60 |
| 14 | 60 |
| 15 | 60 |
| 16 | 60 |
| 17 | 60 |
| 18 | 60 |
| 19 | 60 |
| 20 | 60 |
| 21 | 60 |
| 22 | 60 |
| 23 | 60 |
| 24 | 60 |
| 25 | 60 |
| Mean = | 60 |
| S.D. = | 0 |

**Positional Frequencies:**

| | Position in the BWS choice set | | | | |
|---|---|---|---|---|---|
| Statement | 1 | 2 | 3 | 4 | 5 |
| 1 | 12 | 12 | 12 | 12 | 12 |
| 2 | 11 | 13 | 12 | 12 | 12 |
| 3 | 12 | 13 | 12 | 12 | 11 |
| 4 | 12 | 11 | 13 | 12 | 12 |
| 5 | 12 | 12 | 12 | 13 | 11 |
| 6 | 12 | 13 | 11 | 12 | 12 |
| 7 | 12 | 12 | 12 | 13 | 11 |
| 8 | 12 | 12 | 12 | 11 | 13 |
| 9 | 12 | 11 | 12 | 12 | 13 |
| 10 | 12 | 12 | 12 | 12 | 12 |
| 11 | 12 | 12 | 12 | 12 | 12 |
| 12 | 12 | 13 | 12 | 11 | 12 |
| 13 | 12 | 12 | 12 | 12 | 12 |
| 14 | 12 | 12 | 12 | 12 | 12 |
| 15 | 12 | 11 | 12 | 13 | 12 |
| 16 | 12 | 12 | 12 | 12 | 12 |
| 17 | 12 | 12 | 12 | 12 | 12 |
| 18 | 11 | 12 | 12 | 12 | 13 |
| 19 | 12 | 12 | 12 | 12 | 12 |
| 20 | 13 | 12 | 12 | 11 | 12 |
| 21 | 12 | 12 | 13 | 11 | 12 |
| 22 | 13 | 12 | 11 | 12 | 12 |
| 23 | 12 | 11 | 12 | 13 | 12 |
| 24 | 12 | 12 | 12 | 11 | 13 |
| 25 | 12 | 12 | 12 | 13 | 11 |
| Mean = | 12 | | | | |
| S.D. = | 0.522 | | | | |

**Two Way Frequencies:**

| Statement | 1 | 2 | 3 | 4 | 5 | 6 | 7 | 8 | 9 | 10 | 11 | 12 | 13 | 14 | 15 | 16 | 17 | 18 | 19 | 20 | 21 | 22 | 23 | 24 | 25 |
|---|---|---|---|---|---|---|---|---|---|---|---|---|---|---|---|---|---|---|---|---|---|---|---|---|---|
| 1 | 60 | 10 | 11 | 10 | 10 | 10 | 10 | 10 | 10 | 10 | 10 | 11 | 9 | 11 | 10 | 10 | 10 | 10 | 10 | 10 | 9 | 10 | 9 | 10 | 10 |
| 2 | 10 | 60 | 10 | 10 | 10 | 10 | 10 | 10 | 10 | 10 | 10 | 10 | 9 | 10 | 10 | 11 | 10 | 9 | 10 | 10 | 11 | 10 | 10 | 10 | 10 |
| 3 | 11 | 10 | 60 | 10 | 10 | 9 | 10 | 10 | 10 | 9 | 10 | 10 | 11 | 10 | 10 | 10 | 10 | 10 | 11 | 9 | 10 | 10 | 10 | 10 | 10 |
| 4 | 10 | 10 | 10 | 60 | 10 | 10 | 9 | 10 | 10 | 10 | 9 | 10 | 10 | 10 | 10 | 10 | 11 | 11 | 10 | 10 | 10 | 10 | 10 | 10 | 10 |
| 5 | 10 | 10 | 10 | 10 | 60 | 10 | 10 | 10 | 10 | 10 | 10 | 10 | 10 | 10 | 10 | 10 | 10 | 10 | 10 | 9 | 10 | 11 | 10 | 10 | 10 |
| 6 | 10 | 10 | 9 | 10 | 10 | 60 | 10 | 11 | 10 | 10 | 10 | 10 | 10 | 10 | 9 | 10 | 10 | 10 | 11 | 10 | 10 | 10 | 9 | 10 | 11 |
| 7 | 10 | 10 | 10 | 9 | 10 | 10 | 60 | 10 | 10 | 11 | 11 | 10 | 10 | 11 | 9 | 10 | 9 | 10 | 10 | 10 | 10 | 10 | 10 | 10 | 10 |
| 8 | 10 | 10 | 10 | 10 | 10 | 11 | 10 | 60 | 10 | 9 | 10 | 10 | 11 | 10 | 10 | 10 | 10 | 10 | 11 | 10 | 9 | 10 | 10 | 10 | 9 |
| 9 | 10 | 10 | 10 | 10 | 10 | 10 | 10 | 10 | 60 | 10 | 10 | 10 | 11 | 9 | 11 | 9 | 10 | 11 | 9 | 10 | 10 | 10 | 10 | 10 | 10 |
| 10 | 10 | 10 | 9 | 10 | 10 | 10 | 11 | 9 | 10 | 60 | 10 | 10 | 10 | 10 | 11 | 9 | 11 | 10 | 10 | 10 | 10 | 10 | 10 | 10 | 10 |
| 11 | 10 | 10 | 10 | 9 | 10 | 10 | 11 | 10 | 10 | 10 | 60 | 9 | 11 | 9 | 11 | 11 | 10 | 10 | 10 | 10 | 10 | 10 | 10 | 9 | 10 |
| 12 | 11 | 10 | 10 | 10 | 10 | 10 | 10 | 10 | 10 | 10 | 9 | 60 | 9 | 11 | 11 | 11 | 9 | 9 | 10 | 10 | 10 | 10 | 10 | 10 | 10 |
| 13 | 9 | 9 | 11 | 10 | 10 | 10 | 10 | 11 | 11 | 10 | 11 | 9 | 60 | 10 | 10 | 10 | 10 | 10 | 9 | 10 | 10 | 10 | 10 | 10 | 10 |
| 14 | 11 | 10 | 10 | 10 | 10 | 10 | 11 | 10 | 9 | 10 | 9 | 11 | 10 | 60 | 10 | 9 | 9 | 10 | 10 | 10 | 10 | 10 | 11 | 10 | 10 |
| 15 | 10 | 10 | 10 | 10 | 10 | 9 | 9 | 10 | 11 | 11 | 11 | 11 | 10 | 10 | 60 | 10 | 10 | 10 | 9 | 11 | 10 | 9 | 10 | 10 | 9 |
| 16 | 10 | 11 | 10 | 10 | 10 | 10 | 10 | 10 | 9 | 9 | 11 | 11 | 10 | 9 | 10 | 60 | 10 | 10 | 10 | 10 | 10 | 10 | 10 | 10 | 10 |
| 17 | 10 | 10 | 10 | 11 | 10 | 10 | 9 | 10 | 10 | 11 | 10 | 9 | 10 | 9 | 10 | 10 | 60 | 10 | 10 | 10 | 10 | 10 | 10 | 10 | 11 |
| 18 | 10 | 9 | 10 | 11 | 10 | 10 | 10 | 10 | 11 | 10 | 10 | 9 | 10 | 10 | 10 | 10 | 10 | 60 | 9 | 10 | 10 | 10 | 10 | 11 | 10 |
| 19 | 10 | 10 | 11 | 10 | 10 | 11 | 10 | 11 | 9 | 10 | 10 | 10 | 9 | 10 | 9 | 10 | 10 | 9 | 60 | 9 | 11 | 11 | 10 | 10 | 10 |
| 20 | 10 | 10 | 9 | 10 | 9 | 10 | 10 | 10 | 10 | 10 | 10 | 10 | 10 | 10 | 11 | 10 | 10 | 10 | 9 | 60 | 11 | 10 | 11 | 10 | 10 |
| 21 | 9 | 11 | 10 | 10 | 10 | 10 | 10 | 9 | 10 | 10 | 10 | 10 | 10 | 10 | 10 | 10 | 10 | 10 | 11 | 11 | 60 | 9 | 10 | 10 | 10 |
| 22 | 10 | 10 | 10 | 10 | 11 | 10 | 10 | 10 | 10 | 10 | 10 | 10 | 10 | 10 | 9 | 10 | 10 | 10 | 11 | 10 | 9 | 60 | 10 | 10 | 10 |
| 23 | 9 | 10 | 10 | 10 | 10 | 9 | 10 | 10 | 10 | 10 | 10 | 10 | 10 | 11 | 10 | 10 | 10 | 10 | 10 | 11 | 10 | 10 | 60 | 10 | 10 |
| 24 | 10 | 10 | 10 | 10 | 10 | 10 | 10 | 10 | 10 | 10 | 9 | 10 | 10 | 10 | 10 | 10 | 10 | 11 | 10 | 10 | 10 | 10 | 10 | 60 | 10 |
| 25 | 10 | 10 | 10 | 10 | 10 | 11 | 10 | 9 | 10 | 10 | 10 | 10 | 10 | 10 | 9 | 10 | 11 | 10 | 10 | 10 | 10 | 10 | 10 | 10 | 60 |
| Mean = | | 10 | | | | | | | | | | | | | | | | | | | | | | | |
| S.D. = | | 0.497 | | | | | | | | | | | | | | | | | | | | | | | |

## Appendix B Results of the Hierarchical Bayesian Mixed Logit Model for BWS Data

| | | | |
|---|---|---|---|
| Total respondents | | 459 | |
| Total Best Choices | | 4590 | |
| Total Worst Choices | | 4590 | |
| Root Likelihood statistics (model fit) | | 0.58 | |
| **BWS Statements** | **Rank** a | **Avg. Imprt. Scores** b | **[95% C.I. Lower, Upper]** |
| 1. I feel assured that the Taiwanese institutions do a good job in adequately protecting consumers. | 12 | 4.80 | [4.50–5.10] |
| 2. I feel assured that the Chinese institutions do a good job in adequately protecting consumers. | 24 | 0.08 | [0.06–0.11] |
| 3. I feel assured that the Japanese institutions do a good job in adequately protecting consumers. | 10 | 5.43 | [5.18–5.69] |
| 4. I generally like to consume conventional red sweet peppers produced in Taiwan | 14 | 4.26 | [4.01–4.51] |
| 5. I generally like to consume conventional red sweet peppers produced in China. | 25 | 0.05 | [0.03–0.07] |
| 6. I generally like to consume conventional red sweet peppers produced in Japan. | 18 | 1.98 | [1.84–2.11] |
| 7. I generally like to consume organic red sweet peppers produced in Taiwan. | 4 | 7.30 | [7.09–7.52] |
| 8. I generally like to consume organic red sweet peppers produced in China. | 22 | 0.33 | [0.24–0.42] |
| 9. I generally like to consume organic red sweet peppers produced in Japan. | 13 | 4.68 | [4.38–4.98] |
| 10. Taiwanese organic red sweet peppers are trustworthy. | 2 | 7.73 | [7.52–7.93] |
| 11. Chinese organic red sweet peppers are trustworthy. | 21 | 0.49 | [0.37–0.61] |
| 12. Japanese organic red sweet peppers are trustworthy. | 7 | 5.96 | [5.67–6.24] |
| 13. I feel sure that organic red sweet peppers contain higher vitamin C and anti-cancer substances than conventional ones. | 16 | 3.24 | [2.98–3.51] |
| 14. I feel sure that organic red sweet peppers contain the same vitamin C and anti-cancer substances as conventional ones. | 15 | 3.53 | [3.23–3.83] |
| 15. With purchasing organic red sweet peppers I help preserve the environment and natural resources. | 11 | 5.29 | [5.01–5.57] |
| 16. There are no differences between buying organic red sweet peppers or conventional ones with respect to preserving the environment and natural resources. | 17 | 2.32 | [2.09–2.54] |
| 17. It is more likely that I buy Taiwanese red sweet peppers if information on chemical residue testing is provided. | 1 | 8.42 | [8.21–8.62] |

| Statement | Rank | Avg. imprt. Scores | Range |
|---|---|---|---|
| 18. It is more likely that I buy Chinese red sweet peppers if information on chemical residue testing is provided. | 19 | 0.75 | [0–0.92] |
| 19. It is more likely that I buy Japanese red sweet peppers if information on chemical residue testing is provided. | 5 | 7.27 | [7–7.48] |
| 20. It is more likely that I buy Taiwanese red sweet peppers if they are on special offer. | 6 | 6.52 | [6–6.84] |
| 21. It is more likely that I buy Chinese red sweet peppers if they are on special offer. | 23 | 0.29 | [0–0.39] |
| 22. It is more likely that I buy Japanese red sweet peppers if they are on special offer. | 8 | 5.78 | [5–6.08] |
| 23. Taiwanese organic red sweet peppers have good value for money. | 3 | 7.39 | [7–7.61] |
| 24. Chinese organic red sweet peppers have good value for money. | 20 | 0.50 | [0.37–0.64] |
| 25. Japanese organic red sweet peppers have good value for money. | 9 | 5.61 | [5–5.92] |

[a] Rank indicates the rank position among the 25 statements based on the importance score. [b] Avg. imprt. Scores = Average importance scores.

## References

1. Peng, G.J.; Chang, M.H.; Fang, M.; Liao, C.D.; Tsai, C.F.; Tseng, S.H.; Kao, Y.-M.; Chou, H.-K.; Cheng, H.F. Incidents of major food adulteration in Taiwan between 2011 and 2015. *Food Control* **2017**, *72*, 145–152. [CrossRef]
2. Cheng, W.-C.; Kuo, C.-W.; Chi, T.-Y.; Lin, L.-C.; Lee, C.-H.; Feng, R.-L.; Tsai, S.J. Investigation on the trend of food-borne disease outbreaks in Taiwan (1991–2010). *J. Food Drug Anal.* **2013**, *21*, 261–267. [CrossRef]
3. Chern, W.S.; Hong, J.-P.; Liu, K.E. Comparison of the Vickrey second-price and random nth-price auctions for analyzing country of origin labeling in Taiwan. *Acad. Econ. Pap.* **2013**, *41*, 215–254.
4. Liu, C.Y. Institutional Isomorphism and Food Fraud: A Longitudinal Study of the Mislabeling of Rice in Taiwan. *J. Agric. Environ. Ethics* **2016**, *29*, 607–630. [CrossRef]
5. Yang, Y.-T. The development of Taiwan's healthy agriculture industry. *Agric. Biotechnol. Ind. Q.* **2013**, *32*, 1–6.
6. Epstein, L. Fifty years since silent spring. *Ann. Rev. Phytopathol.* **2014**, *52*, 377–402. [CrossRef]
7. Yen, T.H.; Lin-Tan, D.T.; Lin, J.L. Food safety involving ingestion of foods and beverages prepared with phthalate-plasticizer-containing clouding agents. *J. Formos. Med. Assoc.* **2011**, *110*, 671–684. [CrossRef]
8. Kang, J.J.; Chu, S.F.; Wu, Z.Z.; Chou, S.W.; Tsai, S.J.; Chiu, W.T. Crisis management turns Taiwan's plasticizer nightmare into progressive policy. *J. Formos. Med. Assoc.* **2012**, *111*, 409–411. [CrossRef]
9. Chern, W.S.; Chang, C.Y. Benefit evaluation of the country of origin labeling in Taiwan: Results from an auction experiment. *Food Policy* **2012**, *37*, 511–519. [CrossRef]
10. Teng, C.C.; Wang, Y.M. Decisional factors driving organic food consumption: Generation of consumer purchase intentions. *Br. Food J.* **2015**, *117*, 1066–1081. [CrossRef]
11. Tonkin, E.; Coveney, J.; Meyer, S.B.; Wilson, A.M.; Webb, T. Managing uncertainty about food risks—Consumer use of food labelling. *Appetite* **2016**, *107*, 242–252. [CrossRef] [PubMed]
12. Grunert, K.G. Food quality and safety: Consumer perception and demand. *Eur. Rev. Agric. Econ.* **2005**, *32*, 369–391. [CrossRef]
13. Vandeplas, A.; Minten, B. Food quality in domestic markets of developing economies: A comparative study of two countries. *Agric. Econ.* **2015**, *46*, 617–628. [CrossRef]
14. Tung, S.J.; Shih, C.C.; Wei, S.; Chen, Y.H. Attitudinal inconsistency toward organic food in relation to purchasing intention and behavior: An illustration of Taiwan consumers. *Br. Food J.* **2012**, *114*, 997–1015. [CrossRef]
15. Bravo, C.P.; Cordts, A.; Schulze, B.; Spiller, A. Assessing determinants of organic food consumption using data from the German National Nutrition Survey II. *Food Qual. Prefer.* **2013**, *28*, 60–70. [CrossRef]
16. Moser, A.K. Consumers' purchasing decisions regarding environmentally friendly products: An empirical analysis of German consumers. *J. Retail. Consum. Serv.* **2016**, *31*, 389–397. [CrossRef]
17. Janssen, D.; Langen, N. The bunch of sustainability labels–Do consumers differentiate? *J. Clean. Prod.* **2017**, *143*, 1233–1245. [CrossRef]
18. Klöckner, H.; Langen, N.; Hartmann, M. COO labeling as a tool for pepper differentiation in Germany: Insights into the taste perception of organic food shoppers. *Br. Food J.* **2013**, *115*, 1149–1168. [CrossRef]
19. Vabø, M.; Hansen, H.; Hansen, K.V.; Kraggerud, H. Ethnocentrism and domestic food choice: Insights from an affluent protectionist market. *J. Food Prod. Mark.* **2017**, *23*, 570–590. [CrossRef]
20. Roos, G.M.; Hansen, K.V.; Skuland, A.V. Consumers, Norwegian food and belonging: A qualitative study. *Br. Food J.* **2016**, *118*, 2359–2371. [CrossRef]
21. Aprile, M.C.; Caputo, V.; Nayga, R.M., Jr. Consumers' valuation of food quality labels: The case of the European geographic indication and organic farming labels. *Int. J. Consum. Stud.* **2012**, *36*, 158–165. [CrossRef]
22. Castaldo, S.; Perrini, F.; Misani, N.; Tencati, A. The missing link between corporate social responsibility and consumer trust: The case of fair trade products. *J. Bus. Ethics* **2009**, *84*, 1–15. [CrossRef]
23. Hall, C.; Osses, F. A review to inform understanding of the use of food safety messages on food labels. *Int. J. Consum. Stud.* **2013**, *37*, 422–432. [CrossRef]
24. Brom, F.W. Food, consumer concerns, and trust: Food ethics for a globalizing market. *J. Agric. Environ. Ethics* **2000**, *12*, 127–139. [CrossRef]
25. Einsiedel, E.F. GM food labeling: The interplay of information, social values, and institutional trust. *Sci. Commun.* **2002**, *24*, 209–221. [CrossRef]

26. De Jonge, J.; Van Trijp, J.C.M.; van der Lans, I.A.; Renes, R.J.; Frewer, L.J. How trust in institutions and organizations builds general consumer confidence in the safety of food: A decomposition of effects. *Appetite* **2008**, *51*, 311–317. [CrossRef]
27. Qiu, H.; Huang, J.; Pray, C.; Rozelle, S. Consumers' trust in government and their attitudes towards genetically modified food: Empirical evidence from China. *J. Chin. Econ. Bus. Stud.* **2012**, *10*, 67–87. [CrossRef]
28. Chen, W. The effects of different types of trust on consumer perceptions of food safety: An empirical study of consumers in Beijing Municipality, China. *China Agric. Econ. Rev.* **2013**, *5*, 43–65. [CrossRef]
29. Lang, J.T. Elements of public trust in the American food system: Experts, organizations, and genetically modified food. *Food Policy* **2013**, *41*, 145–154. [CrossRef]
30. Omari, R.; Ruivenkamp, G.T.; Tetteh, E.K. Consumers' trust in government institutions and their perception and concern about safety and healthiness of fast food. *J. Trust Res.* **2017**, *7*, 170–186. [CrossRef]
31. Frewer, L.J.; Scholderer, J.; Bredahl, L. Communicating about the risks and benefits of genetically modified foods: The mediating role of trust. *Risk Anal.* **2003**, *23*, 1117–1133. [CrossRef] [PubMed]
32. Frewer, L.J. Risk perception, communication and food safety. In *Strategies for Achieving Food Security in Central Asia*; Alpas, H., Smith, M., Kulmyrzaev, A., Eds.; Springer: Dordrecht, The Netherlands, 2012; pp. 123–131.
33. De Jonge, J.; Van Trijp, H.; Renes, R.J.; Frewer, L.J. Consumer confidence in the safety of food and newspaper coverage of food safety issues: A longitudinal perspective. *Risk Anal.* **2010**, *30*, 125–142. [CrossRef] [PubMed]
34. Siegrist, M.; Cvetkovich, G. Perception of hazards: The role of social trust and knowledge. *Risk Anal.* **2000**, *20*, 713–720. [CrossRef] [PubMed]
35. Siegrist, M.; Keller, C.; Kiers, H.A. Lay people's perception of food hazards: Comparing aggregated data and individual data. *Appetite* **2006**, *47*, 324–332. [CrossRef]
36. Henson, S. Demand-side constraints on the introduction of new food technologies: The case of food irradiation. *Food Policy* **1995**, *20*, 111–127. [CrossRef]
37. Wandel, M.; Fagerli, R.A. Consumer concern about food related health risks and their trust in experts. *Ecol. Food Nutr.* **2001**, *40*, 253–283. [CrossRef]
38. Sapp, S.G.; Downing-Matibag, T. Consumer acceptance of food irradiation: A test of the recreancy theorem. *Int. J. Consum. Stud.* **2009**, *33*, 417–424. [CrossRef]
39. Bruhn, C.M. United States consumer choice of irradiated food. In *Irradiation for Food Safety and Quality*; Loaharanu, P., Thomas, P., Eds.; Technomic Publishing Company: Lancaster, PA, USA, 2001; pp. 169–173.
40. Siegrist, M.; Cousin, M.E.; Kastenholz, H.; Wiek, A. Public acceptance of nanotechnology foods and food packaging: The influence of affect and trust. *Appetite* **2007**, *49*, 459–466. [CrossRef]
41. Siegrist, M.; Stampfli, N.; Kastenholz, H.; Keller, C. Perceived risks and perceived benefits of different nanotechnology foods and nanotechnology food packaging. *Appetite* **2008**, *51*, 283–290. [CrossRef]
42. Siegrist, M.; Stampfli, N.; Kastenholz, H. Acceptance of nanotechnology foods: A conjoint study examining consumers' willingness to buy. *Br. Food J.* **2009**, *111*, 660–668. [CrossRef]
43. Roosen, J.; Bieberstein, A.; Blanchemanche, S.; Goddard, E.; Marette, S.; Vandermoere, F. Trust and willingness to pay for nanotechnology food. *Food Policy* **2015**, *52*, 75–83. [CrossRef]
44. Poortinga, W.; Pidgeon, N.F. Trust in risk regulation: Cause or consequence of the acceptability of GM food? *Risk Anal.* **2005**, *25*, 199–209. [CrossRef] [PubMed]
45. Fandos Herrera, C.; Flavián Blanco, C. Consequences of consumer trust in PDO food products: The role of familiarity. *J. Prod. Brand Manag.* **2011**, *20*, 282–296. [CrossRef]
46. Lassoued, R.; Hobbs, J.E. Consumer confidence in credence attributes: The role of brand trust. *Food Policy* **2015**, *52*, 99–107. [CrossRef]
47. Lassoued, R.; Hobbs, J.E.; Micheels, E.T.; Zhang, D.D. Consumer trust in chicken brands: A structural equation model. *Can. J. Agric. Econ.* **2015**, *63*, 621–647. [CrossRef]
48. Taylor, A.W.; Coveney, J.; Ward, P.R.; Henderson, J.; Meyer, S.B.; Pilkington, R.; Gill, T.K. Fruit and vegetable consumption–the influence of aspects associated with trust in food and safety and quality of food. *Public Health Nutr.* **2012**, *15*, 208–217. [CrossRef]
49. Viktoria Rampl, L.; Eberhardt, T.; Schütte, R.; Kenning, P. Consumer trust in food retailers: Conceptual framework and empirical evidence. *Int. J. Retail Distrib. Manag.* **2012**, *40*, 254–272. [CrossRef]
50. Banwell, C.; Kelly, M.; Dixon, J.; Seubsman, S.A.; Sleigh, A. Trust: The missing dimension in the food retail transition in Thailand. *Anthropol. Forum* **2016**, *26*, 138–154. [CrossRef]

51. Xie, Y.; Batra, R.; Peng, S. An extended model of preference formation between global and local brands: The roles of identity expressiveness, trust, and affect. *J. Int. Mark.* **2015**, *23*, 50–71. [CrossRef]
52. Chinomona, R.; Mahlangu, D.; Pooe, D. Brand service quality, satisfaction, trust and preference as predictors of consumer brand loyalty in the retailing industry. *Mediterr. J. Soc. Sci.* **2013**, *4*. [CrossRef]
53. Becker, M.; Hanley, M. The mediating effects of privacy and preference management on trust and consumer participation in a mobile marketing initiative: A proposed conceptual model. In *Trust and New Technologies: Marketing and Management on the Internet and Mobile Media*; Kautonen, T., Karjaluoto, H., Eds.; Edward Elgar Publishing Limited: Cheltenham, UK, 2008; pp. 127–145.
54. Hariyanto, E. The Influence of Brand Experience through Brand Trust and Brand Satisfaction toward Brand Loyalty Consumer at Carl's Jr Surabaya. *Petra Bus. Manag. Rev.* **2018**, *4*, 20–29.
55. Earle, T.C. Trust in risk management: A model-based review of empirical research. *Risk Anal.* **2010**, *30*, 541–574. [CrossRef] [PubMed]
56. Jin, N.; Line, N.D.; Merkebu, J. The impact of brand prestige on trust, perceived risk, satisfaction, and loyalty in upscale restaurants. *J. Hosp. Mark. Manag.* **2016**, *25*, 523–546. [CrossRef]
57. Woolley, K.; Fishbach, A. A recipe for friendship: Similar food consumption promotes trust and cooperation. *J. Consum. Psychol.* **2017**, *27*, 1–10. [CrossRef]
58. Yim, D.S. Trust of Agricultural Food and Affective Commitment: A Comparison of Levels of Consumer's Income. *J. Korean Soc. Int. Agric.* **2012**, *24*, 293–298.
59. Mayer, R.C.; Davis, J.H.; Schoorman, F.D. An integrative model of organizational trust. *Acad. Manag. Rev.* **1995**, *20*, 709–734. [CrossRef]
60. McKnight, D.H.; Chervany, N.L. Trust and distrust definitions: One bite at a time. In *Trust in Cyber-Societies*; Falcone, R., Singh, M., Tan, Y.-H., Eds.; Springer: Berlin, Germany, 2001; pp. 27–54.
61. Kożuch, B.; Magala, S.J.; Paliszkiewicz, J. *Managing Public Trust*; Springer: Schoonhoven, The Netherlands, 2018.
62. Siegrist, M.; Gutscher, H.; Earle, T.C. Perception of risk: The influence of general trust, and general confidence. *J. Risk Res.* **2005**, *8*, 145–156. [CrossRef]
63. Ding, Y.; Veeman, M.M.; Adamowicz, W.L. The impact of generalized trust and trust in the food system on choices of a functional GM food. *Agribusiness* **2012**, *28*, 54–66. [CrossRef]
64. McKnight, D.H.; Cummings, L.L.; Chervany, N.L. Initial trust formation in new organizational relationships. *Acad. Manag. Rev.* **1998**, *23*, 473–490. [CrossRef]
65. Love, B.; Mackert, M.; Silk, K. Consumer trust in information sources: Testing an interdisciplinary model. *Sage Open* **2013**, *3*. [CrossRef]
66. Tsai, M.T.; Chin, C.W.; Chen, C.C. The effect of trust belief and salesperson's expertise on consumer's intention to purchase nutraceuticals: Applying the theory of reasoned action. *Soc. Behav. Personal.* **2010**, *38*, 273–287. [CrossRef]
67. Zhang, J.; Liu, H.; Sayogo, D.S.; Picazo-Vela, S.; Luna-Reyes, L. Strengthening institutional-based trust for sustainable consumption: Lessons for smart disclosure. *Gov. Inf. Q.* **2016**, *33*, 552–561. [CrossRef]
68. McKnight, D.H.; Choudhury, V.; Kacmar, C. Developing and validating trust measures for e-commerce: An integrative typology. *Inf. Syst. Res.* **2002**, *13*, 334–359. [CrossRef]
69. Lam, T.; Heales, J.; Hartley, N.; Hodkinson, C. Information Transparency Matters in Relation to Consumer Trust in Food Safety. Presented at the Australasian Conference on Information Systems, Sydney, Australia, 3–5 December 2018.
70. Zhang, X.; Zhang, Q. Online trust forming mechanism: Approaches and an integrated model. In Proceedings of the 7th International Conference on Electronic Commerce, Xi'an, China, 15–17 August 2005; pp. 201–209.
71. McKnight, D.H.; Chervany, N.L. What trust means in e-commerce customer relationships: An interdisciplinary conceptual typology. *Int. J. Electron. Commun.* **2001**, *6*, 35–59. [CrossRef]
72. Bachmann, R.; Inkpen, A.C. Understanding institutional-based trust building processes in inter-organizational relationships. *Organ. Stud.* **2011**, *32*, 281–301. [CrossRef]
73. Morrow, J.L., Jr.; Hansen, M.H.; Pearson, A.W. The cognitive and affective antecedents of general trust within cooperative organizations. *J. Manag. Issues* **2014**, *16*, 48–64.
74. Benson, T.; Lavelle, F.; Spence, M.; Elliott, C.T.; Dean, M. The development and validation of a toolkit to measure consumer trust in food. *Food Control* **2019**, *110*, 106988. [CrossRef]
75. Couch, L.L.; Jones, W.H. Measuring levels of trust. *J. Res. Personal.* **1997**, *31*, 319–336. [CrossRef]

76. Moorman, C.; Deshpande, R.; Zaltman, G. Factors Affecting Trust in Market Research Relationships. *J. Mark.* **1993**, *57*, 81–101. [CrossRef]
77. Yan, Z. Security via trusted communications. In *Handbook of Information and Communication Security*; Stavroulakis, P., Stamp, M., Eds.; Springer: Berlin, Germany, 2010; pp. 719–746.
78. Feldmann, C.; Hamm, U. Consumers' perceptions and preferences for local food: A review. *Food Qual. Prefer.* **2015**, *40*, 152–164. [CrossRef]
79. Mørkbak, M.R.; Christensen, T.; Gyrd-Hansen, D. Valuation of food safety in meat—A review of stated preference studies. *Food Econ. Acta Agric. Scand. C* **2008**, *5*, 63–74.
80. Moser, R.; Raffaelli, R.; Thilmany-McFadden, D. Consumer preferences for fruit and vegetables with credence-based attributes: A review. *Int. Food Agribus. Manag.* **2011**, *14*, 121–142.
81. Tonkin, E.; Wilson, A.M.; Coveney, J.; Webb, T.; Meyer, S.B. Trust in and through labelling—A systematic review and critique. *Br. Food J.* **2015**, *117*, 318–338. [CrossRef]
82. Kim, R. Japanese consumers' use of extrinsic and intrinsic cues to mitigate risky food choices. *Int. J. Consum. Stud.* **2008**, *32*, 49–58. [CrossRef]
83. Zheng, Y.; Li, X.; Peterson, H.H. In pursuit of safe foods: Chinese preferences for soybean attributes in soymilk. *Agribusiness* **2013**, *29*, 377–391. [CrossRef]
84. Charity, N.E.J. Economic Analysis of Consumers' Awareness and Willingness to Pay for Geographical Indicators and Other Quality Attributes of Honey in Kenya. Master's Thesis, University of Nairobi, Nairobi, Kenya, 2016.
85. Bolduc, D.; Alvarez-Daziano, R. On estimation of hybrid choice models. In *Choice Modelling: The State-of-the-Art and the State-of-Practice: Proceedings from the Inaugural International Choice Modelling Conference*; Hess, S., Daly, A., Eds.; Emerald Group Publishing Limited: Bingley, UK, 2010; pp. 259–287.
86. Abou-Zeid, M.; Ben-Akiva, M. Hybrid choice model. In *Handbook of Choice Modelling*; Hess, S., Daly, A., Eds.; Edward Elgar Publishing Limited: Cheltenham, UK, 2014; pp. 283–412.
87. Hensher, D.A.; Rose, J.M.; Greene, W.H. *Applied Choice Analysis*, 2nd ed.; Cambridge University Press: Cambridge, UK, 2015.
88. Langen, N. Are ethical consumption and charitable giving substitutes or not? Insights into consumers' coffee choice. *Food Qual. Prefer.* **2011**, *22*, 412–421. [CrossRef]
89. Yangui, A.; Costa-Font, M.; Gil, J.M. The effect of personality traits on consumers' preferences for extra virgin olive oil. *Food Qual. Prefer.* **2016**, *51*, 27–38. [CrossRef]
90. O'Neill, V.; Hess, S.; Campbell, D. A question of taste: Recognising the role of latent preferences and attitudes in analysing food choices. *Food Qual. Prefer.* **2014**, *32*, 299–310. [CrossRef]
91. Song, F.; Hess, S.; Dekker, T. Comparing and Combining Best-Worst Scaling and Stated Choice Data to Understand Attribute Importance in Mode-Choice Behavior. In Proceedings of the Transportation Research Board 97th Annual Meeting, Washington, DC, USA, 7–11 January 2018.
92. Kikulwe, E.M.; Birol, E.; Wesseler, J.; Falck-Zepeda, J. A latent class approach to investigating demand for genetically modified banana in Uganda. *Agric. Econ.* **2011**, *42*, 547–560. [CrossRef]
93. Owusu-Sekyere, E.; Owusu, V.; Akwetey, W.Y.; Jordaan, H.; Ogundeji, A.A. Economic welfare implications of policy changes regarding food safety and quality in Ghana. *Afr. J. Agric. Resour. Econ.* **2018**, *13*, 357–371.
94. Thurstone, L.L. A law of comparative judgement. *Psychol. Rev.* **1927**, *34*, 273–286. [CrossRef]
95. McFadden, D. Conditional logit analysis of qualitative choice behavior. In *Frontiers in Econometrics*; Zarembka, P., Ed.; Academic Press: New York, NY, USA, 1973; pp. 105–142.
96. Batinic, B.; Reips, U.D.; Bosnjak, M. *Online Social Sciences*; Hogrefe & Huber: Seattle, WA, USA, 2002.
97. Bethlehem, J.; Biffignandi, S. *Handbook of Web Surveys*; John Wiley & Sons: Hoboken, NJ, USA, 2011.
98. Lancaster, K.J. A new approach to consumer theory. *J. Political Econ.* **1966**, *74*, 132–157. [CrossRef]
99. Rachlin, H. Economics and behavioral psychology. In *Limits to Action: The Allocation of Individual Behavior*; Academic Press: New York, NY, USA, 1980; pp. 205–236.
100. McFadden, D. The choice theory approach to market research. *Mark. Sci.* **1986**, *5*, 275–297. [CrossRef]
101. The Committee of Agricultural Extension. Sweet peppers and agricultural facilities. In *Review of Agricultural Extension Science, 68*; National Ilan University: Ilan, Taiwan, 2014.
102. Taichung District Agricultural Research and Extension Station. *Sweet Peppers Production and Cultivation Technology*; Council of Agriculture Executive Yuan: Taipei, Taiwan, 2016. Available online: http://www.tdais.gov.tw (accessed on 4 April 2016).

103. Dekhili, S.; Sirieix, L.; Cohen, E. How consumers choose olive oil: The importance of origin cues. *Food Qual. Prefer.* **2011**, *22*, 757–762. [CrossRef]
104. Fernqvist, F.; Ekelund Axelson, L. Consumer attitudes towards origin and organic-the role of credence labels on consumers' liking of tomatoes. *Eur. J. Hortic. Sci.* **2013**, *78*, 184–190.
105. Choice Metrics. *Ngene 1.1.1 User Manual & Reference Guide*; Choice Metrics, Ltd.: Sydney, Australia, 2012.
106. Zwerina, K.; Huber, J.; Kuhfeld, W.F. *A General Method for Constructing Efficient Choice Designs*; Fuqua School of Business, Duke University: Durham, NC, USA, 1996.
107. Carlsson, F.; Martinsson, P. Design techniques for stated preference methods in health economics. *Health Econ.* **2003**, *12*, 281–294. [CrossRef]
108. Rose, J.M.; Bliemer, M.C. *Constructing Efficient Choice Experiments*; ITLS Working Paper, ITLS-WP-05-07; Institute of Transport and Logistic Studies: Sydney, Australia, 2005.
109. Scarpa, R.; Campbell, D.; Hutchinson, W.G. Benefit estimates for landscape improvements: Sequential Bayesian design and respondents' rationality in a choice experiment. *Land Econ.* **2007**, *83*, 617–634. [CrossRef]
110. Train, K.E. *Discrete Choice Methods with Simulation*, 2nd ed.; Cambridge University Press: New York, NY, USA, 2009.
111. Haughton, D.; Legrand, P.; Woolford, S. Review of three latent class cluster analysis packages: Latent Gold, poLCA, and MCLUST. *Am. Stat.* **2009**, *63*, 81–91. [CrossRef]
112. Boxall, P.C.; Adamowicz, W.L. Understanding heterogeneous preferences in random utility models: A latent class approach. *Environ. Resour. Econ.* **2002**, *23*, 421–446. [CrossRef]
113. Swait, J. A structural equation model of latent segmentation and product choice for cross-sectional revealed preference choice data. *J. Retail. Consum. Serv.* **1994**, *1*, 77–89. [CrossRef]
114. McCutcheon, A.L. *Latent Class Analysis*; Sage Publications: Cheltenham, UK, 1987.
115. McCutcheon, A.L. Basic concepts and procedures in single-and multiple-group latent class analysis. In *Applied Latent Class Analysis*; McCutcheon, A.L., Ed.; Cambridge University Press: Cambridge, UK, 2002; pp. 56–88.
116. Lusk, J.L.; Roosen, J.; Fox, J.A. Demand for beef from cattle administered growth hormones or fed genetically modified corn: A comparison of consumers in France, Germany, the United Kingdom, and the United States. *Am. J. Agric. Econ.* **2003**, *85*, 16–29. [CrossRef]
117. Bech, M.; Gyrd-Hansen, D. Effects coding in discrete choice experiments. *Health Econ.* **2005**, *14*, 1079–1083. [CrossRef] [PubMed]
118. Ortega, D.L.; Wang, H.H.; Wu, L.; Olynk, N.J. Modeling heterogeneity in consumer preferences for select food safety attributes in China. *Food Policy* **2011**, *36*, 318–324. [CrossRef]
119. Hole, A.R. A comparison of approaches to estimating confidence intervals for willingness to pay measures. *Health Econ.* **2007**, *16*, 827–840. [CrossRef]
120. Hein, K.A.; Jaeger, S.R.; Carr, B.T.; Delahunty, C.M. Comparison of five common acceptance and preference methods. *Food Qual. Prefer.* **2008**, *19*, 651–661. [CrossRef]
121. Flynn, T.N.; Marley, A.A.J. Best-worst scaling: Theory and methods. In *Handbook of Choice Modelling*; Hess, S., Daly, A., Eds.; Edward Elgar Publishing: Cheltenham, UK, 2004; pp. 178–201.
122. Finn, A.; Louviere, J.J. Determining the appropriate response to evidence of public concern: The case of food safety. *J. Public Policy Mark.* **1992**, *11*, 12–25. [CrossRef]
123. Cohen, S. Maximum difference scaling: Improved measures of importance and preference for segmentation. In *Sawtooth Software Conference Proceedings*; Sawtooth Software, Inc.: Provo, UT, USA, 2003; Volume 530, pp. 61–74.
124. Louviere, J.J.; Flynn, T.N.; Marley, A.A.J. *Best-Worst Scaling: Theory, Methods and Applications*; Cambridge University Press: Cambridge, UK, 2015.
125. Cohen, S.H.; Markowitz, P. Renewing market segmentation: Some new tools to correct old problems. In *ESOMAR 2002 Congress Proceedings*; ESOMAR: Amsterdam, The Netherlands, 2002; pp. 595–612.
126. Louviere, J.J.; Woodworth, G.G. *Best-Worst Analysis*; Working Paper; Department of Marketing and Economic Analysis, University of Alberta: Edmonton, AB, Canada, 1990.
127. Smith, N.F.; Street, D.J. The use of balanced incomplete block designs in designing randomized response surveys. *Aust. N. Z. J. Stat.* **2003**, *45*, 181–194. [CrossRef]
128. Sawtooth Software. *The MaxDiff v6.0 Technical Paper*; Sawtooth Software Inc.: Provo, UT, USA, 2007.

129. Lagerkvist, C.J.; Okello, J.; Karanja, N. Anchored vs. relative best–worst scaling and latent class vs. hierarchical Bayesian analysis of best–worst choice data: Investigating the importance of food quality attributes in a developing country. *Food Qual. Prefer.* **2012**, *25*, 29–40. [CrossRef]
130. De-Magistris, T.; Royo, A.G. Wine consumers preferences in Spain: An analysis using the best-worst scaling approach. *Span. J. Agric. Res.* **2014**, *12*, 529–541. [CrossRef]
131. Campbell, D.; Erdem, S. Position bias in best-worst scaling surveys: A case study on trust in institutions. *Am. J. Agric. Econ.* **2015**, *97*, 526–545. [CrossRef]
132. Claybaugh, C.C.; Haseman, W.D. Understanding professional connections in LINKEDIN—A question of trust. *J. Comput. Inf. Syst.* **2013**, *54*, 94–105. [CrossRef]
133. Hämäläinen, A. Trust Antecedents in Social Networking Services. Master's Thesis, School of Business, Aalto University, Espoo, Finland, 2015.
134. Kivijärvi, H.; Leppänen, A.; Hallikainen, P. Antecedents of Information Technology Trust and the Effect of Trust on Perceived Performance Improvement. *Int. J. Soc. Org.* **2013**, *3*, 17–32. [CrossRef]
135. Allenby, G.M.; Ginter, J.L. Using extremes to design products and segment markets. *J. Mark. Res.* **1995**, *32*, 392–403. [CrossRef]
136. Lenk, P.J.; DeSarbo, W.S.; Green, P.E.; Young, M.R. Hierarchical Bayes conjoint analysis: Recovery of partworth heterogeneity from reduced experimental designs. *Mark. Sci.* **1996**, *15*, 173–191. [CrossRef]
137. Akinc, D.; Vandebroek, M. Comparing the performances of maximum simulated likelihood and hierarchical Bayesian estimation for mixed logit models. In Proceedings of the International Choice Modelling Conference 2017, Cape Town, South Africa, 3–5 April 2017.
138. Lusk, J.L.; Briggeman, B.C. Food values. *Am. J. Agric. Econ.* **2009**, *91*, 184–196. [CrossRef]
139. Hansson, H.; Lagerkvist, C.J. Dairy farmers' use and non-use values in animal welfare: Determining the empirical content and structure with anchored best-worst scaling. *J. Dairy Sci.* **2016**, *99*, 579–592. [CrossRef]
140. Rodríguez-Delgado, M.Á.; González-Hernández, G.; Conde-González, J.E.; Pérez-Trujillo, J.P. Principal component analysis of the polyphenol content in young red wines. *Food Chem.* **2002**, *78*, 523–532. [CrossRef]
141. Costello, A.B.; Osborne, J.W. Best practices in exploratory factor analysis: Four recommendations for getting the most from your analysis. *Pract. Assess. Res. Eval.* **2005**, *10*, 1–9.
142. Horn, J.L. A rationale and test for the number of factors in factor analysis. *Psychometrika* **1965**, *30*, 179–185. [CrossRef]
143. Fabrigar, L.R.; Wegener, D.T.; MacCallum, R.C.; Strahan, E.J. Evaluating the use of exploratory factor analysis in psychological research. *Psychol. Methods* **1999**, *4*, 272–299. [CrossRef]
144. Hayton, J.C.; Allen, D.G.; Scarpello, V. Factor retention decisions in exploratory factor analysis: A tutorial on parallel analysis. *Organ. Res. Methods* **2004**, *7*, 191–205. [CrossRef]
145. Dinno, A. Implementing Horn's parallel analysis for principal component analysis and factor analysis. *Stata J.* **2009**, *9*, 291–298. [CrossRef]
146. Timmerman, M.E.; Lorenzo-Seva, U. Dimensionality assessment of ordered polytomous items with parallel analysis. *Psychol. Methods* **2011**, *16*, 209. [CrossRef] [PubMed]
147. Orme, B. Sample size issues for conjoint analysis. In *Getting Started with Conjoint Analysis: Strategies for Product Design and Pricing Research*, 2nd ed.; Research Publishers LLC: Madison, WI, USA, 2010.
148. Lu Hsu, J.; Lin, Y.T. Consumption and attribute perception of fluid milk in Taiwan. *Nutr. Food Sci.* **2006**, *36*, 177–182. [CrossRef]
149. Coltman, T.R.; Devinney, T.M.; Midgley, D.F. E-Business strategy and firm performance: A latent class assessment of the drivers and impediments to success. *J. Inf. Technol.* **2007**, *22*, 87–101. [CrossRef]
150. Hauser, J.R. Testing the accuracy, usefulness, and significance of probabilistic choice models: An information-theoretic approach. *Oper. Res.* **1978**, *26*, 406–421. [CrossRef]
151. Ogawa, K. An approach to simultaneous estimation and segmentation in conjoint analysis. *Mark. Sci.* **1987**, *6*, 66–81. [CrossRef]
152. Bozdogan, H. Model selection and Akaike's information criterion (AIC): The general theory and its analytical extensions. *Psychometrika* **1987**, *52*, 345–370. [CrossRef]
153. Schwarz, G. Estimating the dimension of a model. *Ann. Stat.* **1978**, *6*, 461–464. [CrossRef]
154. Hair, J.F.; Black, W.C.; Babin, B.J.; Anderson, R.E. *Multivariate Data Analysis: Pearson New International Edition*; Pearson Education Limited: London, UK, 2014.

155. Huang, S. Nation-branding and transnational consumption: Japan-mania and the Korean wave in Taiwan. *Media Cult. Soc.* **2011**, *33*, 3–18. [CrossRef]
156. Ma, J.; Wang, S.; Hao, W. Does cultural similarity matter? Extending the animosity model from a new perspective. *J. Consum. Mark.* **2012**, *29*, 319–332. [CrossRef]
157. Becker, K. Positioning strategies against nations with perceived quality advantages. *J. Transl. Manag.* **2009**, *14*, 74–99. [CrossRef]
158. Lim, K.H.; Hu, W.; Maynard, L.J.; Goddard, E. US consumers' preference and willingness to pay for country-of-origin-labeled beef steak and food safety enhancements. *Can. J. Agric. Econ.* **2013**, *61*, 93–118. [CrossRef]
159. Rahmawati, N.A.; Muflikhati, I. Effect of Consumer Ethnocentrism and Perceived Quality of Product on Buying Behavior of Domestic and Foreign Food Products: A Case Study in Pekanbaru, Riau, Indonesia. *J. Consum. Sci.* **2016**, *1*, 1–13. [CrossRef]
160. Grunert, K.G.; Sonntag, W.I.; Glanz-Chanos, V.; Forum, S. Consumer interest in environmental impact, safety, health and animal welfare aspects of modern pig production: Results of a cross-national choice experiment. *Meat Sci.* **2018**, *137*, 123–129. [CrossRef]
161. Koistinen, L.; Pouta, E.; Heikkilä, J.; Forsman-Hugg, S.; Kotro, J.; Mäkelä, J.; Niva, M. The impact of fat content, production methods and carbon footprint information on consumer preferences for minced meat. *Food Qual. Prefer.* **2013**, *29*, 126–136. [CrossRef]
162. Boncinelli, F.; Contini, C.; Romano, C.; Scozzafava, G.; Casini, L. Territory, environment, and healthiness in traditional food choices: Insights into consumer heterogeneity. *Int. Food Agribus. Manag. Rev.* **2017**, *20*, 143–157. [CrossRef]
163. Apostolidis, C.; McLeay, F. Should we stop meating like this? Reducing meat consumption through substitution. *Food Policy* **2016**, *65*, 74–89. [CrossRef]
164. Peschel, A.O.; Grebitus, C.; Steiner, B.; Veeman, M. How does consumer knowledge affect environmentally sustainable choices? Evidence from a cross-country latent class analysis of food labels. *Appetite* **2016**, *106*, 78–91. [CrossRef]
165. Chen, M.F. Attitude toward organic foods among Taiwanese as related to health consciousness, environmental attitudes, and the mediating effects of a healthy lifestyle. *Br. Food J.* **2009**, *111*, 165–178. [CrossRef]
166. Yiridoe, E.K.; Bonti-Ankomah, S.; Martin, R.C. Comparison of consumer perceptions and preference toward organic versus conventionally produced foods: A review and update of the literature. *Renew. Agric. Food Syst.* **2005**, *20*, 193–205. [CrossRef]
167. Festinger, L. *A Theory of Cognitive Dissonance*; Stanford University Press: Stanford, CA, USA, 1957; Volume 2.
168. Losch, M.E.; Cacioppo, J.T. Cognitive dissonance may enhance sympathetic tonus, but attitudes are changed to reduce negative affect rather than arousal. *J. Exp. Soc. Psychol.* **1990**, *26*, 289–304. [CrossRef]
169. Hasimu, H.; Marchesini, S.; Canavari, M. A concept mapping study on organic food consumers in Shanghai, China. *Appetite* **2017**, *108*, 191–202. [CrossRef] [PubMed]
170. Xie, B.; Wang, L.; Yang, H.; Wang, Y.; Zhang, M. Consumer perceptions and attitudes of organic food products in Eastern China. *Br. Food J.* **2015**, *117*, 1105–1121. [CrossRef]
171. Liang, R.D. Predicting intentions to purchase organic food: The moderating effects of organic food prices. *Br. Food J.* **2016**, *118*, 183–199. [CrossRef]
172. Jing, B.; Wen, Z. Finitely loyal customers, switchers, and equilibrium price promotion. *J. Econ. Manag. Strategy* **2008**, *17*, 683–707. [CrossRef]
173. Koçaş, C.; Bohlmann, J.D. Segmented switchers and retailer pricing strategies. *J. Mark.* **2008**, *72*, 124–142. [CrossRef]
174. Gao, Z.; Schroeder, T.C. Consumer responses to new food quality information: Are some consumers more sensitive than others? *Agric. Econ.* **2009**, *40*, 339–346. [CrossRef]

# 10

# The Role of Intrinsic and Extrinsic Sensory Factors in Sweetness Perception of Food and Beverages

**Qian Janice Wang [1,2,*], Line Ahm Mielby [1], Jonas Yde Junge [1], Anne Sjoerup Bertelsen [1], Ulla Kidmose [1], Charles Spence [2] and Derek Victor Byrne [1]**

1 Department of Food Science, Faculty of Science and Technology, Aarhus University, 5792 Aarslev, Denmark; lineh.mielby@food.au.dk (L.A.M.); jonas.junge@food.au.dk (J.Y.J.); annesbertelsen@food.au.dk (A.S.B.); ulla.kidmose@food.au.dk (U.K.); derekv.byrne@food.au.dk (D.V.B.)

2 Crossmodal Research Laboratory, Department of Experimental Psychology, University of Oxford, Oxford OX2 6GG, UK; charles.spence@psy.ox.ac.uk

* Correspondence: qianjanice.wang@food.au.dk.

**Abstract:** When it comes to eating and drinking, multiple factors from diverse sensory modalities have been shown to influence multisensory flavour perception and liking. These factors have heretofore been strictly divided into either those that are intrinsic to the food itself (e.g., food colour, aroma, texture), or those that are extrinsic to it (e.g., related to the packaging, receptacle or external environment). Given the obvious public health need for sugar reduction, the present review aims to compare the relative influences of product-intrinsic and product-extrinsic factors on the perception of sweetness. Evidence of intrinsic and extrinsic sensory influences on sweetness are reviewed. Thereafter, we take a cognitive neuroscience perspective and evaluate how differences may occur in the way that food-intrinsic and extrinsic information become integrated with sweetness perception. Based on recent neuroscientific evidence, we propose a new framework of multisensory flavour integration focusing not on the food-intrinsic/extrinsic divide, but rather on whether the sensory information is perceived to originate from within or outside the body. This framework leads to a discussion on the combinability of intrinsic and extrinsic influences, where we refer to some existing examples and address potential theoretical limitations. To conclude, we provide recommendations to those in the food industry and propose directions for future research relating to the need for long-term studies and understanding of individual differences.

**Keywords:** sugar reduction; multisensory integration; intrinsic factors; extrinsic factors; sweetness perception

---

## 1. Introduction

Eating and drinking are amongst the most multisensory of the experiences that we have. When people think about the consumption of food and drink, the senses of taste and smell usually come to mind first. However, a growing body of research conducted over the last decade or two has increasingly demonstrated that *all* of our senses play a role in influencing flavour perception (see References [1–3] for reviews). For instance, recalling the experience of eating an apple will usually evoke not just taste and smell, but also its colour, weight, shape, its firmness, crunchiness, juiciness and even the sound of chewing and perhaps its provenance (e.g., supermarket, organic, local, or the tree in the backyard).

A large body of research now supports the view that both *food-intrinsic* sensory factors (e.g., product colour, aroma, texture, viscosity, etc.) as well as *food-extrinsic* factors (e.g., visual, olfactory, and tactile properties of product packaging or servingware, background music, ambient lighting, temperature and aroma, etc.) play a role in determining whether we accept and how we perceive food and beverages (e.g., for intrinsic factors [2,4,5] and for extrinsic factors [6–12]). What is less clear, however, is how these different factors interact and the relative importance of intrinsic and extrinsic factors to our perception of, not to mention our behaviour towards, food and drink.

In this review, we focus on how intrinsic and extrinsic factors can enhance the perception of sweetness in foods and beverages and address the question of how (and if) they can be combined in order to deliver an enhanced perception of sweetness. The decision to target the perception of sweetness is informed by the growing public health concern over excessive sugar consumption. The consumption of sweet foods has been argued to be one of the major contributors to the current obesity epidemic, with more than 3 million deaths globally each year [13–16].

Moreover, sugar reduction is of critical concern to major food and beverage companies such as PepsiCo, Givaudan, and Arla, who have been engaging in a number of major initiatives in order to reduce added sugars and develop naturally resourced sweeteners [17–19]. Therefore, a multisensory, psychological model of sweetness perception is especially important when it comes to the design of sugar-reduced/replaced foods and beverages.

Hutchings et al. [20] recently outlined four general strategies for sugar reduction. Sugar substitution, altering food structure (e.g., heterogeneously distributing sucrose, modifying tastant release, or reducing particle size), gradual long-term sugar reduction, and using the principles of multisensory integration. However, Hutchings et al. [20] do not address the role of product-extrinsic factors in sweetness perception. Therefore, in the current review, we endeavour to assess the complex interplay between various different sensory factors surrounding the multisensory eating experience, with a cognitive neuroscience view of how the senses combine to shape taste perception (Note that the scope of the current review is limited to the influence of *sensory cues* on sweetness enhancement.

Various cognitive factors, such as, for instance, any information provided by packaging, can also influence consumer flavour perception and acceptance of sugar-reduced foods [21]; however, such cognitive effects fall beyond the scope of the present article. Similarly, sugar substitution schemes with non-caloric sweeteners are also outside of the scope of the present review [22]).

## 2. Food-Intrinsic versus Food-Extrinsic Influences on Sweetness Perception

In the following section, we will target each sensory modality in turn and review the literature on the intrinsic and/or extrinsic cues regarding their influence on sweetness perception. Table 1 provides a representative summary of studies demonstrating sweetness enhancement effects from the influence of different sensory modalities.

**Table 1.** A representative selection of studies demonstrating sweetness enhancement via food-intrinsic and extrinsic sensory cues.

| Study | Sense | Intrinsic or Extrinsic | Sweet Enhancing Stimuli | Control/Comparison Stimuli | Taste Stimuli | Scale | % Difference |
|---|---|---|---|---|---|---|---|
| Crisinel et al. (2012) [7] | Hearing | Extrinsic | Sweet soundtrack | Bitter soundtrack | Cinder toffee | 1–9 rating (bitter–sweet) | 15% |
| Höchenberger et al. (2018) [23] | Hearing | Extrinsic | Sweet soundtrack | Bitter soundtrack | Toffee | 0–100 rating (bitter–sweet) | 8% |
| Höchenberger et al. (2018) [23] | Hearing | Extrinsic | Sweet soundtrack | Bitter soundtrack | Toffee | 0–100 rating (sweet, bitter, salt, sour) | No significant difference |
| Reinoso Carvalho et al. (2016) [9] | Hearing | Extrinsic | Sweet soundtrack | Bitter soundtrack | Belgian beer | 1–7 rating sweetness | 20% |
| Reinoso Carvalho et al. (2016) [9] | Hearing | Extrinsic | Sweet soundtrack | Sour soundtrack | Belgian beer | 1–7 rating sweetness | 20% |
| Reinoso Carvalho et al. (2017) [24] | Hearing | Extrinsic | Legato soundtrack | Staccato soundtrack | Dark chocolate | 1–7 rating sweetness | 11% |
| Wang and Spence, (2016) [25] | Hearing | Extrinsic | Consonant soundtrack | Dissonant soundtrack | Fruit juice (apple, orange, grapefruit) | 1–10 rating (sour–sweet) | 19% |
| Wang and Spence (2017) [26] | Hearing | Extrinsic | Consonant soundtrack | Dissonant soundtrack | Fruit juice (apple, orange, grapefruit) | 0–10 rating (sour–sweet) | 17% |
| Wang and Spence, (2017) [27] | Hearing | Extrinsic | Sweet soundtrack | Sour soundtrack | Off-dry white wine | 0–10 rating sweetness | 19% |
| Wang et al. (2019) [28] | Hearing | Extrinsic | Sweet soundtrack | Bitter soundtrack | Apple elderflower juice | 1–9 rating sweetness | 8% |
| Carvalho and Spence (2019) [29] | Sight | Extrinsic | Pink coffee cup | White coffee cup | Espresso | 0–10 rating (sweetness) | 30% |
| Clydesdale et al. (1992) [30] | Sight | Intrinsic | More red colouring | Less red colouring | Dry beverage base and sugar solution | 1–7 rating sweetness | 14% |
| Fairhurst et al. (2015) [31] | Sight | Both | Round plate and round food presentation | Angular plate and angular food presentation | Beetroot salad | 0–10 rating sweetness | 17% |
| Frank et al. (1989) [32] | Sight | Intrinsic | Red colouring | No colour | Sucrose solution | Rating sweetness | No effect |
| Hidaka and Shimoda (2014) [33] | Sight | Intrinsic | Pink solution | No colouring | Sucrose solution 4% and 6% | 10 cm visual analogue scale (VAS) less–sweeter | 40% |

**Table 1.** *Cont.*

| Study | Sense | Intrinsic or Extrinsic | Sweet Enhancing Stimuli | Control/Comparison Stimuli | Taste Stimuli | Scale | % Difference |
|---|---|---|---|---|---|---|---|
| Johnson and Clydesdale (1982) [34] | Sight | Intrinsic | Darker red coloured solution | Lighter red reference solution | Sucrose solutions 2.7–5.3% | Magnitude estimation sweetness | 2–10% |
| Lavin and Lawless (1998) [35] | Sight | Intrinsic | Darker red solution | Lighter red solution | Fruit beverage + aspartame to 9% sucrose level | 1–9 category scale sweetness | 10% |
| Lavin and Lawless (1998) [35] | Sight | Intrinsic | Lighter green solution | Darker green solution | Fruit beverage + aspartame to 9% sucrose level | 1–9 category scale sweetness | 8% |
| Maga (1974) [36] | Sight | Intrinsic | Red colouring | Green, yellow, uncoloured solutions | Sucrose solution | Recognition threshold | No effect |
| Pangborn and Hansen (1963) [37] | Sight | Intrinsic | Red solution | Green, yellow, uncoloured solutions | Pear nectar | Rating sweetness | No effect |
| Pangborn et al. (1963) [38] | Sight | Intrinsic | Pink colouring | Yellow, brown, light red, dark red colouring | White wine | Rating sweetness | Rose sweetest |
| Pangborn (1960) [39] | Sight | Intrinsic | Red colouring | Green, yellow, uncoloured solutions | Sucrose solution | 2-AFC (alternative forced choice) which one sweeter | No effect |
| Pangborn (1960) [39] | Sight | Intrinsic | Red colouring | Green, yellow, uncoloured solutions | Pear nectar | 2-AFC which one sweeter | No effect |
| Piqueras–Fiszman et al. (2012) [8] | Sight | Extrinsic | White plate | Black plate | Strawberry mousse | 10 cm sweetness scale | 15% |
| Stewart and Goss (2013) [40] | Sight | Extrinsic | White plate | Black plate | Cheesecake | 10 cm sweetness scale | 28% |
| Wang and Spence (2017) [26] | Sight | Extrinsic | Image of happy child | Image of sad child | Fruit juice (apple, orange, grapefruit) | 0–10 rating (sour–sweet) | 20% |
| Wang et al. (2017) [41] | Sight | Intrinsic | Round shape | Angular shape | Dark chocolate | 1–9 rating expected sweetness | 30% |
| Dalton et al. (2000) [42] | Smell | Extrinsic (Orthonasal) | Benzaldehyde odour (cherry almond aroma) | No odour | Saccharin solution | Threshold test | 29% increase in benzaldehyde threshold in benz + saccharin condition |

**Table 1.** *Cont.*

| Study | Sense | Intrinsic or Extrinsic | Sweet Enhancing Stimuli | Control/Comparison Stimuli | Taste Stimuli | Scale | % Difference |
|---|---|---|---|---|---|---|---|
| Delwiche and Heffelfinger (2005) [43] | Smell | Intrinsic (Retronasal) | Pineapple odour, high concentration | Pineapple odour, lower concentration | Aspartame/acesulfame potassium solution | 2-AFC threshold detection | Additive taste-odour |
| Frank and Byram (1988) [44] | Smell | Intrinsic (Retronasal) | Strawberry odour | No odour | Sweetened whipped cream | 0–20 rating sweetness | 13% at 0.6 M and 1.2 M; 40% at 0.25 M |
| Frank et al., 1989 [32] | Smell | Intrinsic (Retronasal) | Strawberry odour | No odour | Sucrose solution | 0–20 rating sweetness | ~18% at 0.3 M, 7% at 0.5 M concentration |
| Schifferstein and Verlegh (1996) [45] | Smell | Intrinsic (Retronasal) | Strawberry odour, lemon odour | No odour | Sucrose solution | 150 mm sweetness scale | 25% |
| Wang et al. (2019) [28] | Smell | Intrinsic | Pomegranate aroma | No added aroma | Apple elderflower juice | 1–9 rating sweetness | 5% |
| Biggs et al. (2016) [46] | Touch | Extrinsic | Rough plate | Smooth plate | Biscuits | How did the biscuits taste? | Biscuits in smooth plate 3 times more likely to be rated as sweet compared to those in rough plate |
| van Rompay et al. (2016) [47] | Touch | Extrinsic | Rounded cup surface pattern | Angular cup surface pattern | Hot coffee and chocolate | 1–7 rating sweetness | 20% |
| Wang and Spence (2018) [48] | Touch | Extrinsic | Velvet swatch | Sandpaper swatch | Off-dry white wine (10 g/L) | 1–9 rating sweetness | 13% |
| Wang and Spence (2018) [48] | Touch | Extrinsic | Velvet swatch | Sandpaper swatch | Fortified red dessert wine (110 g/L) | 1–7 rating sweetness | 14% |

## 2.1. *Vision*

### 2.1.1. Colour

A growing body of scientific research shows that people systematically associate different colours of foods and beverages (regardless of whether they are found in the food itself or in the food presentation/packaging), with specific basic tastes (see References [49–51] for reviews). In one early study, O'Mahony [52] reported that North American participants consistently matched the colour red to sweet taste, yellow to sour tastes, and white to salty tastes. The impact of particular colours on the perception of specific tastes has been repeatedly demonstrated over the years. Specifically, in terms of sweetness, red-coloured drinks have been found to enhance sweetness detection [34], expectations of sweetness [53], and perceived intensity [30,33,35,39,50]. However, in terms of the sensitivity to sweet taste, Maga [36] did not observe any effect of the colour red on taste detection thresholds. Rather, this colour was found to decrease bitter taste sensitivity (see Reference [49] for a review).

What is perhaps more surprising, is the role of extrinsic colour cues—that are not themselves part of the product itself—on perceived taste. So, for instance, Harrar et al. [54] reported that salty popcorn was perceived as slightly, but significantly, sweeter when served in a red or blue bowl, as compared to when presented in a white bowl. Meanwhile, in another study, a pinkish-red strawberry mousse was rated as tasting roughly 15% sweeter when served on a white plate than when served on a black plate [8,55]. Harrar and Spence [56] replicated these results using black versus white spoons with different coloured yoghurts. Furthermore, they posited that the contrast between the colour of the cutlery and the colour of the food may help to explain differences in consumer liking and perceived value of the food [57], even if the effect on sweetness was based on the colour of the cutlery alone. Additionally, it may be the combination of colours that especially signal sweetness; Woods et al [58] examined how single colours compared to colour combinations were rated in terms of sweetness, and found that the combination of pink and white was rated significantly sweeter than white and pink represented individually as single colours.

When it comes to beverages, hot chocolate was rated as slightly sweeter when served in a dark cream coloured cup as compared to orange, red, or white cups [59]. Elsewhere, it has been demonstrated that the same café latte tasted less sweet in a transparent glass cup with a white sleeve as compared to cups having a blue sleeve or no sleeve at all [60]. Furthermore, the same espresso coffee is both expected to be sweeter in a pink cup compared to a white one and, in fact, does taste sweeter too [29] (also see Reference [61] for a review of the literature on background colour).

### 2.1.2. Food Shape

Needless to say, shape is not solely a visual attribute but also related to touch, as discussed below. People associate round shapes with sweetness, whereas angular shapes tend to be associated with sourness, saltiness, and bitterness instead [62–65]. This association has been observed in food shapes [41,66]. That is, food shape can influence expectations. For instance, Wang et al. [41] demonstrated that people expected round-shaped chocolates to taste sweeter and less bitter than angular-shaped chocolates. Alternatively, it is also possible that food with a round shapes feel smoother and creamier in the mouth, and creaminess may be associated with sweetness [24].

Now, beyond the association with abstract shapes and taste, there is also a more semantic association. In a recent study, Spence, Corujo, and Youssef [67] demonstrated that a candy-floss like shape/texture was strongly associated with sweetness in both British and Spanish participants. Like food shape, food-extrinsic shape can also evoke taste similar associations when it comes to plate shape [31,57] (though see Reference [8]) and even packaging shape [68] and typeface/font [69] (see Reference [11] for a review of the typeface of taste).

### 2.2. *Audition*

#### 2.2.1. Intrinsic-Food Sounds

The sounds of mastication (i.e., the sounds that we hear when eating) can contribute to our perception of crispness, freshness and pleasantness for foods such as crisps, biscuits and fruit [70–73] (see Reference [74] for a review). For instance, an apple is rated as less crispy and softer when the biting sound is attenuated in either just the high-frequency range, or over the entire sound envelope [70]. However, as far as we are aware, there has not yet been a study published that has investigated whether changing mastication sounds influences sweetness perception.

#### 2.2.2. Extrinsic-Food Sounds

Beyond the sound of eating, incidental sounds playing in the background can also influence our taste perception. Woods et al. [75] conducted studies assessing the impact of loud versus quiet background white noise (75–85 dB versus 45–55 dB) on the perception of sweetness, sourness and liking for a variety of foods. Sweet foods (biscuits and flapjacks) were rated as tasting significantly less sweet in the loud background noise condition compared to the quiet background noise condition. As far as sweetness is concerned, Yan and Dando [76] found similar results when they assessed intensity ratings for a variety of pure taste solutions in silence and with simulated airplane noise (presented at 80–85 dB over headphones). Once again, sweetness ratings were suppressed in the noise condition compared to silence.

"Sonic seasoning", the idea that certain auditory stimuli can be deliberately used to alter people's taste perception, is becoming an increasingly popular topic in both the academic literature as well as the popular press [9,74,77–79]. Previously, it has been shown that specific auditory attributes are associated with basic tastes, both when presented as taste words (for instance, "sweet"), and in the form of tasting solutions [80–88] (see Reference [89] for a review). For instance, both sweet and sour tastes are mapped to high pitch, whereas bitterness is mapped to low pitch [81,85–87]. Crisinel et al. [7] first demonstrated that beyond any cross-modal associations between sounds and taste words, auditory stimuli could also affect people's taste evaluations. The participants in their study were given samples of bittersweet cinder toffee to evaluate while listening to one of two soundtracks that had been specifically composed to correspond to either sweet or bitter tastes. Crucially, the participants rated the cinder toffee samples higher on the sweet–bitter scale (i.e., sweeter and less bitter) while listening to the sweet soundtrack than while listening to the bitter soundtrack. Similar sonic seasoning effects, specifically involving sweetness, have since been found in a range of food and beverage stimuli ranging from juice to beer, and from chocolate to wine [9,24,25,27,90,91]. Moreover, technology is beginning to be incorporated into smart food packaging, such as the "sonic sweetener" prototype coffee container which can deliver sweet- or bitter-sounding music as people consume the beverage inside [92].

### 2.3. *Olfaction*

#### 2.3.1. Retronasal Olfaction

In terms of the effect of aroma on sweetness perception, the body of research supporting the modifying effects of aromas on sweetness perception is certainly not new. Aromas such as caramel, vanilla, and berry flavours have been found to increase the perception of sweetness, at least in Western participants (e.g., [2,44,45,93–100], see References [1,101] for reviews). Here, it is important to note that previous co-exposure is key to taste enhancement, which can differ with participants from different cultural backgrounds. For instance, in one study, the aroma of vanilla was found to enhance perceived sweetness in French participants more than in Vietnamese participants, but the reverse was true for lemon aroma [101,102]. Furthermore, Frank and Byram [44] studied the perception of sweetness and saltiness in different food matrices: Sucrose-sweetened whipped cream and strawberry aroma, sucrose-sweetened whipped cream and peanut butter aroma, sodium chloride salted whipped cream and strawberry aroma and finally sucrose-sweetened whipped cream and strawberry aroma evaluated with the nose pinched. The strawberry aroma tended to enhance the perception of sweetness; the results also showed that an aroma's ability to enhance sweetness was aroma-dependent, that an aroma's ability to enhance taste was tastant-dependent, and that the influence of the strawberry aroma on sweetness was olfactory rather than gustatory in origin.

Attempts have been made to predict the level of sweetness enhancement from different aromas. Schifferstein and Verlegh [45] investigated the effect of congruency in the form of co-occurring aroma-taste pairs. They studied strawberry, lemon and ham aromas together with sucrose and thereby created different aroma-tastant pairs varying in levels of congruency. As expected, they found that strawberry and lemon aromas increased sweetness ratings, but that the aroma of ham did not. However, the ratings of congruency did not contribute to the predictive power of the regression predicting sweetness enhancement. This indicates that congruency is a necessary condition for aroma-induced sweetness enhancement to occur, but that the degree of congruency does not relate to the level of sweetness enhancement. The degree of congruency did, however, affect the pleasantness of the mixtures, but as for congruency, the pleasantness scores could not predict the level of sweetness enhancement.

Stevenson et al. [103] investigated how to best predict aroma-induced sweetness enhancement by screening 12 different aromas, including aromas that smelled "sweet", "acidic", or non-food-like. They found that the degree to which an aroma smelled sweet was the best predictor of an aroma's ability to enhance sweetness of sucrose, but that the degree to which an aroma was perceived to smell sweet was significantly correlated to whether the aroma was regarded to be food or non-food-like. The more an aroma was judged to be food-like, the sweeter-smelling it was perceived.

Besides being aroma- and tastant-dependent, aroma–sweetness interactions are also dependent on the context. Indeed, comparing samples to a sucrose reference have been shown to not only reduce, but also change the cross-modal effect of aromas on sweet taste (unpublished data). Another important fact to keep in mind is the possible interaction between aroma and sweetener. Several studies have investigated the sweetness-enhancing effect of aromas at varying sucrose levels. While not all studies have shown bigger effects at lower rather than higher sucrose concentrations [44], most studies have [45,104,105]. Indeed, a greater degree of sensory integration has been shown for weaker compared to stronger stimuli [106].

#### 2.3.2. Orthonasal Olfaction

In the majority of studies investigating specific odour-induced changes in taste perception including that of sweetness, odorants and tastants have been presented together intrinsically in liquid mixtures which have been sipped (see References [107,108] for reviews). This means that it has not been possible to study the origin of this effect—in other words, whether it is peripheral or central. However, in a study on aspartame solutions and vanilla aroma, Sakai et al. [109] demonstrated that independent of whether vanilla and aspartame were mixed or delivered separately, sweetness enhancement occurred. These early results therefore suggested, surprisingly, that there is little functional difference between retro- and orthonasal-olfaction when it comes to sweetness enhancement. In a study of strawberry and soy sauce odours on perceived sweetness and saltiness, where odorants and tastants were delivered separately, Djordjevic et al. [110] found taste–smell interactions for congruent odour–taste pairs, thus leading them to conclude that odour-induced changes in taste perception are centrally mediated.

In a study using saccharin solutions and benzaldehyde, Pfeiffer et al. [111] found that out of a panel of 16 people, 12 showed perceptual integration while the remaining four did not. This suggested that there are individual differences in how odour-induced changes in taste perception occurs. While evaluating the intensity of fruit flavour, Hort and Hollowood [112] further found that a subgroup of trained assessors was unaffected by cross-modal aroma-sweetness interactions. Additionally, Pfeiffer et al. [111] found that while retro- and the orthonasal-presentation both produced perceptual integration, the threshold values for the different presentation methods were different.

### 2.4. Taste

Taste–taste interactions are not well understood, partly since research has reported contradictory results. In their review on binary taste interactions at suprathreshold levels, Keast and Breslin [113] suggested that the position of the individual taste stimulus on the concentration-intensity psychophysical curve (expansive, linear, or the compressive phase of the curve) can predict important interactions of taste mixtures, and that these interactions are thus concentration-dependent.

In relation to the perception of sweetness, many compounds can elicit a sweet taste. As sugar reduction is a highly interesting topic for industry, academia, as well as for society in general, several studies have investigated sweet tasting compounds and perceptual interactions between them [114–118]. For instance, the perception of sweetness in mixtures of aspartame and acesulfame K has been extensively investigated and the two sweeteners have demonstrated a perceptual synergy when mixed [32,114,115]. In general, Keast and Breslin [113] suggested that, at low intensity/concentrations (corresponding to the expansive phase of the psychophysical curve), synergy is reported when mixing sweet tasting compounds. On the other hand, at higher intensities/concentrations (the linear of compressive phase of the psychophysical curve) sweetness enhancement is less common and suppression is reported instead.

The perception of sweetness is, however, also modified by other taste qualities. In their review, Keast and Breslin [113] generalised on the effect of other taste qualities as follows: At low intensities/concentrations (the expansive phase of the psychophysical curve), the perception of sweetness is increased by salty taste, while results on the effect of adding bitterness and sourness was inconclusive. At medium intensities/concentrations (the linear phase), sweetness was suppressed by bitterness while results were inconclusive for saltiness and sourness. At high intensities/concentrations (the compressive phase), both bitterness and sourness suppressed the perception of sweetness while saltiness either suppressed or else had no effect. Similar results were found in the review on heterogeneous binary taste interactions by Wilkie and Capaldi Phillips [119]. However, they found that salt generally suppresses sweet taste, but that, as with Keast and Breslin [113], the effect of salt was concentration-dependent. Only a few studies have investigated the interaction between sweetness and umami, and the results are, thus far, inconclusive [113,119].

With respect to more complex taste mixtures such as ternary and quaternary mixtures including sweetness, Green et al. [120] followed a sip-and-spit procedure to investigate if asymmetries in suppression between stimuli in binary mixtures could predict taste perception in ternary and quaternary mixtures. The authors found a consistent pattern of mixture suppression in which sucrose sweetness was both the least suppressed quality as well as the strongest suppressor of other tastes (sourness, saltiness, and bitterness). Further, they concluded that, the overall intensity of mixtures was found to be predicted best by perceptual additivity. In other words, the overall taste intensity was derived from the sum of the tastes perceived within a mixture, rather than the sum of the perceived intensities of the individual stimuli (stimulus additivity).

### 2.5. Touch

#### 2.5.1. Oral-Somatosensation

The oral texture or mouthfeel of food and drink can influence the way in which multisensory flavours are experienced (e.g., [121–123]). Christensen [124] first suggested that increased viscosity can reduce taste perception in sweet and salty solutions. In a similar study, this time involving viscosity and flavour intensity, Bult et al. [121] presented a creamy odour either orthonasally or retronasally using an olfactomer. At the same time, milk-like solutions with different viscosities were delivered to the mouth of the participants. The intensity of perceived flavour decreased as the viscosity of the liquid increased, regardless of whether the odour was presented orthonasally or retronasally. Similarly, Weel et al. [123] studied the perception of ethylbutyrate (pineapple like, fruity) and diacetyl (buttery) flavours in whey protein gels, detected by 10 trained panellists. They found that perceived flavour intensity decreased linearly with increasing gel hardness.

Of course, it should be kept in mind that flavour compounds can interact with various ingredients in the food matrix through either chemical or physical interactions, possibly leading to either increased or decreased flavour release (e.g., [125–127]). However, Tournier et al. [100] found that physico-chemical interactions could not explain all of the sensory interactions that they observed, suggesting that cognitive mechanisms were involved after all.

Beyond mouthfeel, the temperature of food in the mouth can also influence the perception of taste. In general, the threshold for detection of sweet tastes (as well as bitter, salty and sour) shows a U-shaped response as a function of temperature, with the lowest threshold, i.e., highest sensitivity, in the 20–30 °C temperature range [128–130]. Furthermore, some individuals experience a phantom taste when small areas of the tongue are rapidly heated or cooled, with most such thermal tastes experiencing sweetness when the anterior edge of the tongue is warmed [131]. Therefore, it is possible that for these temperature tasting individuals, sweetness may be enhanced when they consume warm foods.

#### 2.5.2. Tactile Feedback

A recent body of empirical research has demonstrated that the taste and hedonic evaluation of food and beverages can indeed be influenced by the surface texture of packaging materials [59,132], servingware [46,47] and even the food itself [133]. For instance, Biggs et al. [46] served caramelised biscuits on two plates of the same shape, one with a rough and grainy surface texture, the other with a smooth and shiny texture instead. Biscuits taken from the smoother of the two plates were rated as sweeter and less crunchy than those from the rough plate. In another study examining larger-scale surface textures, van Rompay and colleagues [47] demonstrated that 3D-printed surface patterns influenced, amongst other attributes, the sweetness intensity of beverages. Namely, hot chocolate and coffee tasted from a cup with a rounded macrogeometric outer surface pattern were rated as tasting sweeter and less intense than the same beverages when tasted from a cup that had a much more angular outer pattern instead. Conversely, the angular surface pattern increased perceived bitterness and taste intensity when compared to the rounded pattern. Slocombe et al. [133] examined the influence of both food-intrinsic and extrinsic surface texture on sweetness, bitterness, and sourness of citrus-flavoured fondant sugar. While these researchers observed no influence of plate texture (rough versus smooth) on the taste of these food samples, they did observe that, when it came to the surface texture of the food itself, samples were rated as tasting significantly sourer when it had a rough surface texture as compared to a smooth one. Recently, Wang and Spence [48] demonstrated for the first time that haptic feedback of surface texture can apply to orthonasal olfaction as well as to the taste and hedonics of foodstuffs. The authors found that haptic sensations from touching different surface textures (velvet versus sandpaper) can (at least when attended) influence the wine-tasting experience, both in terms of olfaction and in terms of in-mouth sensations. The wine was rated as smelling fruitier, as tasting sweeter and more pleasant, when the participants touched the velvet as compared to when they touched the sandpaper.

## 3. A Neuroscientific Perspective on Sensory Interactions

### *3.1. The Role of Multisensory Flavour Perception*

When it comes to rationalising multisensory integration, Gibson [134] proposed an ecological model whereby information about an object is processed and interpreted via different sensory channels, as part of an active process to acquire information about the environment (see Reference [1] for a review). Flavour perception, then, can be considered as a system that controls ingestion, with the goal of picking up all available information about the food that is about to enter the body in order to secure an adequate supply of nutrients and avoid poisons [135]. Moreover, this process can be considered in multiple stages: first, there is the pre-ingestion period when food is identified and expectations are formed—this is probably most naturally gathered via visual information, together with some degree of tactile (e.g., weight, surface texture, hardness), orthonasal olfactory, and auditory information (e.g., sizzling, fizzing, bubbling). Then, there is the actual eating/mastication period where additional properties of the food—such as its taste, retronasal aroma, texture, temperature and piquancy—are detected by various taste and oral-somatosensory receptors. These receptors serve to detect nutrients and poisons in the food [136,137]. At the same time, hedonic judgments are made continuously during ingestion as a way of motivating and curtailing ingestion (e.g., [138]). Finally, learned associations are formed between different sensory stimuli as a result of the eating process (e.g., many red-coloured fruits are ripe and sweet [49]).

Just as the tactile system combines disparate information from various parts of the body and various different classes of receptors to register invariant stimuli, this proposed flavour system combines information from all the senses in order to form flavour percepts that ultimately optimise nutrient intake. Viewed from this perspective, extrinsic information such as packaging colour or background sound can act to provide extra information about the food that one is about to taste or is currently tasting. According to Bayesian decision theory, the brain uses prior knowledge about what sensory signals go together—whether inborn or explicitly learned—to integrate appropriate sensory stimuli with the goal of maximising the reliability of perceived information [139–141] and, presumably, to reduce cognitive load by combining disparate sensory cues into a single object. Cross-modal correspondences involving sweetness (such as with round shapes or consonant harmonies), could act as a conduit (i.e., in the form of Bayesian priors) to help the brain interpret multisensory cues in order to help form taste/flavour evaluations.

### 3.2. *Evidence of Multisensory Flavour Perception in the Brain*

In humans, taste is first projected from the tongue and oral cavity to the primary taste cortex in an area of the anterior insula and frontal operculum (see References [142,143] for reviews), along with oral texture and temperature [144,145].

#### 3.2.1. Aroma + Taste Binding

In support of the idea that the different channels of information contributing to flavour are bound together in a unified entity, neuroimaging evidence has demonstrated that convergence for taste and odour occur in the orbitofrontal and anterior cingulate cortex [146–148]. Indeed, increased blood oxygen level dependent (BOLD) response was observed in the orbitofrontal cortex (OFC) and amygdala when taste and odour stimuli are presented in combination, compared to the summed activity of taste and smell when pretended alone [149].

The role of olfaction, however, is made more complicated by the comparison of orthonasal versus retronasal olfaction. Whereas retronasally perceived odours are referred to the oral cavity (see a further discussion on oral referral in Section 4.1), orthonasal odours are perceived to come from outside the oral cavity. For instance, there are differences in neural responses evoked by orthonasal versus retronasal odours, with orthonasal perception leading to preferential activation in the insula, thalamus, hippocampus, amygdala and caudolateral orbitofrontal cortex and retronasal perception leading to preferential activation in the perigenual cingulate and medial orbitofrontal cortex [150]. Interestingly, these different patterns of activation were only found in the odorants associated with food (i.e., chocolate compared to lavender).

#### 3.2.2. Vision + Taste Binding

Vision is crucial in the prediction of food flavour. By altering the expectation of flavour, colour can enhance orthonasal olfactory intensity and reduce retronasal intensity (relative to colourless solutions, see Reference [151]). Furthermore, there is evidence to suggest that visual inputs associated with food also converge with olfactory and taste information in the OFC [152–154] (see References [155,156] for reviews). Multisensory neurons in the OFC have parallel sensitivities to the input quality from the different modalities; neurons which are most responsive to sweet tastants (glucose) also have a stronger response to the visual representations of sweet fruit juice or olfactory stimuli of fruit odours [157].

#### 3.2.3. Sound + Taste Binding

The influence of sound on taste and flavour perception can be considered in terms of expectations, emotion mediation, attentional direction, physiological reaction or just bias [79,158]. Of course, there is the possibility that any cross-modal effects of music on taste perception may have, at least in part, a direct, low-level, physiological basis (although see Reference [159] for one reported null-effect of a specific sour soundtrack on increasing salivation levels). Such a suggestion is inspired by Wesson and Wilson's [160] surprising discovery of direct connections between the ear and the olfactory tubercule in mice (see Reference [161] for a review of the multiple functions of the olfactory tubercule). In another example of rodent olfactory–auditory integration, single-cell recordings in the primary auditory cortex of female mice revealed that exposure to a pup's body odour can reshape neuronal response to pure tones and natural auditory stimuli [162]. It is certainly feasible that such a cross-modal connection might exist in humans, especially as some exciting preliminary neuroimaging results have recently started to appear demonstrating activation in the primary taste cortex by taste-congruent soundtracks [163].

A different possible neural connection was suggested by Brown et al. [164] who demonstrated that the processing of aesthetic stimuli—be they paintings, music or food—overlap within the primary gustatory cortex. The authors propose that the aesthetics system evolved first for the appraisal of objects necessary for survival, such as evaluating the suitability of food/energy sources; over time, according to their suggestion, the same neural circuitry was co-opted for the appreciation of artworks. Therefore, the fact that our evaluation of music and food are processed in overlapping brain areas might potentially account for the associations that we make between them, not to mention how the evaluation (especially hedonic) of stimuli in one sensory modality might influence the evaluation of another.

#### 3.2.4. Touch + Taste Binding

Oral-somatosensory information regarding the food we happen to have in our mouths is transferred to the brain via the trigeminal nerve [165]. The triminal nerve then carries information concerning touch, texture, temperature, proprioception, nociception and chemical irritation from the receptors in the mouth directly to the primary somatic sensory cortex of the brain [166,167]. The importance of somatosensory perception is demonstrated in the large areas of the sensory cortex dedicated to the lips and tongue [168]. Moreover, oral texture (including the perception of fattiness in food) is also represented in the OFC [169,170]. Congruent combinations of colour and orthonasally presented odours have been shown to lead to enhanced activation in the OFC [171]. To the best of our knowledge, neuroimaging techniques have not yet been applied to the case of somatosensory–taste interactions.

## 4. A Framework for How Intrinsic and Extrinsic Factors Influence Multisensory Flavour Perception

### *4.1. Differences between Exteroceptive and Interoceptive Senses*

When thinking about the senses and their role in multisensory flavour perception, it can be helpful to distinguish between two categories: the exteroceptive sense of vision, audition, and orthonasal olfaction are typically stimulated prior to (and sometimes during) the consumption of food, and the interoceptive senses (retronasal olfaction, oral-somatosensation and gustation) are those that are stimulated during eating [172]. In the latter case, the relevant senses are taste, retronasal smell, oral-somatosensation and the sounds associated with the consumption of food. Different brain mechanisms may be involved in these two cases. Small et al. [173] found different and overlapping neurological representations of anticipatory and consummatory phases of eating; specifically, the amygdala and mediodorsal thalamus respond preferentially to odours associated with a nutritive drink, whereas the left insula/operculum responds preferentially to the consumption of the drink itself. The right insula/operculum and left OFC responded preferentially to both anticipatory and consumptive phases. Overall, it would seem likely that the multisensory integration of interoceptive flavour cues is more automatic than the combination of cues that is involved in interpreting exteroceptive food-related signals [1,174,175].

One of the most important means by which exteroceptive cues influence food perception relates to expectancy effects [176–179]. That is, visual appearance cues, orthonasal olfactory cues, and distal food sounds can all set up powerful expectations regarding the food that someone is about to eat. When the food or drink is then evaluated, assimilation may occur if there is only a small discrepancy between what was expected and what was provided. However, if the discrepancy between expectations and the actual interoceptive information is too large, then contrast may occur instead. Human neuroimaging and animal electrophysiology has shown that expectations can modulate sensory processing at both early and late stages, and the response modulation can be either dampened or enhanced (see References [180–182] for reviews).

Another example of differences between interoceptive and exteroceptive senses come from Koza et al. [151]. These researchers demonstrated that colour had a qualitatively different effect on the perception of orthonasally (interoceptive) versus retronasally (exteroceptive) presented odours associated with a commercial fruit-flavoured water drink (see also References [124,183]). In particular, they found that colouring the solutions red led to odour enhancement in those participants who sniffed the odour orthonasally, while leading to a reduction in perceived odour intensity when it was presented retronasally. The authors suggested that this surprising result may be accounted for by the fact that it may be more important for us to correctly evaluate foods once they have entered our mouths, since that is when they pose a greater risk of poisoning. By contrast, the threat of poisoning from foodstuffs located outside the mouth is less severe. Alternatively, however, it may well be that people simply attend more to the stimuli within their bodies as compared to those stimuli that are situated externally [141], and that this influence biased the pattern of sensory dominance that was reported.

Given the above considerations, rather than a food-intrinsic versus food-extrinsic divide, it may be more appropriate, with neuroscience and physiology in mind, to divide sensory cues depending on where it is referred. In other words, the key question to consider here is, is the sensory stimulus localised (or perceived to be) coming from within or outside the mouth?

### 4.2. *Oral Referral*

The importance of the oral cavity can be seen through the observation that flavours appear to originate from the oral cavity, even if olfactory stimuli are detected in the nose (e.g., [184–186], see Reference [187] for a review). In addition, the phenomenon of oral referral appears to go beyond merely changing the perceived location of olfactory stimuli; in fact, they are combined with taste information from the tongue to form integrated flavour percepts that cannot be attended to separately [184,188]. Notably, people find it difficult to attend selectively to olfactory stimuli after the stimuli have been localised in the mouth [188,189]. The loss of the source of olfactory information is most likely a result of gustatory attention capture (according to Reference [187]), where the most intense stimulus (normally taste) directs one's attention to the spatial location where that stimulus comes from. This is supported by studies indicating that the degree of oral referral is proportional to the intensity of the tastants, and inversely proportional to the intensity of olfactory stimuli [186].

Intriguingly, the occurrence of oral referral also seems to be related to the degree of congruency between the oral and taste stimuli. Lim and Johnson [190] demonstrated that, when participants were introduced to a simultaneous retronasal odour (soy sauce, vanilla) and a taste solution (sweet, salty, water), they rated the odours as coming from the mouth significantly more often when the odour–taste combination was congruent (vanilla–sweet, soy sauce–salty) than when the solution was neutral or when the combination was incongruent. Further studies conducted with solid gelatine disks instead of liquid solutions [191], and with more ecologically valid stimulus combinations (citral aroma with sweet or sour tastants, coffee aroma with sweet or bitter tastants) revealed similar results where oral referral was enhanced proportional to the degree of self-reported smell–taste congruency [192]. In addition, more recent research supports the hypothesis that retronasal enhancement of odour by taste is dictated by the nutritive value of the tastants in addition to odour–taste congruency; sweet, salt, and umami tastes—which signal the presence of elements essential for survival—presented evidence of enhancing retronasal odour, but no such effect was seen for sour or bitter tastes [193]. In the context of sweetness perception, then, it certainly seems that multisensory cues localised in the mouth (such as food-intrinsic aroma or textural cues) would be more effective in enhancing sweetness perception than those cues localised elsewhere.

## 5. Combining Intrinsic and Extrinsic Influences

There has been relatively little research on the interaction between food-intrinsic and food-extrinsic factors. The available cognitive neuroscience research suggests that the biggest impact on our experiences and behaviours occur when several sensory attributes are changed at once, and when they complement one another [172]. This is precisely the sort of situation in which one might expect to see an additive response (both in the brain and in behaviour), a response that is far bigger than that which can be achieved by manipulating a single sense individually at a time [106,194].

### 5.1. *Intrinsic–Intrinsic Interactions*

Several studies have examined interactions between intrinsic factors in relation to sweetness. For instance, Zampini and his colleagues [195] found that an orange-coloured water solution only was found significantly sweeter than other colours in combination with orange flavour. Similarly, in several studies, Junge [196] investigated interactions between aroma and colour in relation to sweetness, as well as between aroma and viscosity, in an apple/elderflower fruit drink. In one study conducted with trained panellists, Junge [196] found that both pomegranate aroma and red (different from the original yellowish) colour increased sweetness and showed normal additivity in mixture. In another study [196] conducted with trained panellists, increasing the viscosity of the fruit drink with pectin resulted in a tendency to an increase in rated sweetness. This tendency was maintained even if vanilla and pomegranate aromas were added. On the contrary, Knoop et al. [197] found that both a modification of texture and of aroma had a sweetness enhancing effect on an apple juice gel, but in combination, these showed the same sweetness enhancement as aroma and texture had individually. Thus, the interaction effect was smaller than the sum of the two individually and thereby not additive.

Besides cross-modal additive effects, it is also worth considering intramodal additive effects between different stimuli. For instance, Woods et al. [58] found that the pairwise combinations of colours had stronger associations with sweetness as compared to single colours. Along a similar line of thought, given the efficacy of sweet-smelling aromas in enhancing perceived sweetness, it would be worth testing whether combinations of sweet smells might have a bigger overall impact compared to just individual smells.

In terms of interactions between food-intrinsic factors, it has also been demonstrated that taste–aroma interactions are moderated by the nature of the food matrix in question. Labbe et al. [95] tested the taste enhancement effects of cocoa and vanilla flavouring in cocoa and caffeinated milk. They found that in the cocoa beverage, cocoa flavouring led to an enhancement of bitterness and vanilla flavouring enhanced sweetness. However, when it came to the relatively less familiar caffeinated milk product, the addition of vanilla flavouring unexpectedly enhanced bitterness instead of sweetness. Elsewhere, Alcaire et al. [198] found that while an increase in vanilla flavour in a dairy dessert product had a minor effect on sweetness enhancement, the combination of increased vanilla concentration together with higher starch concentration led to an increase in vanilla flavour intensity as well as an increase in perceived sweetness. This was presumably due to the thickened viscosity of the dessert product from the addition of starch.

### *5.2. Extrinsic–Extrinsic Interactions*

A few studies have also examined the interaction between various food-extrinsic factors. For instance, Spence et al. [194] conducted a study in which the participants tasted the same red wine in an opaque black glass under different illumination and music conditions. In Experiment 1, the wine was liked more (by approximately 5%) in the combined sweet music and red lighting condition, compared to red lighting alone without music and compared to the control condition (white light, silence). However, there were no changes in rated flavour intensity or freshness (compared to fruitiness). In Experiment 2, the wine was rated as being significantly fresher (an increase of approximately 14%) when tasted under green lighting while listening to sour music, as compared to green lighting alone and to the control condition. What the studies did not assess, however, was the individual effect of music, which leaves the question of whether both visual and auditory cues are required for a noticeable effect on wine perception.

Elsewhere, Fairhurst et al. [31] manipulated both plate shape (extrinsic) and the arrangement and shape of the food (intrinsic) in a restaurant setting. Participants rated the sweetness, sourness, intensity and pleasantness of two beetroot salads served on either round or angular plates, with the beetroots cut into either angular or round shapes. One group were served beetroot salads only from round plates, and the other group only from angular plates. Serving the salad in the round plate plus round food shape led to significantly sweeter ratings than the congruent angular plate plus angular food shape condition, with the incongruent conditions (where plate shape did not match food shape) rating somewhere in-between.

### *5.3. Intrinsic-Extrinsic Interactions*

Recently, Wang et al. [28] conducted a study measuring the relative influence of both extrinsic and intrinsic factors. In a mixed model design, participants tasted samples of apple-elderflower nectars under different visual (looking at a pink or pale-yellow rectangle on an iPad screen) and auditory (listening to bitter or sweet soundtrack) conditions. Three levels of added pomegranate aroma (none, medium level, high level) were also included as an additional between-participant food-intrinsic manipulation. The results revealed that both added aroma and the presence of the soundtrack exhibited significant sweetness enhancement effects. An interaction analysis revealed that the enhancement effect of aroma and soundtrack were independent of each other, and the combined influence of aroma and soundtrack appeared to be greater than the influence of aroma or soundtrack when assessed individually. In other words, there appeared to be a linearly additive relationship between the enhancement effect of aroma and the enhancement effect of soundtrack, even when one factor was intrinsic (aroma) and the other extrinsic (soundtrack). This result gives additional support to the theory that changing multiple sensory stimuli at once could result in a greater sweetness enhancement effect than manipulating single factors alone.

### *5.4. Theoretical Limitations*

In terms of sugar reduction, the goal is typically to maintain consumer satisfaction even if the consumers can detect that sweetness has been altered. Intriguingly, while a reduction in added sugar of 20% (from 9% to 7.2% sugar) for chocolate-flavoured milks was detected by trained assessors and consumers, consumer liking was unaffected by the reduction in sugar, for not only 20% but even up to a 40% reduction [105]. Similarly, for orange nectar drinks, lowering added sugar from 10% to 8.5% did not influence sensory attributes or acceptance [199]. In fact, the authors [199] suggested a strategy of long-term gradual reduction in added sugar, from 8.5% to 7.2% and eventually 5.5%. One challenge in working with sugar reduction, as shown in Section 5.1, is that it is food-matrix-dependent, therefore, it is difficult to compare acceptable levels of sugar reduction across the board.

It is important to keep in mind that exteroceptive sensory cues mostly operate by shaping sensory expectations (see Reference [182] for a review). However, expectancy effects have their limits. Yeomans et al. [200] found that a salmon-based mousse that looked like strawberry ice cream evoked a negative hedonic reaction when participants wrongly assumed it was strawberry ice cream. This shows that there is an envelope in which expectations for a food or drink, such as that driven by food colour, can affect food evaluation via assimilation. Assimilation can only go so far before the disconfirmation of expectations become too great and contrast sets in. The size for such an envelope remains unclear to this day, although it is doubtlessly food and participant-dependent. Table 1 shows that the degree of sweetness enhancement by extrinsic factors (at least, those due to the expectation effects) is usually around 10–20%.

Finally, it is also important to keep in mind that besides adding a sweet taste to a food or beverage matrix, sugar plays multiple roles in foods and beverages including being a bulking agent as well as contributing to the mouthfeel and texture (see Reference [201] for a review). Therefore, when reducing and/or replacing the sugar, it is necessary to find a substitute which can stand in its place from a structural perspective. However, given the many functionalities sugars have in foods and beverages, reducing and/or replacing it can be a rather complicated process.

## 6. Conclusions, Future Directions, and Open Questions

The results summarised here are of relevance for those working on understanding human sweetness perception, as well as those working to design healthier, sugar-reduced food products. Indeed, the knowledge that multiple sensory cues can, at least under the appropriate conditions, work in conjunction for a greater taste modulation effect will allow designers to come up with more effective sugar-reduced products without taking away consumer satisfaction. It also presents a convincing argument for food researchers and packaging designers to work together, in the philosophy of designing the "Total Product Experience" [202] in order to optimally balance product-intrinsic and extrinsic cues for maximal sweetness enhancement.

### *6.1. Future Directions and Open Questions*

#### 6.1.1. Long-Term Studies

As mentioned by Pineli et al. [199], one potential strategy going forward is to gradually lower added sugar % in the long term, in order for consumers to adapt to and eventually accept reduced sugar beverages. Unfortunately, very few long-term studies have been performed in this area. One such study was conducted by Chung and Vickers [203] using sweetened teas. Participants initially tasted three types of tea, each at a low and optimum level of sweetness. Then, over each of the next 20 consumption sessions, participants selected a tea, tasted it, and rated their liking, tiredness, and satisfaction with their choice. The authors observed that liking for the low sweetness levels did increase over time. Moreover, there was a trend for participants to become more tired of the tea at optimum sweetness while becoming less tired of the lower sweetness teas. That said, given the choice of all six teas at each session, participants did not choose the less sweet teas more frequently, as the authors hypothesised. This implies that long-term strategies should focus on improving consumer acceptability for only the reduced sweetness product at any one time, without offering the full-sweetness product as an alternative. As a testament to the difficulty of gradual sugar reduction, while UK retailers have successfully reduced the salt content of breakfast cereals by 47% between 1992 and 2015 via incremental salt reduction targets, sugar content has not changed significantly [204]. The authors speculate that sugar reduction is much more difficult because sugar influences not only the flavour, but also the texture and structure of breakfast cereals.

#### 6.1.2. Individual Differences

Given that we all live in different taste worlds, it is important to consider whether our taster status influences the extent to which sensory factors influence the way we perceive sweetness. One particularly interesting result to have emerged in this regard comes from Zampini et al. [195].

These researchers found that supertasters showed less visual dominance over their perception of flavour than medium tasters who, in turn, showed less visual dominance than non-tasters. The participants in this particular study had to identify the flavour of a large number of fruit-flavoured drinks presented among other flavour less drinks. The drinks could be coloured red, orange, yellow, grey or else presented as clear and colourless solutions. Overall, the non-tasters correctly identified 19% of the solutions, the medium tasters 31% and the supertasters 67% of the drinks that they were given to taste. Interestingly, the addition of colouring had the largest effect on the performance of the non-tasters, less of an effect on the medium-tasters, and very little effect on the colour identification responses of the supertasters.

Genetic differences in terms of supertaster status have also been demonstrated to play a role in terms of sonic seasoning effects, whereby listening to specific soundtracks congruent to specific tastes/flavours can alter the perception of food. Using a mixed-model design, Wang tested 27 participants who tasted 70% and 85% cacao chocolates while listening to sweet and bitter soundtracks [158]. All participants then took a Phenylthiocarbamide (PTC) taste strip test at the end of the study. Results revealed an intriguing difference when it came to the influence of music. While there were no differences between the two taste sensitivity groups for 70% chocolate, when it came to the more bitter 85% chocolate, the high taste sensitivity group appeared to be more influenced by the different soundtracks than the low sensitivity group (i.e., they found a bigger difference in the taste of the 85% chocolate between the bitter and sweet soundtrack; see [205]).

Moreover, any sugar reduction strategies likely also need to consider individual differences in sweetness liking. Looy et al. [206] categorised people into sweet likers and dislikers based on their hedonic response to sugar solutions of increasing sweetness. Notably, the authors found that the sweetness liker–disliker distinction held across different sugar types (sucrose, glucose, fructose), and even when the solution was coloured red and strawberry-flavoured. Moreover, sweet liking has been demonstrated to be partly inherited [207] and linked with alcoholism [208]. It is important to realise that sweetness-enhancing multisensory cues reviewed earlier may have different outcomes when it comes to consumer acceptance of sugar-reduced foods. For instance, odours which were associated with sweetness via training became more pleasant for sweet likers but more unpleasant in sweet dislikers [209].

### *6.2. Industry Implications*

The current review on the role of intrinsic and extrinsic sensory factors in sweetness perception of food and beverages suggests that, changing multiple sensory stimuli, intrinsic as well as extrinsic, at once could result in a greater sweetness enhancement effect than manipulating single factors alone such as working with intrinsic sensory factors only. This has industry implications since in research as well as in academia, work on intrinsic and extrinsic factors are considered as belonging to different disciplines. In industry, this is often reflected in the organizational structure. Namely, Research and Development (R&D), which is in charge of food-intrinsic properties, often sits far away from, and actually has little interaction with the marketing department, who may be responsible for food-extrinsic decisions, such as those involving product packaging.

However, in order to fully maximize sweetness enhancement in foods and beverages, R&D and the marketing department should work closely together in order to optimally balance product-intrinsic and extrinsic cues. Efforts and resources are unarguably needed to create successful collaborations, but they are crucial to get the full potential of sweetness enhancement. The current review highlights the importance of understanding how food-intrinsic and extrinsic factors work together to form our overall perception of sweetness. Of course, while the topic of this review is focused on sugar reduction, similar strategies can be considered for the reduction of salt/fat content in food to promote healthy eating behaviour.

**Author Contributions:** Writing—original draft preparation, Q.J.W.; L.A.M.; J.Y.J.; A.S.B.; C.S.; writing—review and editing, Q.J.W.; L.A.M.; U.K.; C.S.; D.V.B.; funding acquisition, L.A.M.; U.K.; D.V.B.

**Acknowledgments:** In this section you can acknowledge any support given which is not covered by the author contribution or funding sections. This may include administrative and technical support, or donations in kind (e.g., materials used for experiments).

## References


1. Auvray, M.; Spence, C. The multisensory perception of flavor. *Conscious Cog.* **2018**, *17*, 1016–1031. [CrossRef] [PubMed]
2. Delwiche, J. The impact of perceptual interactions on perceived flavor. *Food Qual. Pref.* **2004**, *15*, 137–146. [CrossRef]
3. Stevenson, R.J. Attention and flavor binding. In *Multisensory Flavor Perception: From Fundamental Neuroscience through to the Marketplace*; Piqueras-Fiszman, B., Spence, C., Eds.; Elsevier: Duxford, UK, 2016; pp. 15–35.
4. Mielby, L.H.; Andersen, B.V.; Jensen, S.; Kildegaard, H.; Kuznetsova, A.; Eggers, N.; Brockhoff, P.B.; Byrne, D.V. Changes in sensory characteristics and their relation with consumers' liking, wanting and sensory satisfaction: Using dietary fibre and lime flavour in Stevia rebaudiana sweetened fruit beverages. *Food Res. Int.* **2016**, *82*, 14–21. [CrossRef]
5. Schifferstein, H.N.J. The drinking experience: Cup or content? *Food Qual. Pref.* **2009**, *20*, 268–276. [CrossRef]
6. Ares, G.; Deliza, R. Studying the influence of package shape and colour on consumer expectations of milk desserts using word association and conjoint analysis. *Food Qual. Pref.* **2010**, *21*, 930–937. [CrossRef]
7. Crisinel, A.-S.; Cosser, S.; King, S.; Jones, R.; Petrie, J.; Spence, C. A bittersweet symphony: Systematically modulating the taste of food by changing the sonic properties of the soundtrack playing in the background. *Food Qual. Pref.* **2012**, *24*, 201–204. [CrossRef]
8. Piqueras-Fiszman, B.; Alcaide, J.; Roura, E.; Spence, C. Is it the plate or is it the food? Assessing the influence of the color (black or white) and shape of the plate on the perception of the food placed on it. *Food Qual. Pref.* **2012**, *24*, 205–208. [CrossRef]
9. Carvalho, F.R.; Wang, Q.J.; Van Ee, R.; Spence, C. The influence of soundscapes on the perception and evaluation of beers. *Food Qual. Pref.* **2016**, *52*, 32–41. [CrossRef]
10. Spence, C. *Gastrophysics: The New Science of Eating*; Viking Penguin: London, UK, 2017.
11. Velasco, C.; Spence, C. The role of typeface in packaging design. In *Multisensory Packaging: Designing New Product Experiences*; Velasco, C., Spence, C., Eds.; Palgrave MacMillan: Cham, Switzerland, 2019; pp. 79–101.
12. Wang, Q.J.; Keller, S.; Spence, C. The sound of spiciness: Enhancing the evaluation of piquancy by means of a customized crossmodally congruent soundtrack. *Food Qual. Pref.* **2017**, *58*, 1–9. [CrossRef]
13. Johnson, R.J.; Segal, M.S.; Sautin, Y.; Nakagawa, T.; Feig, D.I.; Kang, D.; Gersch, M.S.; Benner, S.; Sánchez-Lozada, L.G. Potential role of sugar (fructose) in the epidemic of hypertension, obesity and the metabolic syndrome, diabetes, kidney disease, and cardiovascular disease. *Am. J. Clin. Nutr.* **2007**, *86*, 899–906.
14. Malik, V.S.; Schulze, M.B.; Hu, F.B. Intake of sugar-sweetened beverages and weight gain: A systematic review. *Am. J. Clin. Nutr.* **2006**, *84*, 274–288. [CrossRef] [PubMed]
15. Maier, J.X.; Wachowiak, M.; Katz, D.B. Chemosensory convergence on primary olfactory cortex. *J. Neurosci.* **2012**, *32*, 17037–17047. [CrossRef] [PubMed]
16. Vartanian, L.R.; Schwartz, M.B.; Brownell, K.D. Effects of soft drink consumption on nutrition and health: A systematic review and meta-analysis. *Am. J. Public Health* **2007**, *97*, 667–675. [CrossRef] [PubMed]
17. Sugar Reduction: Blend the Trends with Functional Formulations. Available online: https://www.foodnavigator.com/Article/2019/05/15/Sugar-reduction-Blend-the-trends-with-functional-formulations (accessed on 21 May 2019).
18. Why Sugar Is on Everyone's Lips. Available online: https://www.arla.com/food-for-thought/health/why-sugar-is-on-everyones-lips/ (accessed on 21 May 2019).
19. Khan, M. PepsiCo R&D: A catalyst for change in the food and beverage industry. *New Food* **2015**, *16*, 10–13.
20. Hutchings, S.C.; Low, J.Y.Q.; Keast, S.J. Sugar reduction without compromising sensory perception. An impossible dream? *Crit. Rev. Food Sci. Nutr.* **2018**. [CrossRef] [PubMed]

21. Skaczkowski, G.; Durkin, S.; Kashima, Y.; Wakefield, M. The effect of packaging, branding and labeling on the experience of unhealthy food and drink: A review. *Appetite* **2016**, *99*, 219–234. [CrossRef] [PubMed]
22. Shankar, P.; Ahuja, S.; Sriram, K. Non-nutritive sweeteners: Review and update. *Nutrition* **2013**, *29*, 1293–1299. [CrossRef] [PubMed]
23. Höchenberger, R.; Ohla, K. A bittersweet symphony: Evidence for taste-sound correspondences without effects on taste quality-specific perception. *J. Neurosci. Res.* **2019**, *97*, 267–275.
24. Carvalho, F.R.; Wang, Q.J.; Van Ee, R.; Persoone, D.; Spence, C. "Smooth operator": Music modulates the perceived creaminess, sweetness, and bitterness of chocolate. *Appetite* **2017**, *108*, 383–390.
25. Wang, Q.J.; Spence, C. 'Striking a sour note': Assessing the influence of consonant and dissonant music on taste perception. *Multisens. Res.* **2016**, *29*, 195–208. [CrossRef]
26. Wang, Q.J.; Spence, C. "A sweet smile": The modulatory role of emotion in how extrinsic factors influence taste evaluation. *Cogn. Emot.* **2017**, *32*, 1052–1061. [CrossRef]
27. Wang, Q.J.; Spence, C. Assessing the influence of music on wine perception amongst wine professionals. *Food Sci. Nutr.* **2017**, *52*, 211–217.
28. Wang, Q.J.; Mielby, L.A.; Thybo, A.K.; Bertelsen, A.S.; Kidmose, U.; Spence, C.; Byrne, D.V. Sweeter together? Assessing the combined influence of product intrinsic and extrinsic factors on perceived sweetness of fruit beverages. *J. Sens. Stud.* **2019**, *34*, e12492. [CrossRef]
29. Carvalho, F.; Spence, C. Cup colour influences consumers' expectations and experience on tasting specialty coffee. *Food Qual. Pref.* **2019**, *75*, 157–169. [CrossRef]
30. Clydesdale, F.M.; Gover, R.; Philipsen, D.H.; Fugardi, C. The effect of color on thirst quenching, sweetness, acceptability and flavour intensity in fruit punch flavored beverages. *J. Food Qual.* **1992**, *15*, 19–38. [CrossRef]
31. Fairhurst, M.; Pritchard, D.; Ospina, D.; Deroy, O. Bouba-Kiki in the plate: Combining crossmodal correspondences to change flavour experience. *Flavour* **2015**, *4*, 22. [CrossRef]
32. Frank, R.A.; Ducheny, K.; Mize, S.J.S. Strawberry odor, but not red color, enhances the sweetness of sucrose solutions. *Chem. Senses* **1989**, *14*, 371–377. [CrossRef]
33. Hidaka, S.; Shimoda, K. Investigation of the effects of color on judgments of sweetness using a taste adaptation method. *Multisens. Res.* **2014**, *27*, 189–205. [CrossRef] [PubMed]
34. Johnson, J.; Clydesdale, F.M. Perceived sweetness and redness in colored sucrose solutions. *J. Food Sci.* **1982**, *47*, 747–752. [CrossRef]
35. Lavin, J.G.; Lawless, H.T. Effects of color and odor on judgments of sweetness among children and adults. *Food Qual. Pref.* **1998**, *9*, 283–289. [CrossRef]
36. Maga, J.A. Influence of color on taste thresholds. *Chem. Senses Flavor* **1974**, *1*, 115–119. [CrossRef]
37. Pangborn, R.M.; Hansen, B. The influence of color on discrimination of sweetness and sourness in pear-nectar. *Am. J. Psychol.* **1963**, *76*, 315. [CrossRef]
38. Pangborn, R.M.; Berg, H.W.; Hansen, B. The influence of color on discrimination of sweetness in dry table-wine. *Am. J. Psychol.* **1963**, *76*, 492. [CrossRef]
39. Pangborn, R.M. Influence of color on the discrimination of sweetness. *Am. J. Psychol.* **1960**, *73*, 229–238. [CrossRef]
40. Stewart, P.C.; Goss, E. Plate shape and colour interact to influence taste and quality judgments. *Flavour* **2013**, *2*, 27. [CrossRef]
41. Wang, Q.J.; Reinoso Carvalho, F.; Persoone, D.; Spence, C. Assessing the effect of shape on expected and actual chocolate flavor. *Flavour* **2017**, *6*, 2. [CrossRef]
42. Dalton, P.; Doolittle, N.; Nagata, H.; Breslin, P.A.S. The merging of the senses: Integration of subthreshold taste and smell. *Nat. Neurosci.* **2000**, *3*, 431–432. [CrossRef] [PubMed]
43. Delwiche, J.; Heffelfinger, A.L. Cross-modal additivity of taste and smell. *J. Sens. Stud.* **2005**, *20*, 137–146. [CrossRef]
44. Frank, R.A.; Byram, J. Taste-smell interactions are tastant and odorant dependent. *Chem. Senses* **1988**, *13*, 445–455. [CrossRef]
45. Schifferstein, H.N.J.; Verlegh, P.W.J. The role of congruency and pleasantness in odor-induced taste enhancement. *Acta Psychol.* **1996**, *94*, 87–105. [CrossRef]
46. Biggs, L.; Juravle, G.; Spence, C. Haptic exploration of plateware alters the perceived texture and taste of food. *Food Qual. Pref.* **2016**, *50*, 129–134. [CrossRef]

47. Van Rompay, T.J.L.; Finger, F.; Saakes, D.; Fenko, A. "See me, feel me": Effects of 3D-printed surface patterns on beverage evaluation. *Food Qual. Pref.* **2017**, *62*, 332–339. [CrossRef]
48. Wang, Q.J.; Spence, C. A smooth wine? Haptic influences on wine evaluation. *Int. J. Gastron. Food Sci.* **2018**, *14*, 9–13. [CrossRef]
49. Spence, C.; Levitan, C.; Shankar, M.U.; Zampini, M. Does food color influence taste and flavor perception in humans? *Chemosens. Percept.* **2010**, *3*, 68–84. [CrossRef]
50. Spence, C. On the psychological impact of food colour. *Flavour* **2015**, *4*, 21. [CrossRef]
51. Spence, C.; Velasco, C. On the multiple effects of packaging colour on consumer behaviour and product experience in the 'food and beverage' and 'home and personal care' categories. *Food Qual. Pref.* **2018**, *68*, 226–237. [CrossRef]
52. O'Mahony, M. Gustatory responses to nongustatory stimuli. *Perception* **1983**, *12*, 627–633. [CrossRef]
53. Wei, S.-T.T.; Ou, L.-C.C.; Luo, M.R.R.; Hutchings, J.B. Optimisation of food expectations using product colour and appearance. *Food Qual. Pref.* **2012**, *23*, 49–62. [CrossRef]
54. Harrar, V.; Piqueras-Fiszman, B.; Spence, C. There's more to taste in a coloured bowl. *Perception* **2011**, *40*, 880–882. [CrossRef]
55. Piqueras-Fiszman, B.; Giboreau, A.; Spence, C. Assessing the influence of the color of the plate on the perception of a complex food in a restaurant setting. *Flavour* **2013**, *2*, 24. [CrossRef]
56. Harrar, V.; Spence, C. The taste of cutlery: How the taste of food is affected by the weight, size, shape, and colour of the cutlery used to eat it. *Flavour* **2013**, *2*, 21. [CrossRef]
57. Chen, Y.-C.; Woods, A.; Spence, C. Sensation transference from plateware to food: The sounds and tastes of plates. *Int. J. Food Des.* **2018**, *3*, 41–62. [CrossRef]
58. Woods, A.T.; Marmolejo-Ramos, F.; Velasco, C.; Spence, C. Using single colors and color pairs to communicate basic tastes II: Foreground-background color combinations. *i-Perception* **2016**, *7*. [CrossRef] [PubMed]
59. Piqueras-Fiszman, B.; Laughlin, Z.; Miodownik, M.; Spence, C. Tasting spoons: Assessing how the material of a spoon affects the taste of the food. *Food Qual. Pref.* **2012**, *24*, 24–29. [CrossRef]
60. Van Doorn, G.; Wuillemin, D.; Spence, C. Does the colour of the mug influence the taste of the coffee? *Flavour* **2014**, *3*, 10. [CrossRef]
61. Spence, C. Background colour & its impact on food perception & behaviour. *Food Qual. Pref.* **2018**, *68*, 156–166.
62. Spence, C.; Deroy, O. Tasting shapes: A review of four hypotheses. *Theor. Hist. Sci.* **2013**, *10*, 207–238.
63. Spence, C.; Ngo, M.K. Assessing the shape symbolism of the taste, flavor, and texture of foods and beverages. *Flavour* **2012**, *1*, 12. [CrossRef]
64. Spence, C. Assessing the influence of shape and sound symbolism on the consumer's response to chocolate. *New Food* **2014**, *17*, 59–62.
65. Velasco, C.; Woods, A.T.; Petit, O.; Cheok, A.D.; Spence, C. Crossmodal correspondences between taste and shape, and their implications for product packaging: A review. *Food Qual. Pref.* **2016**, *52*, 17–26. [CrossRef]
66. Lenfant, F.; Hartmann, C.; Watzke, B.; Breton, O.; Loret, C.; Martin, N. Impact of the shape on sensory properties of individual dark chocolate pieces. *LWT Food Sci. Technol.* **2013**, *51*, 545–552. [CrossRef]
67. Spence, C.; Corujo, A.; Youssef, J. Cotton candy: A gastrophysical investigation. *Int. J. Gastron. Food Sci.* **2019**, *16*, 100146. [CrossRef]
68. Velasco, C.; Salgado-Montejo, A.; Marmolejo-Ramos, F.; Spence, C. Predictive packaging design: Tasting shapes, typefaces, names, and sounds. *Food Qual. Pref.* **2013**, *34*, 88–95. [CrossRef]
69. Velasco, C.; Woods, A.T.; Hyndman, S.; Spence, C. The taste of typeface. *i-Perception* **2015**, *6*. [CrossRef] [PubMed]
70. Demattè, M.L.; Pojer, N.; Endrizzi, I.; Corollaro, M.L.; Betta, E.; Aprea, E.; Charles, M.; Biasioli, F.; Zampini, M.; Gasperia, F. Effects of the sound of the bite on apple perceived crispness and hardness. *Food Qual. Pref.* **2014**, *38*, 58–64. [CrossRef]
71. Masuda, M.; Okajima, K. Added mastication sound affects food texture and pleasantness. Presented at the 12th International Multisensory Research Forum Meeting, Fukuoka, Japan, 17–20 October 2011.
72. Zampini, M.; Spence, C. The role of auditory cues in modulating the perceived crispness and staleness of potato chips. *J. Sens. Stud.* **2004**, *19*, 347–363. [CrossRef]
73. Zampini, M.; Spence, C. Modifying the multisensory perception of a carbonated beverage using auditory cues. *Food Qual. Pref.* **2005**, *16*, 632–641. [CrossRef]

74. Spence, C. Eating with our ears: Assessing the importance of the sounds of consumption to our perception and enjoyment of multisensory flavour experiences. *Flavour* **2015**, *4*, 3. [CrossRef]
75. Woods, A.T.; Poliakoff, E.; Lloyd, D.M.; Kuenzel, J.; Hodson, R.; Gonda, H.; Batchelor, J.; Dijksterhuis, G.B.; Thomas, A. Effect of background noise on food perception. *Food Qual. Pref.* **2011**, *22*, 42–47. [CrossRef]
76. Yan, K.S.; Dando, R. A crossmodal role for audition in taste perception. *J. Exp. Psychol. Hum. Percept. Perform.* **2015**, *41*, 590–596. [CrossRef]
77. Spence, C. Sonic seasoning. In *Audio Branding: Using Sound to Build Your Brand*; Minsky, L., Fahey, C., Eds.; Kogan Page: London, UK, 2017; pp. 52–58.
78. Spence, C.; Wang, Q.J. Wine & music (II): Can you taste the music? Modulating the experience of wine through music and sound. *Flavour* **2015**, *4*, 33.
79. Spence, C.; Reinoso-Carvalho, F.; Velasco, C.; Wang, Q.J. Extrinsic auditory contributions to food perception & consumer behaviour: An interdisciplinary review. *Multisens. Res.* **2019**, 1–44. [CrossRef]
80. Crisinel, A.-S.; Spence, C. Implicit association between basic tastes and pitch. *Neurosci. Lett.* **2009**, *464*, 39–42. [CrossRef] [PubMed]
81. Crisinel, A.-S.; Spence, C. A sweet sound? Food names reveal implicit associations between taste and pitch. *Perception* **2010**, *39*, 417–425. [CrossRef] [PubMed]
82. Crisinel, A.-S.; Spence, C. As bitter as a trombone: Synesthetic correspondences in nonsynesthetes between tastes/flavors and musical notes. *Atten. Percept. Psychophys.* **2010**, *72*, 1994–2002. [CrossRef] [PubMed]
83. Crisinel, A.-S.; Spence, C. A fruity note: Crossmodal associations between odours and musical notes. *Chem. Senses* **2011**, *37*, 151–158. [CrossRef] [PubMed]
84. Mesz, B.; Sigman, M.; Trevisan, M.A. A composition algorithm based on crossmodal taste-music correspondences. *Front. Hum. Neurosci.* **2012**, *6*, 71. [CrossRef] [PubMed]
85. Mesz, B.; Trevisan, M.A.; Sigman, M. The taste of music. *Perception* **2011**, *40*, 209–219. [CrossRef] [PubMed]
86. Wang, Q.J.; Wang, S.; Spence, C. "Turn up the taste": Assessing the role of taste intensity and emotion in mediating crossmodal correspondences between basic tastes and pitch. *Chem. Senses* **2016**, *29*, 345–356. [CrossRef] [PubMed]
87. Knoeferle, K.M.; Woods, A.; Käppler, F.; Spence, C. That sounds sweet: Using crossmodal correspondences to communicate gustatory attributes. *Psychol. Mark.* **2015**, *32*, 107–120. [CrossRef]
88. Simner, J.; Cuskley, C.; Kirby, S. What sound does that taste? Cross-modal mappings across gustation and audition. *Perception* **2010**, *39*, 553–569. [CrossRef] [PubMed]
89. Knöferle, K.M.; Spence, C. Crossmodal correspondences between sounds and tastes. *Psychon. Bull. Rev.* **2012**, *19*, 992–1006. [CrossRef] [PubMed]
90. Reinoso Carvalho, F.; Van Ee, R.; Rychtarikova, M.; Touhafi, A.; Steenhaut, K.; Persoone, D.; Spence, C.; Leman, M. Does music influence the multisensory tasting experience? *J. Sens. Stud.* **2015**, *30*, 404–412. [CrossRef]
91. Wang, Q.J.; Spence, C. Assessing the effect of musical congruency on wine tasting in a live performance setting. *i-Perception* **2015**, *6*. [CrossRef] [PubMed]
92. Hold the Sugar: A Chinese Café Brand Is Offering Audio Sweeteners. Available online: https://www.campaignasia.com/video/hold-the-sugar-a-chinese-cafe-brand-is-offering-audio-sweeteners/433757 (accessed on 20 May 2019).
93. Boakes, R.A.; Hemberger, H. Odour-modulation of taste ratings by chefs. *Food Qual. Pref.* **2012**, *25*, 81–86. [CrossRef]
94. Burseg, K.M.M.; Camacho, S.; Knoop, J.; Bult, J.H.F. Sweet taste intensity is enhanced by temporal fluctuation of aroma and taste, and depends on phase shift. *Physiol. Behav.* **2010**, *101*, 726–730. [CrossRef] [PubMed]
95. Labbe, D.; Damevin, L.; Vaccher, C.; Morgenegg, C.; Martin, N. Modulation of perceived taste by olfaction in familiar and unfamiliar beverages. *Food Qual. Pref.* **2006**, *17*, 582–589. [CrossRef]
96. Labbe, D.; Martin, N. Impact of novel olfactory stimuli at supra and subthreshold concentrations on the perceived sweetness of sucrose after associative learning. *Chem. Senses* **2009**, *34*, 645–651. [CrossRef]
97. Labbe, D.; Rytz, A.; Morgenegg, C.; Ali, S.; Martin, N. Subthreshold olfactory stimulation can enhance sweetness. *Chem. Senses* **2007**, *32*, 205–214. [CrossRef]
98. Murphy, C.; Cain, W.S. Taste and olfaction: Independence vs interaction. *Physiol. Behav.* **1980**, *24*, 601–605. [CrossRef]

99. Stevenson, R.J.; Boakes, R.A. Sweet and sour smells: Learned synaesthesia between the senses of taste and smell. In *The Handbook of Multisensory Processing*; Calvert, G.A., Spence, C., Stein, B.E., Eds.; MIT Press: Cambridge, MA, USA, 2004; pp. 69–83.
100. Tournier, C.; Sulmont-Rossé, C.; Sémon, E.; Vignon, A.; Issanchou, S.; Guichard, E. A study on texture-taste-aroma interactions: Physico-chemical and cognitive mechanisms. *Int. Dairy J.* **2009**, *19*, 450–458. [CrossRef]
101. Valentin, D.; Chrea, C.; Nguyen, D.H. Taste-odour interactions in sweet taste perception. In *Optimising Sweet Taste in Foods*; Spillane, W., Ed.; Woodhead Publishing Limited: Cambridge, UK, 2006; pp. 66–84.
102. Nguyen, G.D.; Valentin, D.; Ly, M.H.; Chrea, C.; Sauvageot, F. When does smell enhance taste? The effect of culture and odorant/tastants relationship. Presented at the 4th Pangborn Sensory Science Symposium, Dijon, France, 22–26 July 2001.
103. Stevenson, R.J.; Prescott, J.; Boakes, R.A. Confusing tastes and smells: How odours can influence the perception of sweet and sour tastes. *Chem. Senses* **1999**, *24*, 627–635. [CrossRef] [PubMed]
104. Charles, M.; Endrizzi, I.; Aprea, E.; Zambanini, J.; Betta, E.; Gasperi, F. Dynamic and static sensory methods to study the role of aroma on taste and texture: A multisensory approach to apple perception. *Food Qual. Pref.* **2017**, *62*, 17–30. [CrossRef]
105. Oliveira, D.; Antúnez, L.; Giménez, A.; Castura, J.C.; Deliza, R.; Ares, G. Sugar reduction in probiotic chocolate-flavored milk: Impact on dynamic sensory profile and liking. *Food Res. Int.* **2015**, *75*, 148–156. [CrossRef] [PubMed]
106. Meredith, M.A.; Stein, B.E. Visual, auditory, and somatosensory convergence on cells in superior colliculus results in multisensory integration. *J. Neurophysiol.* **1986**, *56*, 640–662. [CrossRef] [PubMed]
107. Spence, C. Multi-sensory integration and the psychophysics of flavour perception. In *Food Oral Processing: Fundamentals of Eating and Sensory Perception*; Chen, J., Engelen, L., Eds.; Wiley-Blackwell: Oxford, UK, 2012; pp. 203–223.
108. Poinot, P.; Arvisenet, G.; Ledauphin, J.; Gaillard, J.-L.; Prost, C. How can aroma-related cross-modal interactions be analysed? A review of current methodologies. *Food Qual. Pref.* **2012**, *28*, 304–316. [CrossRef]
109. Sakai, N.; Kobayakawa, T.; Gotow, N.; Saito, S.; Imada, S. Enhancement of sweetness ratings of aspartame by a vanilla odor presented either by orthonasal or retronasal routes. *Percept. Mot. Skills* **2001**, *92*, 1002–1008. [CrossRef] [PubMed]
110. Djordjevic, J.; Zatorre, R.J.; Jones-Gotman, M. Odor-induced changes in taste perception. *Exp. Brain Res.* **2004**, *159*, 405–408. [CrossRef]
111. Pfeiffer, J.C.; Hollowood, T.A.; Hort, J.; Taylor, A.J. Temporal synchrony and integration of sub-threshold taste and smell signals. *Chem. Senses* **2005**, *30*, 539–545. [CrossRef]
112. Hort, J.; Hollowood, T.A. Controlled continuous flow delivery system for investigating taste-aroma interactions. *J. Agric. Food Chem.* **2004**, *52*, 4834–4843. [CrossRef]
113. Keast, R.S.J.; Breslin, P.A.S. An overview of binary taste-taste interactions. *Food Qual. Pref.* **2002**, *24*, 111–124. [CrossRef]
114. Ayya, N.; Lawless, H.T. Quantitative and qualitative evaluation of high-intensity sweeteners and sweetener mixtures. *Chem. Senses* **1992**, *17*, 245–259. [CrossRef]
115. Frank, R.A.; Mize, S.J.S.; Carter, R. An assessment of binary mixture interactions for nine sweeteners. *Chem. Senses* **1989**, *14*, 621–632. [CrossRef]
116. Hanger, L.Y.; Lotz, A.; Lepeniotis, S. Descriptive profiles of selected High Intensity Sweeteners (HIS), HIS blends, and sucrose. *J. Food Sci.* **1996**, *61*, 456–459. [CrossRef]
117. Hutteau, F.; Mathlouthi, M.; Portmann, M.O.; Kilcast, D. Physicochemical and psychophysical characteristics of binary mixtures of bulk and intense sweeteners. *Food Chem.* **1998**, *63*, 9–16. [CrossRef]
118. Schiffman, S.S.; Booth, B.J.; Carr, B.T.; Losee, M.L.; Sattely-Miller, E.A.; Graham, B.G. Investigation of synergism in binary mixtures of sweeteners. *Brain Res. Bull.* **1995**, *38*, 105–120. [CrossRef]
119. Wilkie, L.M.; Capaldi Phillips, E.D. Heterogeneous binary interactions of taste primaries: Perceptual outcomes, physiology, and future directions. *Neurosci. Biobehav. Rev.* **2014**, *47*, 70–86. [CrossRef] [PubMed]
120. Green, B.G.; Lim, J.; Osterhoff, F.; Blacher, K.; Nachtigal, D. Taste mixture interactions: Suppression, additivity, and the predominance of sweetness. *Physiol. Behav.* **2010**, *101*, 731–737. [CrossRef]
121. Bult, J.H.F.; de Wijk, R.A.; Hummel, T. Investigations on multimodal sensory integration: Texture, taste, and ortho- and retronasal olfactory stimuli in concert. *Neurosci. Lett.* **2007**, *411*, 6–10. [CrossRef]

122. Szczesniak, A.S. Texture is a sensory property. *Food Qual. Pref.* **2002**, *13*, 215–225. [CrossRef]
123. Weel, K.G.C.; Boelrijk, A.C.; Alting, P.J.J.M.; van Mil, J.J.; Burger, H.; Gruppen, H.; Voragen, A.G.J.; Smit, G. Flavour release and perception of flavoured whey protein gels: Perception is determined by texture rather than by release. *J. Agric. Food Chem.* **2002**, *50*, 5149–5155. [CrossRef]
124. Christensen, C.M. Effects of solution viscosity on perceived saltiness and sweetness. *Percept. Psychophys.* **1980**, *28*, 347–353. [CrossRef] [PubMed]
125. Guichard, E. Interactions between flavor compounds and food ingredients and their influence on flavor perception. *Food Rev. Int.* **2002**, *18*, 49–70. [CrossRef]
126. Siefarth, C.; Tyapkova, O.; Beauchamp, J.; Schweiggert, U.; Buettner, A.; Bader, S. Influence of polyols and bulking agents on flavor release from low-viscosity solutions. *Food Chem.* **2011**, *129*, 1462–1468. [CrossRef]
127. Yang, Z.Y.; Fan, Y.G.; Xu, M.; Ren, J.N.; Liu, Y.L.; Zhang, L.L.; Li, J.J.; Zhang, Y.; Dong, M.; Fan, G. Effects of xanthan and sugar on the release of aroma compounds in model solution. *Flavour Fragr. J.* **2017**, *32*, 112–118. [CrossRef]
128. McBurney, D.H.; Collins, V.B.; Glanz, L.M. Temperature dependence of human taste responses. *Physiol. Behav.* **1973**, *11*, 89–94. [CrossRef]
129. Sato, M. Gustatory response as a temperature-dependent process. *Contrib. Sens. Physiol.* **1967**, *2*, 223–251. [PubMed]
130. Talavera, K.; Ninomiya, Y.; Winkel, C.; Voets, T.; Nilius, B. Influence of temperature on taste perception. *Cell. Mol. Life Sci.* **2007**, *64*, 377–381. [CrossRef]
131. Cruz, A.; Green, B.G. Thermal stimulation of taste. *Nature* **2000**, *403*, 889–892. [CrossRef]
132. Tu, Y.; Yang, Z.; Ma, C. Touching tastes: The haptic perception transfer of liquid food packaging materials. *Food Qual. Pref.* **2015**, *39*, 124–130. [CrossRef]
133. Slocombe, B.G.; Carmichael, D.A.; Simner, J. Cross-modal tactile-taste interactions in food evaluations. *Neuropsychologia* **2015**, *88*, 58–64. [CrossRef]
134. Gibson, J.J. *The Senses Considered as Perceptual Systems*; Houghton Mifflin: Boston, MA, USA, 1966.
135. Stevenson, R.J. *The Psychology of Flavour*; Oxford University Press: Oxford, UK, 2009.
136. Chalé-Rush, A.; Burgess, J.R.; Mattes, R.D. Multiple routes of chemosensitivity to free fatty acids in humans. *Am. J. Physiol.-Gastrointest. Liver Physiol.* **2007**, *292*, G1206–G1212. [CrossRef] [PubMed]
137. Green, B.G.; Lawless, H.T. The psychophysics of somatosensory chemoreception in the nose and mouth. In *Smell and Taste in Health and Disease*; Getchell, T.V., Bartoshuk, L.M., Doty, R.L., Snow, J.B., Eds.; Raven Press: New York, NY, USA, 1991; pp. 235–253.
138. Hetherington, M.M. Sensory-specific satiety and its importance in meal termination. *Neurosci. Biobehav. Rev.* **1996**, *20*, 113–117. [CrossRef]
139. Ernst, M.O. Learning to integrate arbitrary signals from vision and touch. *J. Vis.* **2007**, *7*, 7. [CrossRef] [PubMed]
140. Parise, C.V.; Spence, C. "When birds of a feather flock together": Synesthetic correspondences modulate audiovisual integration in non-synesthetes. *PLoS ONE* **2009**, *4*, e5664. [CrossRef] [PubMed]
141. Spence, C. Crossmodal correspondences: A tutorial review. *Atten. Percept. Psychophys.* **2011**, *73*, 971–995. [CrossRef] [PubMed]
142. Small, D.M. Taste representation in the human insula. *Brain Struct. Funct.* **2010**, *214*, 551–561. [CrossRef] [PubMed]
143. Small, D.M.; Zald, D.H.; Jones-Gotman, M.; Zatorre, R.J.; Pardo, J.V.; Frey, S.; Petrides, M. Human cortical gustatory areas: A review of functional neuroimaging data. *NeuroReport* **1999**, *10*, 7–14. [CrossRef]
144. De Araujo, I.E.T.; Rolls, E.T. The representation in the human brain of food texture and oral fat. *J. Neurosci.* **2004**, *24*, 3086–3093. [CrossRef]
145. Guest, S.; Grabenhorst, F.; Essick, G.; Chen, Y.; Young, M.; McGlone, F.; de Araujo, I.; Rolls, E.T. Human cortical representation of oral temperature. *Physiol. Behav.* **2007**, *92*, 975–984. [CrossRef]
146. De Araujo, I.E.T.; Rolls, E.T.; Kringelbach, M.L.; McGlone, F.; Phillips, N. Taste-olfactory convergence, and the representation of the pleasantness of flavour, in the human brain. *Eur. J. Neurosci.* **2003**, *18*, 2059–2068. [CrossRef]
147. Small, D.M.; Prescott, J. Odor/taste integration and the perception of flavor. *Exp. Brain Res.* **2005**, *166*, 345–357. [CrossRef] [PubMed]

148. Small, D.M.; Voss, J.; Mak, Y.E.; Simmons, K.B.; Parrish, T.; Gitelman, D. Experience-dependent neural integration of taste and smell in the human brain. *J. Neurophysiol.* **2004**, *92*, 1892–1903. [CrossRef] [PubMed]
149. Small, D.M.; Jones-Gotman, M.; Zatorre, R.J.; Petrides, M.; Evans, A.C. Flavor processing: More than the sum of its parts. *NeuroReport* **1997**, *8*, 3913–3917. [CrossRef] [PubMed]
150. Small, D.M.; Gerber, J.C.; Mak, Y.E.; Hummel, T. Differential neural responses evoked by orthonasal versus retronasal odorant perception in humans. *Neuron* **2005**, *47*, 593–605. [CrossRef] [PubMed]
151. Koza, B.J.; Cilmi, A.; Dolese, M.; Zellner, D.A. Color enhances orthonasal olfactory intensity and reduces retronasal olfactory intensity. *Chem. Senses* **2005**, *30*, 643–649. [CrossRef] [PubMed]
152. O'Doherty, J.P.; Deichmann, R.; Critchley, H.D.; Dolan, J.D. Neural responses during anticipation of a primary taste reward. *Neuron* **2002**, *33*, 815–826. [CrossRef]
153. Simmons, W.K.; Martin, A.; Barsalou, L.W. Pictures of appetizing foods activate gustatory cortices for taste and reward. *Cereb. Cortex* **2005**, *25*, 1602–1608. [CrossRef]
154. Wang, G.J.; Volkow, N.D.; Telang, F.; Jayne, M.; Ma, J.; Rao, M.; Zhu, W.; Wong, C.T.; Pappas, N.R.; Geliebter, A.; et al. Exposure to appetitive food stimuli markedly activates the human brain. *NeuroImage* **2004**, *21*, 1790–1797. [CrossRef]
155. Spence, C. The neuroscience of flavor. In *Multisensory Flavor Perception: From Fundamental Neuroscience Through to the Marketplace*; Piqueras-Fiszman, B., Spence, C., Eds.; Elsevier: Oxford, UK, 2016; pp. 235–248.
156. Spence, C. The neuroscience of flavour. In *Food and Museums*; Levent, N., Mihalache, I.D., Eds.; Bloomsbury Academic: London, UK, 2017; pp. 57–70.
157. Rolls, E.T.; Baylis, L.L. Gustatory, olfactory, and visual convergence within the primate orbitofrontal cortex. *J. Neurosci.* **1994**, *14*, 5437–5452. [CrossRef]
158. Wang, Q.J. Assessing the Mechanisms behind Sound-Taste Correspondences and Their Impact on Multisensory Flavour Perception and Evaluation. Ph.D. Thesis, University of Oxford, Oxford, UK, 2017.
159. Wang, Q.J.; Knoeferle, K.; Spence, C. Music to make your mouth water? Assessing the potential influence of sour music on salivation. *Front. Psychol.* **2017**, *8*, 638. [CrossRef]
160. Wesson, D.W.; Wilson, D.A. Smelling sounds: Olfactory-auditory sensory convergence in the olfactory tubercle. *J. Neurosci.* **2010**, *30*, 3013–3021. [CrossRef] [PubMed]
161. Wesson, D.W.; Wilson, D.A. Sniffing out the contributions of the olfactory tubercle to the sense of smell: Hedonics, sensory integration, and more? *Neurosci. Biobehav. Rev.* **2011**, *35*, 655–668. [CrossRef] [PubMed]
162. Cohen, L.; Rothschild, G.; Mizrahi, A. Multisensory integration of natural odors and sounds in the auditory cortex. *Neuron* **2011**, *72*, 357–369. [CrossRef] [PubMed]
163. Callan, A.; Callan, D.; Ando, H. Differential effects of music and pictures on taste perception—An fMRI study. Presented at the Annual Meeting of the International Multisensory Research Forum, Toronto, ON, Canada, 14–17 June 2018.
164. Brown, S.; Gao, X.; Tisdelle, L.; Eickhoff, S.B.; Liotti, M. Naturalizing aesthetics: Brain areas for aesthetic appraisal across sensory modalities. *NeuroImage* **2011**, *58*, 250–258. [CrossRef] [PubMed]
165. de Araujo, I.E.; Simon, S.A. The gustatory cortex and multisensory integration. *Int. J. Obes.* **2009**, *33*, S34–S43. [CrossRef]
166. Simon, S.A.; de Araujo, I.E.; Gutierrez, R.; Nicolelis, M.A.L. The neural mechanisms of gustation: A distributed processing code. *Nat. Rev. Neurosci.* **2006**, *7*, 890–901. [CrossRef]
167. Viana, F. Chemosensory properties of the trigeminal system. *ACS Chem. Neurosci.* **2011**, *2*, 38–50. [CrossRef]
168. Kringelbach, M.L. The pleasure of food: Underlying brain mechanisms of eating and other pleasures. *Flavour* **2015**, *4*, 20. [CrossRef]
169. Cerf-Ducastel, B.; Van de Moortele, P.-F.; Macleod, P.; Le Bihan, D.; Faurion, A. Interaction of gustatory and lingual somatosensory perceptions at the cortical level in the human: A functional magnetic resonance imaging study. *Chem. Senses* **2001**, *26*, 371–383. [CrossRef]
170. Eldeghaidy, S.; Marciani, L.; McGlone, F.; Hollowood, T.; Hort, J.; Head, K.; Taylor, A.J.; Busch, J.; Spiller, R.C.; Gowland, P.A.; et al. The cortical response to the oral perception of fat emulsions and the effect of taster status. *J. Neurophysiol.* **2011**, *105*, 2572–2581. [CrossRef]
171. Österbauer, R.A.; Matthews, P.M.; Jenkinson, M.; Beckmann, C.F.; Hansen, P.C.; Calvert, G.A. Colour of scents: Chromatic stimuli modulate odor responses in the human brain. *J. Neurophysiol.* **2005**, *93*, 3434–3441. [CrossRef] [PubMed]

172. Spence, C.; Piqueras-Fiszman, B. *The Perfect Meal: The Multisensory Science of Food and Dining*; Wiley-Blackwell: Oxford, UK, 2014.
173. Small, D.M.; Veldhuizen, M.G.; Felsted, J.; Mak, Y.E.; McGlone, F. Separable substrates for anticipatory and consummatory food chemosensation. *Neuron* **2008**, *57*, 786–797. [CrossRef] [PubMed]
174. Stevenson, R.J. Flavor binding: Its nature and cause. *Psychol. Bull.* **2014**, *140*, 487–510. [CrossRef] [PubMed]
175. Van der Klaauw, N.J.; Frank, R.A. Scaling component intensities of complex stimuli: The influence of response alternatives. *Environ. Int.* **1996**, *22*, 21–31. [CrossRef]
176. Deliza, R.; MacFie, H.J.H. The generation of sensory expectation by external cues and its effect on sensory perception and hedonic ratings: A review. *J. Sens. Stud.* **1996**, *11*, 103–128. [CrossRef]
177. Hutchings, J.B. *Expectations and the Food Industry: The Impact of Color and Appearance*; Kluwer Academic/Plenum Publishers: New York, NY, USA, 2003.
178. Cardello, A.V.; Sawyer, F. Effects of disconfirmed consumer expectations of food acceptability. *J. Sens. Stud.* **1992**, *7*, 253–277. [CrossRef]
179. Cardello, A.V. Measuring consumer expectations to improve food product development. In *Consumer-Led Food Product Development*; MacFie, H.J.H., Ed.; Woodhead Publishing: Cambridge, UK, 2007; pp. 223–261.
180. Shankar, M.U.; Levitan, C.; Spence, C. Grape expectations: The role of cognitive influences in color-flavor interactions. *Conscious Cogn.* **2010**, *19*, 380–390. [CrossRef] [PubMed]
181. De Lange, F.P.; Heilbron, M.; Kok, P. How do expectations shape perception? *Trends Cogn. Sci.* **2018**, *1811*, 1–16. [CrossRef]
182. Piqueras-Fiszman, B.; Spence, C. Sensory expectations based on product-extrinsic food cues: An interdisciplinary review of the empirical evidence and theoretical accounts. *Food Qual. Pref.* **2015**, *40*, 165–179. [CrossRef]
183. Zellner, D.A.; Kautz, M.A. Color affects perceived odor intensity. *J. Exp. Psychol. Hum. Percept. Perform.* **1990**, *16*, 391–397. [CrossRef]
184. Hollingworth, H.L.; Poffenberger, A.T. *The Sense of Taste*; Moffat Yard: New York, NY, USA, 1917.
185. Rozin, P. "Taste-smell confusions" and the duality of the olfactory sense. *Percept. Psychophys.* **1982**, *31*, 397–401. [CrossRef] [PubMed]
186. Stevenson, R.J.; Oaten, M.J.; Mahmut, M.K. The role of attention in the localization of odors to the mouth. *Atten. Percept. Psychophys.* **2011**, *73*, 247–258. [CrossRef] [PubMed]
187. Spence, C. Oral referral: On the mislocalization of odours to the mouth. *Food Qual. Pref.* **2016**, *50*, 117–128. [CrossRef]
188. Spence, C.; Smith, B.; Auvray, M. Confusing tastes and flavours. In *Perception and Its Modalities*; Stokes, D., Matthen, M., Biggs, S., Eds.; Oxford University Press: Oxford, UK, 2015; pp. 247–274.
189. Ashkenazi, A.; Marks, L.E. Effect of endogenous attention on detection of weak gustatory and olfactory flavors. *Percept. Psychophys.* **2004**, *66*, 596–608. [CrossRef] [PubMed]
190. Lim, J.; Johnson, M.B. Potential mechanisms of retronasal odor referral to the mouth. *Chem. Senses* **2011**, *36*, 283–289. [CrossRef] [PubMed]
191. Lim, J.; Johnson, M. The role of congruency in retronasal odor referral to the mouth. *Chem. Senses* **2012**, *37*, 515–521. [CrossRef] [PubMed]
192. Lim, J.; Fujimaru, T.; Linscott, T.D. The role of congruency in taste-odor interactions. *Food Qual. Pref.* **2014**, *34*, 5–13. [CrossRef]
193. Linscott, T.D.; Lim, J. Retronasal odor enhancement by salty and umami tastes. *Food Qual. Pref.* **2016**, *48*, 1–10. [CrossRef]
194. Spence, C.; Velasco, C.; Knoeferle, K. A large sample study on the influence of the multisensory environment on the wine drinking experience. *Flavour* **2014**, *3*, 8. [CrossRef]
195. Zampini, M.; Wantling, E.; Phillips, N.; Spence, C. Multisensory flavor perception: Assessing the influence of fruit acids and color cues on the perception of fruit-flavored beverages. *Food Qual. Pref.* **2008**, *19*, 335–343. [CrossRef]
196. Junge, J.Y. Bi- and Ternary Sensory Cross Modal Interactions: The Effect of Aroma, Colour, and Viscosity on Perceived Sweetness in a Beverage. Master's Thesis, Department of Food Science, Aarhus University, Årslev, Denmark, 2019.
197. Knoop, J.E.; Sala, G.; Smit, G.; Stieger, M. Combinatory effects of texture and aroma modification on taste perception of model gels. *Chemosens. Percept.* **2013**, *6*, 60–69. [CrossRef]

198. Alcaire, F.; Antúnez, L.; Vidal, L.; Giménez, A.; Ares, G. Aroma-related cross-modal interactions for sugar reduction in milk desserts: Influence on consumer perception. *Food Res. Int.* **2017**, *97*, 45–50. [CrossRef] [PubMed]
199. Pineli, L.L.O.; de Aguiar, L.A.; Fiusa, A.; de Assunção Botelho, R.B.; Zandonadi, R.P.; Melo, L. Sensory impact of lowering sugar content in orange nectars to design healthier, low-sugar industrialized beverages. *Appetite* **2016**, *96*, 239–244. [CrossRef] [PubMed]
200. Yeomans, M.R.; Chambers, L.; Blumenthal, H.; Blake, A. The role of expectancy in sensory and hedonic evaluation: The case of smoked salmon ice-cream. *Food Qual. Pref.* **2008**, *19*, 565–573. [CrossRef]
201. Clemens, R.A.; Jones, J.M.; Kern, M.; Lee, S.; Mayhew, E.J.; Slavin, J.L.; Zivanovic, S. Functionality of sugars in foods and health. *Compr. Rev. Food Sci. Food Saf.* **2016**, *15*, 433–470. [CrossRef]
202. Dijksterhuis, G. The total product experience and the position of the sensory and consumer sciences: More than meets the tongue. *New Food Mag.* **2012**, *15*, 38–41.
203. Chung, S.-J.; Vickers, Z. Long-term acceptability and choice of teas differing in sweetness. *Food Qual. Pref.* **2007**, *18*, 963–974. [CrossRef]
204. Pombo-Rodrigues, S.; Hashem, K.M.; He, F.J.; MacGregor, G.A. Salt and sugars content of breakfast cereals in the UK from 1992 to 2015. *Public Health Nutr.* **2017**, *20*, 1500–1512. [CrossRef]
205. Crisinel, A.-S.; Spence, C. The impact of pleasantness ratings on crossmodal associations between food samples and musical notes. *Food Qual. Pref.* **2012**, *24*, 136–140. [CrossRef]
206. Looy, H.; Callaghan, S.; Weingarten, H.P. Hedonic response of sucrose likers and dislikers to other gustatory stimuli. *Physiol. Behav.* **1992**, *52*, 219–225. [CrossRef]
207. Keskitalo, K.; Knaapila, A.; Kallela, M.; Palotie, A.; Wessman, M.; Sammalisto, S.; Peltonen, L.; Tuorila, H.; Perola, M. Sweet taste preferences are partly genetically determined: Identification of a trait locus on chromosome 16. *Am. J. Clin. Nutr.* **2007**, *86*, 55–63. [CrossRef] [PubMed]
208. Kampov-Polevoy, A.B.; Garbutt, J.C.; Davis, C.E.; Janowsky, D.S. Preference for higher sugar concentrations and tridimensional personality questionnaire scores in alcoholic and nonalcoholic men. *Alcohol. Clin. Exp. Res.* **1998**, *22*, 610–614. [CrossRef] [PubMed]
209. Yeomans, M.R.; Prescott, J.; Gould, N.J. Acquired hedonic and sensory characteristics of odours: Influence of sweet liker and propylthiouracil taster status. *Q. J. Exp. Psychol.* **2009**, *62*, 1648–1664. [CrossRef] [PubMed]

# 11

# The Effect of Sleep Curtailment on Hedonic Responses to Liquid and Solid Food

**Edward J. Szczygiel, Sungeun Cho and Robin M. Tucker ***

Department of Food Science and Human Nutrition, Michigan State University, East Lansing, MI 48824, USA; szczygi7@msu.edu (E.J.S.); chosunge@msu.edu (S.C.)
* Correspondence: tucker98@msu.edu.

**Abstract:** It is currently unclear whether changes in sweet taste perception of model systems after sleep curtailment extend to complex food matrices. Therefore, the primary objective of this study was to use a novel solid oat-based food (crisps) and oat-based beverage stimulus sweetened with sucralose to assess changes in taste perception after sleep curtailment. Forty-one participants recorded a habitual and curtailed night of sleep using a single-channel electroencephalograph. The next morning, overall sweetness, flavor, and texture liking responses to energy- and nutrient-matched oat products across five concentrations of sweetness were measured. Overall ($p = 0.047$) and flavor ($p = 0.017$) liking slopes across measured concentrations were steeper after curtailment, suggesting that sweeter versions of the oat products were liked more after sleep curtailment. Additionally, a hierarchical cluster analysis was used to classify sweet likers and non-likers. While the effect of sleep curtailment on sweet liking did not differ between sweet liking classification categories, sleep curtailment resulted in decreased texture liking in the solid oat crisps for sweet non-likers ($p < 0.001$), but not in the oat beverage. These findings illustrate the varied effects of sleep on hedonic response in complex food matrices and possible mechanisms by which insufficient sleep can lead to sensory-moderated increases in energy intake.

**Keywords:** sleep curtailment; hedonics; complex food matrices; sweet liking phenotype; sweet taste; texture

---

## 1. Introduction

There is a growing body of evidence that insufficient sleep can alter taste perception. Several recent psychophysical studies have reported that short sleep duration is associated with increased preferred sucrose concentration [1–3] and increased perceived intensity of sour and umami taste [4]. Insufficient sleep-induced changes in taste perception may partially moderate the well-supported relationship between short sleep duration, increased dietary intake of highly palatable high-carbohydrate and high-fat foods, and weight gain [5–7]. Brain imaging research suggests that insufficient sleep results in increased neural sensitivity to the reward properties of food [8–12]. This heightened sensitivity may increase the consumption of palatable food for pleasure, also known as hedonic eating. Hedonic eating is thought to promote weight gain, as highly palatable food tends to be energy-dense [13]. Sweetness is commonly associated with the palatability of food [14] and when tasted, initiates brain reward processes [15]; therefore, sweet taste is of particular interest when exploring relationships between insufficient sleep and hedonic eating. Nearly 40% of the US adult population is reported to sleep less than the recommended 7 h per night [16] and nearly 40% of American adults suffer from obesity [17]. A similar prevalence of insufficient sleep has been documented globally [18,19]. Therefore, understanding the mechanisms by which insufficient sleep can lead to weight gain is of importance to scientists. Furthermore, these mechanisms may provide deeper insight regarding food choices and food acceptance and therefore, are of particular interest to both the food industry and public health advocates.

Very few studies utilize complex food when examining the effect of insufficient sleep on taste function [4,20]. Instead, nearly all existing sleep-taste research has been conducted using model systems—prototypical tastants dissolved in deionized water—and evaluated while wearing nose clips [1–4,21,22]. Results from previous psychophysical studies examining the effects of sleep on taste perception need to be replicated in more complex food matrices as findings in model systems do not always align with findings using complex foods [23–26]. The simplicity of model systems allows participants to evaluate taste with minimal distraction from other sensory inputs like texture or aroma, but affective judgments of foods and beverages are determined using all senses, including appearance, mouthfeel, auditory characteristics, geometry, and the physical state of food [27]. Thus, further efforts are needed to assess the generalizability of taste-related findings from psychophysical studies to complex food matrices.

In addition to the general issues discussed above regarding translating findings from model systems to food, there are particular reasons to believe that the generalizability of findings from model stimuli to complex foods under conditions of insufficient sleep could be especially problematic. In the context of complex food, research suggests two important effects of insufficient sleep that could alter perception: impaired sensory neural processing [28,29] and increased somatosensory sensitivity [30]. First, given that the orbitofrontal cortex (OFC), often described as the neural control center for appetite [31], is impaired after sleep curtailment, the ability to interpret multimodal information may be compromised [28,29]. Under normal conditions, processing of specific attributes within multimodal sensory information is already limited. For example, when consuming complex foods, the ability of participants to separate perceived sweet taste liking from perceived flavor or overall liking may be diminished due to sensory interactions [32]. Thus, after sleep curtailment, impairment of OFC activity may result in further differences between perception in controlled systems and complex food systems [32]. The second concern about the generalizability of findings from model systems to more complex food matrices under insufficient sleep conditions stems from documented changes in somatosensory perception. Sleep curtailment has been implicated in acute reward system-mediated hyperalgesia—an increased sensitivity to pain [33]—and increased oro-facial somatosensory sensitivity, particularly the tongue [30]. While speculative, increased hyperalgesia and increased oro-facial sensitivity might decrease the acceptability of the texture of crispy or crunchy solid foods and increase preference for softer foods, semisolids, or beverages that require less oral processing. In summary, processing of sensory information, reward processing of that information, and changes in oral sensory sensitivity all represent opportunities for insufficient sleep to affect hedonic food perception.

Individual differences in hedonic response to taste make it challenging to study the relationship between insufficient sleep and gustatory perception. Despite being an innately palatable taste at birth [34], liking responses to sweet taste as the concentration of sweetness increases differ across individuals. Three fundamental patterns of liking over a range of sweetness levels have been identified previously: sweet likers, who display a rise in liking as sweetener concentration increases; inverted U-shape responders, who show an increasing liking pattern up until a certain concentration before beginning to show a decrease; and dislikers, who display a reduction in liking as concentration increases [35–37]. Additionally, a fourth pattern where hedonic response to sweetness is the same regardless of sweetness concentration has been reported [36], but others have reported not observing these phenotypes [38,39]. These fundamental patterns of liking are partially determined by genetic factors [40,41], and thus, they are commonly described as "sweet liking phenotypes" (SLP). Sweet likers differ in expressed behaviors compared to the other phenotypes, including increased intake of sugar and sugar-sweetened beverages [42,43]. These behavioral traits suggest that sweet liking phenotypes are heritable indicators of general brain reward processing alteration. Given that the central hypothesis of this research is that insufficient sleep-induced reward processing alterations may influence hedonic perception of food, individual differences in response to sweet taste are an important factor to consider, as these baseline differences in reward processing may reduce the effects of insufficient sleep on the brain reward processing. While our previous work found that preferred sweetener concentration was similarly increased across sweet liking phenotypes after sleep curtailment [21], this relationship has not been evaluated in complex foods. Therefore, the question of whether SLP is an important factor moderating the effect of sleep curtailment merits further investigation in the context of complex foods.

The main objective of this study was to evaluate changes in hedonic response to two complex sucralose-sweetened foods across a range of sweetness levels after a habitual and curtailed night of sleep and to compare these responses to a model system consisting of sucralose solutions. It was hypothesized that hedonic perception in the model system would change in accordance with our previous findings [21], that is, after a night of sleep curtailment, preferred sucralose solution concentration would be increased relative to hedonic response after a night of habitual sleep and a non-significant increase in the steepness of the slope of liking over a range of concentration would be observed. While it was expected that changes in patterns of sweetness liking would agree with our previous findings that showed that sleep curtailment increased the rate of liking as sweetener concentration increased, it was hypothesized that the change in pattern would be more pronounced when tasting "real" foods instead of sweetener solutions due to altered processing of multi-modal sensory information. It was also expected that broader hedonic measures, such as a flavor and overall liking, would show increases corresponding with increasing sweetness after sleep curtailment, as these terms have a greater potential to capture changes in multisensory perception unique to complex foods. Furthermore, it was hypothesized that sleep curtailment would result in decreased texture liking in a solid food and increased liking in a liquid food. A secondary objective was to assess if food form and SLP interact with sleep curtailment to alter sensory perception of complex foods. Although SLP was not found to differentially moderate changes in hedonic perception in model systems under conditions of insufficient sleep, we sought to confirm this finding in complex foods. It was expected that SLP would not moderate changes in hedonic perception of food after sleep curtailment, in accordance with previous work [21].

## 2. Materials and Methods

The protocol for this study was approved by the Michigan State University Human Research Protection Program (East Lansing, MI, USA). Written informed consent was obtained from all participants and a cash incentive was provided for study completion.

### *2.1. Participants*

Participants between the ages of 18 and 45, without obesity (BMI < 30.0 kg/m$^2$) or diagnosed sleep conditions, who typically slept 7–9 h per weeknight, and who had a consistent weekday bedtime were eligible to participate in the study. Participants were pre-screened using two criteria. First, each participant sampled both the oat "beverage" and oat "crisp" products (see Development of Stimuli section, below) evaluated in the study (sweetened with sucralose at the middle 0.032% *w*/*v* level) and were asked to rate their overall liking of each on a 9-point hedonic scale (extremely dislike (1) to extremely like (9)). The mid-range sweetness product was selected to avoid disproportionally recruiting sweet likers. Participants who rated either sample <6 (like slightly) were not eligible for the study to ensure that products could be considered generally palatable. Secondly, each participant sampled the highest concentration of sucralose in water (0.094% *w*/*v*) used in the study and was asked to report if they tasted any bitterness. Sucralose does not ordinarily display high levels of bitterness [44]. Nevertheless, participants who are extremely sensitive to bitterness [40] may find it challenging to evaluate sweetness in sucralose solutions. To avoid this, participants who tasted bitterness in the highly sweet sucralose sample were excluded from the study. Three individuals who were otherwise eligible were excluded due to tasting bitterness, and two were excluded due to dislike of the oat beverage.

## 2.2. *Development of Stimuli*

Sucralose was selected as the sweetener for this study due to its sensory and functional properties. Sucralose has a taste profile with a similar character to sucrose and has low bitter and off-tastes compared to other high-intensity sweeteners [44]. Sucralose requires very small amounts to achieve the same sweetness as sucrose [45]. This property of sucralose enabled formulation of complex food products that varied in sweetness while minimizing changes in other sensory attributes, such as texture. The iso-sweet concentration and sweetness range were based on a similar study [21], which found these concentrations sufficient to observe changes in liking patterns after sleep curtailment. Briefly, a preliminary study was carried out to assess iso-sweet concentrations of sucralose compared to sucrose using the magnitude estimation methods of Reis et al. 2016 [46]. Sucralose concentrations of 0.004%, 0.011%, 0.032%, 0.060%, and 0.094% *w/v* were selected based on the magnitude estimation data power functions developed using data collected from fifty naive participants. These concentrations are equal in sweetness to 3%, 6%, 12%, 18%, and 24% *w/v* sucrose, respectively. The selected concentrations were found to be iso-sweet when used in a previous sleep-taste study using sweetener solutions [21]. To assess the effect of sleep curtailment on patterns of liking of complex food matrices, two energy and macronutrient-matched oat-based products were developed. Oats were selected as a versatile base ingredient because they can be used to produce similarly nutritious and palatable, solid, semi-solid, and liquid foods consistently across batches. The first product, an oat "beverage", was developed to assess the effect of sleep curtailment on hedonic perceptions of liquid food, and the second product, an oat "crisp", was developed to assess the effect of sleep curtailment on hedonic perceptions of solid food. The two products contained the same ingredients: whole grain rolled quick oats (Quaker Oats Company, Chicago, IL, USA), pure sucralose powder (Sweet Solutions, Edison, NJ, USA), and filtered water (Besco, Battle Creek, MI, USA). In both products, sucralose was added to water at the concentrations discussed previously and used to produce five differently sweet versions of both products. A proximate analysis was performed by Great Lakes Scientific (Stevensville, MI, USA) using the Association of Analytical Chemists (AOAC) method. A breakdown of the macronutrient content per 100 kcal is displayed in Table 1. The two products were matched on macronutrients per kcal. The only major difference between the two products was the moisture content, as designed.

**Table 1.** Macronutrient composition of oat products.

| | **Oat Beverage** | **Oat Crisp** |
|---|---|---|
| **Macronutrient** | **100 kcal** | **100 kcal** |
| Fat | 2 g | 2 g |
| Carbohydrates | 18 g | 17 g |
| Protein | 3 g | 3 g |
| Crude Fiber | <1 g | <1 g |
| Moisture | 189 g | 1 g |
| Ash | <1 g | <1 g |

Stimuli were matched for energy and macronutrient composition. Moisture content differed due to the physical state of the stimuli.

Oat beverages were produced by creating an oat slurry by blending (Nutribullet, NutriLiving, Northridge, CA, USA) 240 g of sucralose-sweetened water and 50 g of whole grain rolled quick oats (Quaker Oat Company, Chicago, IL, USA) for 10 s. The slurry was filtered through a 100 μm steel mesh to produce a smooth milk-like beverage. The oat beverage was stored in glass bottles at 4 °C for no more than 48 h after production.

Oat crisps were prepared using a 1200 W microwave (General Electric, Boston, MA, USA) to dehydrate an oat slurry, which was produced by mixing oats and sucralose-sweetened water in the same procedure as the oat beverage. Differently sweetened oat slurries were microwaved in a 200 mm × 200 mm glass pan for 15 min. The semi-dry oat sheet was then flipped and a 12.7 mm circular cutter was used to cut crisps out of the sheet. The cut crisps were then microwaved for an additional 2 min. The oat crisps were weighed to ensure that each crisp weighed 1.2 ± 0.1 g. The crisps were then cooled in air for 15 min before being vacuum-sealed in plastic and stored at room temperature until served.

### 2.3. Study Timeline

After an initial consent visit to confirm eligibility for the study, participants visited the sensory lab twice: once after a habitual night and once after a curtailed night of sleep. The lab visits occurred at least one week apart on the same weekday and time (±30 min). Sensory testing transpired during 1 h timeslots between the hours of 7:00–10:00 on weekdays. Participants were assigned a timeslot as close to their habitual wake-time as possible. Participants were not directed to adhere to specific protocols during the time gain by sleep curtailment in order to allow for free-living conditions. The sleep condition sequence was randomly assigned during the consent visit. A sleep curtailment of 33% was determined by centering the self-reported habitual sleep duration and equally reducing bed and wake time in order to minimize circadian rhythm effects while still inducing sleepiness [47]. For example, if the curtailment was 2 h, the participant was required to go to bed 1 h later and wake up 1 h earlier. The study was designed to assess change in hedonic perception under free-living conditions, and therefore, partial sleep curtailment was utilized in place of total sleep deprivation [47].

### 2.4. Consent Visit

Participants completed several validated questionnaires during the consent visit. The Pittsburgh Sleep Quality Index (PSQI) [48], Perceived Stress Scale (PSS) [49], and the General Food Craving Questionnaire—Trait version (G-FCQ-T) [50] were used to determine subjective sleep, perceived stress, and general food craving traits, respectively. A relationship between food cravings and reward sensitivity has been reported previously [51] and, thus, food cravings were measured to aid in interpretation of findings. Participants may not have been aware that their sleep habits were abnormal, and thus, the PSQI scores were used to confirm that participants met the criteria for the study. Anthropometrics were also measured for use as covariates. Body mass index (BMI) and percent body fat (% BF) were measured using bioelectrical impedance (TBF-400, Tanita, Arlington Heights, IL, USA).

Objective sleep measures were collected using the Zmachine (General Sleep, Columbus, OH, USA). Participants were trained on how to use the Zmachine at the consent visit. The Zmachine records a single channel ($A_1$–$A_2$) of electroencephalography (EEG) and uses a scoring algorithm to discriminate between light sleep (LS), slow wave sleep (SWS), REM sleep, and waking states. The Zmachine has been reported to have significant agreement with polysomnography (PSG) [52]. To ensure that participants complied with the assigned protocol, they were instructed to wear the Zmachine 30 min before the predetermined bedtime assigned to them.

To ensure that participants would be fasted after both sleep conditions, they were told to not eat or drink anything other than water between their wake time and their laboratory visit. Additionally, they were told that they would be required to take a "Hydrogen Breath Test", the results of which would inform the study administrator if they did not follow the fasting instructions. This deceptive procedure was employed to increase compliance with the fasting instructions. The samples were discarded after testing and participants were made aware of the deceit during debriefing.

### 2.5. Laboratory Visits

The testing procedure used was the same for both laboratory visits. EEG data from the previous night's sleep was promptly uploaded to the Zmachine data viewer upon arrival to the lab. The participant was asked to confirm that the data matched their own recollection of the previous night. If any evidence of machine malfunction, such as significant data loss or disagreement >30 min between participant recollection and machine data readout, then participants returned to the lab no fewer than seven days later with a new sleep recording. Participants then completed the "Hydrogen Breath Test".

Before tasting any stimuli, participants completed several validated questionnaires, including the Karolinska Sleepiness Scale (KSS) [53], the Positive Affect-Negative Affect Schedule (PANAS) [54], and the General Food Craving Questionnaire-State version (G-FCQ-S) [50]. These tools were used to measure sleepiness, affect, and food craving state, respectively. Additionally, a 100 mm visual analog scale (VAS) was used to measure hunger with "Extremely Hungry "(0) and "Extremely Full" (100) serving as anchors [55]. The KSS was used along with objective sleep measure to determine the efficacy of the sleep curtailment. The PANAS was used to measure affect changes between the habitual and curtailed sleep conditions to help interpret findings, as changes in affect have been reported to change with sleep curtailment [56] and influence taste perception [57]. Craving states have been found to be associated with sleep duration [4]; therefore, G-FCQ-S data was collected to help aid in interpretation of findings in case cravings were significantly increased by curtailment. Hunger was measured to confirm whether the fasting protocol was effective.

To assess self-perception of the previous night's sleep quality, participants answered four questions regarding their recollection of the previous night's sleep [21]. The four questions were: "How much sleep did you obtain last night?", "How deeply did you sleep last night?", "How would you rate the quality of your sleep last night?", and "Compared to an average night of sleep, how comfortable were you when sleeping last night?" The sum of the scores from each of these four questions was used as a measure of overall subjective sleep quality.

### *2.6. Sensory Evaluation*

RedJade Sensory Software (RedJade, Redwood Shores, CA, USA) was used to manage sensory data collection. All data collection took place at the Michigan State University sensory laboratory. Participants were required to wear nose-clips during sucralose solution tastings but not when consuming oat products. For the sucralose-in-water tasting, participants were instructed to taste the whole cup (10 mL of sample) and expectorate all samples. For the oat product evaluation, the amount served was normalized to 5 kcal, that is, oat crisps were always served in 1.2 ± 0.1 g quantities (5 kcal) and oat beverage was always served in 10 mL quantities (5 kcal). Oat beverage was served cool at 7 °C and while oat crisps and sucralose solutions were served at room temperature (23 °C). Participants did not expectorate oat products. The sensory evaluation consisted of hedonic evaluation of the sucralose solutions and sweet preference testing followed by evaluation of the oat beverage and oat crisps in a random order.

The solutions and products were assessed by presenting a range of five different concentrations of sweetness of each product identified with three-digit blinding codes in a random order. For the sucralose solutions, participants rated their liking of each solution on a 15 cm VAS scale with anchors at 0 (dislike extremely), 7.5 (neutral) and 15 (like extremely). For the oat products, participants rated their overall liking, sweetness liking, flavor liking and texture liking on an identical 15 cm line scale, in that order. In food acceptance tests, it is typical for participants to rate the overall liking of a product, followed by rating a series of product attributes, such as flavor and texture [58]. Additionally, participants were asked to rate how intensely they perceived the sweetness to be on a 15 cm VAS scale with anchors at 0 (not at all intense), 7.5 (no label) and 15 (extremely intense) for both sucralose solutions and oat products. Following the tasting of a sample, there was a 45 s forced wait period in which the participant was required to rinse three times with filtered water. There were three two-minute breaks after every five samples. In total, participants tasted between 10 and 15 sucralose solutions, 5 oat beverages, and 5 oat crisps at each testing visit.

A modified version of the Monell forced-choice paired comparison protocol [59] was used for preference testing per the methods previously described in Szczygiel et al. 2019 [21]. This version of the protocol reduces the two highest concentrations from the Monell protocol—24% and 36% *w/v*—to 18% and 24% *w/v*. respectively. The modification to the original protocol was made in order to reduce the possibility of off-tastes when high concentrations of sucralose were used. This modification was used to determine preferred sucralose concentration previously [21].

### 2.7. Statistical Analysis

SAS version 9.4 (SAS Institute, Cary, NC, USA.) was used to analyze data. In all analyses, findings were treated as statistically significant if $p < 0.05$ and data are presented as the mean ± standard deviation unless stated otherwise. Overall and attribute liking and intensity scores were plotted against sweetener concentration. The best fit linear functions for each plot were calculated in Excel (Microsoft, Redmond, WA, USA) and the slope of that function became the "Slope" variables used in several analyses.

A hierarchical cluster analysis (HCA) was conducted in XLstat (Addinsoft, Paris, France) using the five liking scores for each concentration of sucralose in water in order to classify participants into sweet liking phenotypes [35]. HCA is recommended as an objective strategy for classifying study participants into sweet liking phenotypes [36]. Three clusters were identified. Due to the limited sample size, the inverted U-shape responders and sucralose dislikers were grouped into a single "non-liker" group to be used as a fixed factor in further analysis.

A mixed model was used to determine differences in liking and intensity responses. Sucralose concentration ($n = 5$, 0.004% *w/v* −0.094% *w/v*), sleep treatment ($n = 2$, curtailed and habitual) food form ($n = 2$, oat beverage and oat crisp), and SLP ($n = 2$, likers and non-likers) were the main fixed factors used throughout the analysis. Participant and interactions between the main fixed factors were included as random factors in all the models. No significant sex, sequence, or period effects were observed in the initial models for sweet taste preference ($p > 0.05$). Therefore, the data for both sexes were pooled and neither sequence nor period were used in any further analysis. Data collected after both nights of sleep, such as PANAS scores or hunger rating, were analyzed using paired $t$-tests and corrected for multiple comparisons using false discovery rate (FDR) with a threshold of $q = 0.05$, which is a strategy used to minimize the risk of type-1 error [2,60].

## 3. Results

### 3.1. Participants

Demographics and anthropometrics for the participants are reported in Table 2. Forty-one non-obese participants finished the study. Participants were primarily white ($n = 27$) and female ($n = 26$). Anthropometric measures as well as G-FCQ-T, PSS, and PSQI scores were not correlated with sucralose preference and therefore, were not used in any further analysis ($p > 0.05$).

**Table 2.** Anthropometric and demographic summary.

| Sex | $n$ | % |
|---|---|---|
| Male | 15 | 37% |
| Female | 26 | 63% |
| Race | | |
| White | 27 | 66% |
| Asian | 13 | 32% |
| Other/More than 1 | 1 | 2% |
| Anthropometrics | Mean ± SD | Range |
| BMI (kg/m$^2$) | 23.1 ± 3.0 | 16.4–29.2 |
| BF (%) | 24.8 ± 11.8 | 9.1–35.5 |
| Age (y) | 24.1 ± 5.0 | 18–41 |
| Traits/Habits | | |
| G-FCQ-T (Score) | 52.5 ± 18.5 | 23–117 |
| PSS (Score) | 12.1 ± 4.6 | 3–23 |
| PSQI (Score | 3.9 ± 1.1 | 1–5 |

Abbreviations: BMI: body mass index, BF: body fat, G-FCQ-T: General Food Craving Questionnaire Trait version, PSS: Perceived Stress Scale, PSQI: Pittsburgh Sleep Quality Index, SD: standard deviation.

### 3.2. *Summary of Curtailment*

A 34.9% reduction in TIB resulted in restriction of TST, LS, and REM ($p \leq 0.001$ for all) but not SWS (Table 3). Sleepiness was significantly increased after sleep curtailment, as evidenced by the KSS score increase ($p < 0.001$). Participants reported that the previous night's sleep was shorter than needed and of a reduced quality ($p < 0.001$). Curtailment reduced perceived sleep quality ($p < 0.001$), but sleep was rated "about average" or higher after both sleep treatments. Participants did not perceive a difference in "deepness" or "comfort" between the two nights.

**Table 3.** Summary of objective and subjective sleep measures.

| | | Habitual | Curtailed | % Reduction | $p$-Value | $q$-Value |
|---|---|---|---|---|---|---|
| Objective Sleep Measures (h) | Time in Bed | 8.3 ± 0.7 | 5.4 ± 0.7 | 34.90% | <0.001 | <0.001 |
| | Total Sleep Time | 7.2 ± 0.7 | 4.5 ± 1.0 | 37.50% | <0.001 | <0.001 |
| | Light Sleep | 3.8 ± 0.5 | 2.0 ± 0.8 | 47.40% | <0.001 | <0.001 |
| | REM Sleep | 1.9 ± 0.5 | 1.2 ± 0.4 | 36.90% | <0.001 | <0.001 |
| | Slow Wave Sleep | 1.5 ± 0.4 | 1.4 ± 0.4 | 6.70% | 0.043 [a] | 0.053 |
| Sleepiness (10 pt) | Karolinska Sleepiness Scale | 3.5 ± 1.4 | 5.7 ± 1.6 | | <0.001 | <0.001 |
| Subjective Previous Night's Sleep Measures (5 pt) | Subjective Sleep Total | 13.5 ± 2.0 | 10.3 ± 2.4 | | <0.001 | <0.001 |
| | How much sleep did you obtain last night? | 3.1 ± 0.4 | 1.5 ± 0.5 | | <0.001 | <0.001 |
| | How deeply did you sleep? | 3.6 ± 0.9 | 3.3 ± 1.0 | | 0.243 | 0.268 |
| | How would you rate the quality of your sleep | 3.8 ± 0.8 | 2.6 ± 1.0 | | <0.001 | <0.001 |
| | Compared to an average night, how comfortable were you when sleeping last night? | 3.0 ± 0.7 | 2.9 ± 1.0 | | 0.593 | 0.593 |

All objective sleep measures were significantly reduced after sleep curtailment. The Karolinska Sleepiness Scale measures sleepiness on a 10-point scale where 1 is "extremely alert" and 10 is "extremely sleepy". Sleepiness was significantly higher after sleep curtailment. Subjective previous night's sleep quality was measured using four questions, and the total score was used to represent general subjective sleep quality. Curtailment resulted in a significantly lower total subjective sleep score. $p$-values were obtained from paired $t$-tests, and $q$-values were obtained by correcting $p$-values for false discovery rate. [a] After false discovery rate correction, the difference between SWS after a habitual and curtailed night was no longer significant.

### 3.3. *Summary of Affect, Cravings and Hunger*

Curtailment did not result in changes in hunger, negative affect, or food cravings—neither the composite score nor any of the five factors (Table 4). However, curtailment resulted in a decrease in positive affect ($p < 0.001$).

**Table 4.** Summary of state-dependent measures.

| Measure | Factor | Habitual | Curtailed | *p*-Value | *q*-Value |
|---|---|---|---|---|---|
| Hunger | Hunger (100 mm VAS) | 67.1 ± 10.24 | 65.5 ± 10.3 | 0.916 | 0.916 |
| G-FCQ-S (0–15 per factor) | Total | 44.2 ± 9.7 | 46.2 ± 12.3 | 0.429 | 0.687 |
| | F1-Desire to Eat | 6.1 ± 2.0 | 6.1 ± 2.2 | 0.948 | 0.916 |
| | F2-Anticipation to positive reinforcement | 8.9 ± 2.0 | 9.5 ± 2.7 | 0.232 | 0.618 |
| | F3-Anticipation to negative reinforcement | 11.2 ± 1.8 | 11.1 ± 2.6 | 0.859 | 0.916 |
| | F4-Obsessive preoccupation | 6.6 ± 2.4 | 7.4 ± 3.0 | 0.124 | 0.496 |
| | F5-Craving as a physiological state | 9.1 ± 2.0 | 9.4 ± 2.7 | 0.405 | 0.687 |
| PANAS | Positive Affect | 23.6 ± 2.0 | 17.6 ± 6.4 | <0.001 | <0.001 |
| | Negative Affect | 12.8 ± 3.9 | 13.2 ± 4.3 | 0.539 | 0.719 |

Positive affect was significantly decreased after sleep curtailment, whereas, hunger, food craving, and negative affect were not. Larger numbers indicate a greater response. For example, positive affect is higher after a habitual night compared to a curtailed night. FDR correction, shown as *q*-values, did not change the significance of any comparisons. Abbreviations: VAS: Visual Analog Scale, G-FCQ-S: General Food Craving Questionnaire State Version, PANAS: Positive Affect Negative Affect Schedule, F1-5: General Food Craving Questionnaire State Version Factors 1–5.

### 3.4. *Sweet Liking Phenotypes*

Three sweet liking phenotypes (SLP) were identified by hierarchical cluster analysis (HCA) using the hedonic response to five concentrations of the model system (sucralose-sweetened water) after the habitual night. Members of cluster 1 ($n$ = 24), the largest cluster, increasingly liked the stimuli as concentration increased until leveling off at 0.032% *w/v* ("likers"). Members of cluster 2 ($n$ = 10) displayed an inverted U-shape of liking ratings, which began to decrease after 0.032% *w/v* ("inverse U-shape"). Members of cluster 3 ($n$ = 8), the smallest cluster, liked solutions less as concentration increased ("dislikers"). After curtailment, there were 26 likers, 11 inverse U-shape, and four dislikers. The number of members in each cluster did not significantly differ after sleep curtailment (Kolmogorov–Smirnov, $p > 0.05$); however, this finding obscures the fact that the SLPs were not entirely stable, as nine participants (22%) changed cluster after sleep curtailment. Seven participants moved from either the inverse U-shape or disliker to the liker cluster and two moved from the liker to the disliker cluster. Due to the small number of participants belonging to clusters 2 and 3 based on the model sucralose solutions after the habitual night, these clusters were combined and will henceforth be referred to as "non-likers" ($n$ = 17).

### 3.5. *Sweetness Perception in the Model System*

Sucralose solution data was analyzed separately from the oat products using a mixed model containing sleep condition, SLP, and the interaction term between the two factors.

#### 3.5.1. Model System Sweet Preference

The preferred concentration from the model system was analyzed to confirm the previously reported SLP-independent increase in preferred sucralose concentration after sleep curtailment and to assess whether the SLPs showed differences in preferred concentration. No interaction was observed between the two factors ($F(1,39) = 3.08$, $p = 0.087$), confirming that SLP and preferred concentration were independent. A significant main effect of the sleep condition on the preferred concentration of sucralose in solution was observed ($F(1,39) = 42.24$, $p < 0.001$), signifying an increase in preferred concentration after sleep curtailment regardless of SLP, (0.042 ± 0.028% *w/v* after the habitual night and 0.063 ± 0.025% *w/v* after the curtailed night). Regardless of sleep condition, sweet likers had a higher preferred concentration (M: 0.067% *w/v* SD: 0.022) compared to non-likers (M: 0.031% *w/v* SD: 0.021) (main effect for SLP on the preferred concentration of sucralose in solution ($F(1,39) = 43.53$, $p < 0.001$)).

#### 3.5.2. Model System Sweet Liking Slopes

Model system sweet liking slopes were analyzed to assess whether sleep curtailment resulted in a change in slope of sucralose liking across the sweetener concentrations and whether changes were independent of SLP. For sucralose liking slope, neither the sleep condition by SLP interaction (F(1,39) = 0.0, $p = 0.953$) nor the main effect of sleep condition were significant (F(1,39) = 2.6, $p = 0.115$), indicating that sucralose slope did not significantly increase in steepness after sleep curtailment, regardless of SLP (habitual slope M: 2.4 liking score/0.1% *w/v* sucralose, curtailed slope M: 3.6 liking score/0.1% *w/v* sucralose). A main effect for SLP was observed (F(1,39) = 89.84, $p < 0.001$), confirming the difference in slopes between sweet likers (M: 6.9 liking score/0.1% *w/v* sucralose) and sweet non-likers (M: −2.5 liking score/0.1% *w/v* sucralose).

### 3.5.3. Model System Sweet Liking by Concentration

To assess whether liking varied at specific concentrations or overall (across all concentrations) after sleep curtailment, sucralose concentration was added as a five-level fixed factor to the model. No tertiary interactions were observed ($p > 0.05$). Sleep curtailment did not result in significant changes in sweet liking by concentration for sucralose solutions, as evidenced by neither the interaction terms nor the main effects for sleep condition showing significance in the model ($p > 0.05$). Differences in sweetness liking between the SLPs depended on sucralose concentration (sucralose concentration by SLP interaction, F(4,156) = 37.09, $p < 0.001$) (Figure 1). Regardless of the sleep condition, sweet likers reported lower sweet liking ratings for the two lowest concentrations (0.004% *w/w*, 0.011% *w/v*, $p < 0.001$ for both) and higher sweet liking ratings for the two highest concentrations of model sucralose solutions (0.06% *w/v*, 0.094% *w/v* $p < 0.001$ for both), with no difference in liking ratings for the middle concentration (0.032% *w/v*), compared to sweet non-likers, confirming significant differences in hedonic responses between likers and non-likers at low and high concentrations.

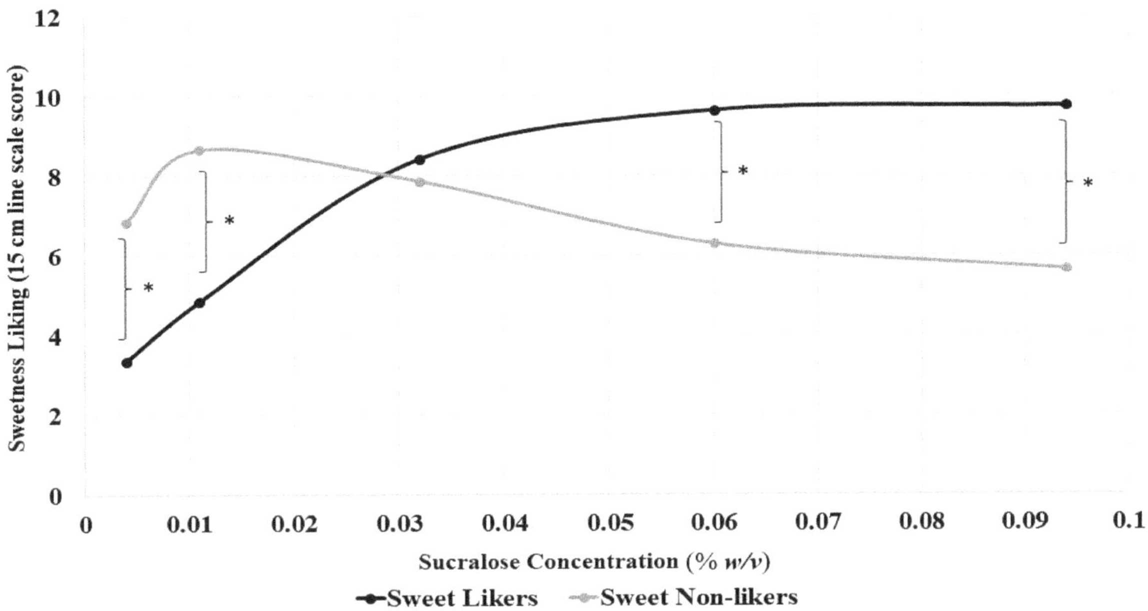


**Figure 1.** Comparison of sweet liking response, averaged across both sleep conditions, by sweet liking phenotype (sweet likers and non-likers) determined using hierarchical cluster analysis based on liking scores over the range of sucralose solutions after a habitual night of sleep. Likers and non-likers showed distinct patterns of liking with sweet likers showing higher sweetness liking at 0.06% and 0.094% *w/v* sucralose and lower sweetness liking at 0.004% and 0.011% *w/v* sucralose, regardless of sleep condition (* $p < 0.001$ for all).

### *3.6. Hedonic Response in the Oat Product Systems*

A four-factor mixed model containing sleep condition, food form, sucralose concentration, and SLP and interactions up to the tertiary level was used to test the primary hypotheses. No tertiary interactions were observed for any oat product models ($p > 0.05$).

3.6.1. Oat Product Sweetness Intensity

Sweetness intensity was measured to confirm previous findings that sleep curtailment does not increase sweet taste intensity perception and to assess whether the products were perceived as iso-sweet at each sucralose concentration across the systems used. It was confirmed that sweet intensity perception was not altered after sleep curtailment, as evidenced by neither the interaction terms nor the main effects for sleep condition showing significance in the model ($p > 0.05$). The second concern, whether iso-sweetness between the products was achieved, was assessed by adding sucralose solution intensity scores to the food form factor and testing the sucralose concentration by food form interaction term in the mixed model. This term was not significant ($F(4,156) = 1.8$, $p = 0.126$), confirming that differences in intensity were similar across the sweetener levels for the food forms and the sucralose (Figure 2). Further, intensity perception did not differ between the SLPs at each sucralose concentration (SLP by sucralose concentration, $F(4,12) = 0.69$, $p = 0.614$), regardless of sleep condition and food form. However, there was a significant main effect of food form effect on sweetness intensity ($F(1,40) = 75.1$, $p < 0.001$), signifying that sweetness was more intense for oat beverage compared to oat crisps regardless of sucralose concentration (Figure 2).

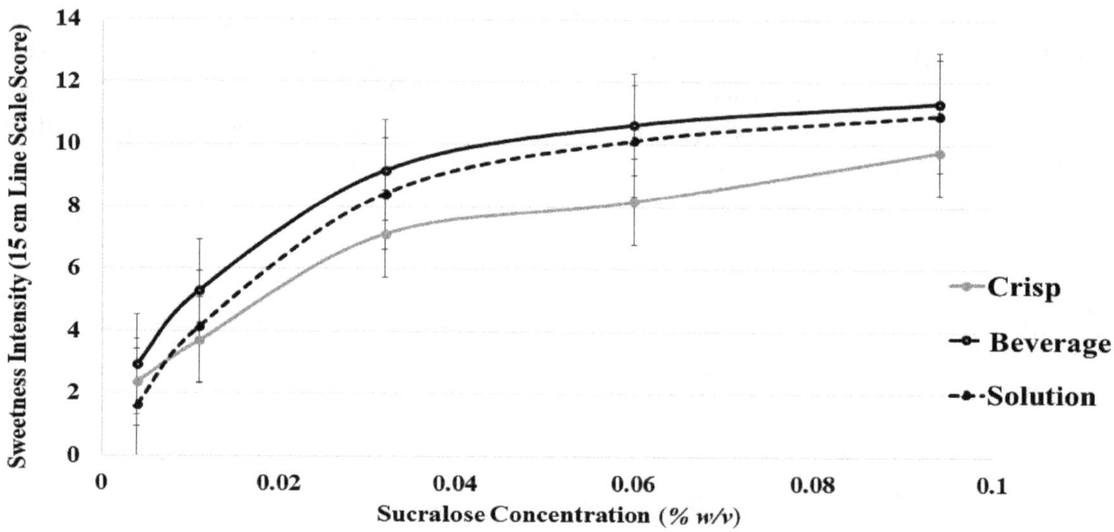


**Figure 2.** Sweet intensity perception over the range of sucralose concentrations for sucralose solutions, oat beverage, and oat crisps. Sweetness intensity was perceived as higher in oat milk compared to oat crisps, regardless of degree of sweetness ($p < 0.001$). The error bars represent the standard error of the mean.

3.6.2. Oat Product Liking Slopes

Oat product liking slopes were analyzed to assess whether sleep curtailment resulted in changes in patterns of hedonic response across a range of sweetness levels. Liking slopes for sweetness liking, flavor liking, texture liking, and overall liking were analyzed using a three-factor mixed model containing sleep condition, food form, and SLP and interactions up to the tertiary level. No tertiary interactions were observed ($p > 0.05$). No significant binary interactions were observed between the factors for overall, sweetness, or flavor slopes ($p > 0.05$). The lack of interactions indicates that the three main effects are independent of one another.

Several main effects were observed across the hedonic attributes measured. First, a main effect of sleep condition was present for the flavor liking slope ($F(1,39) = 11.38$, $p = 0.017$) and the overall liking slope ($F(1,39) = 4.21$, $p = 0.047$), but not for the sweetness or texture liking slope ($p > 0.05$), which demonstrates that overall and flavor liking slopes were steeper after sleep curtailment (Figure 3). The main effect of food form on the slope for overall ($F(1,40) = 5.34$, $p = 0.026$) and sweetness liking $F(1,40) = 9.72$, $p = 0.003$) indicate steeper overall and sweetness liking slopes for the oat crisps compared to the oat beverage regardless of sleep condition. A main effect of food form on slope of flavor liking was not observed. The main effect of SLP on liking slopes was significant for slopes of overall ($F(1,39) = 9.9, p = 0.003$), sweetness liking ($F(1,39) = 12.7, p = 0.001$), and flavor ($F(1,39) = 7.78, p = 0.008$), meaning that positive and negative sweet liking slopes for sweet likers and non-likers, respectively extended to both flavor and overall liking for the oat products.

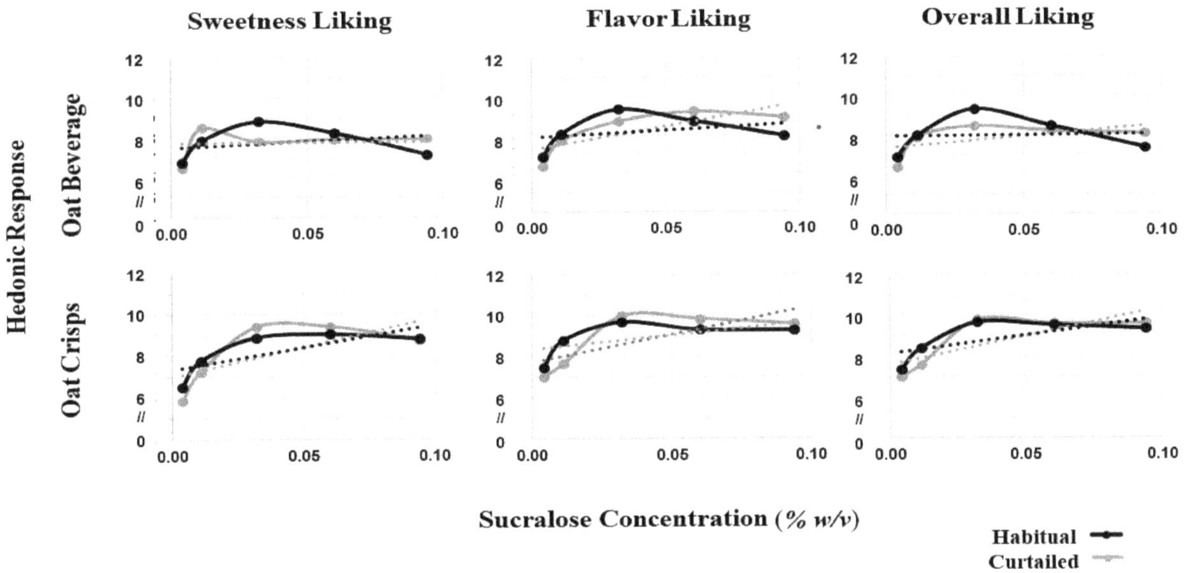


**Figure 3.** Comparisons between liking responses for different hedonic measurements assessed with a 15 cm visual analog scale for the oat crisp and oat beverage. A significant main effect of sleep was observed for both flavor ($p = 0.017$) and overall liking slopes ($p = 0.047$), indicating overall and flavor liking slopes were significantly steeper after curtailment for both oat products. No effect was observed for sweetness. No interaction between sleep condition and food form was observed. A significant food form effect on the overall ($p = 0.026$) and sweetness liking ($p = 0.003$) slopes was observed, indicating a steeper slope for oat crisps compared to oat beverage regardless of sleep condition.

### 3.6.3. Oat Product Liking by Concentration

Similarly to the model system "liking by concentration" analysis, sucralose concentration was added as a five-level fixed factor to the model to assess whether liking varied at specific concentrations or overall (across all concentrations). Sleep condition did not interact with concentration ($p > 0.05$), indicating that sleep curtailment did not produce statistically significant differences in the immediate hedonic response at each concentration. Further, the sleep condition by SLP interaction was not significant for any independent variable in this model, providing further evidence that the effect of sleep curtailment on hedonic response did not depend on SLP.

For sweetness, flavor, and overall liking, no sleep condition by food form interactions was observed ($p > 0.05$). However, a sleep condition by food form interaction was observed for texture ($F(4,156) = 7.5, p = 0.006$), indicating that sleep curtailment resulted in a decrease in texture liking for oat crisps only, regardless of concentration and SLP. Texture liking data for the two oat products were separated and texture liking after a curtailed and habitual night were compared using a reduced mixed model containing only sleep condition and SLP as factors. For the oat beverage, no significant effects of sleep curtailment were observed. For oat crisps, an interaction between sleep condition and SLP was observed ($F(1,39) = 21.16, p < 0.001$), signifying that sweet non-likers were driving differences in texture liking after sleep curtailment. A further analysis revealed a decrease in texture liking in sweet non-likers after sleep curtailment (Habitual: M: 10.5, SD 3.2, Curtailed: M: 9.1, SD: 3.4, $p = 0.021$), but not for sweet likers (Habitual: M: 8.4 SD 2.9, Curtailed: M: 8.8, SD: 3.0, $p > 0.05$).

## 4. Discussion

The primary objective of this study was to determine whether sleep curtailment influences hedonic perception of complex foods, with a focus on sweet taste. The secondary objective was to assess whether these changes are moderated by food form or SLP. Hedonic responses to multiple dimensions of sucralose solutions and sucralose-sweetened liquid and solid oat products were assessed after both a night of habitual sleep and of curtailed sleep. The results from the model solution system are in agreement with our previous findings [21]; preferred sucralose concentration increased and a non-significant increase in liking slope was observed. For the oat products, it was hypothesized that sleep curtailment would result in more pronounced changes in sweet liking and that broader terms such as flavor and overall liking would show significant increases, corresponding with sweetness level. The data partially supports this hypothesis; in oat products, while changes in sweet liking slope were equivalent to model systems, flavor liking and overall liking showed an increase in slope steepness, corresponding with increasing sucralose concentration after sleep curtailment. This finding suggests that participants felt that the products with greater sweetness were holistically preferable to less sweet oat products. Finally, sleep curtailment reduced texture liking of the oat crisps, but not the oat beverage, for sweet non-likers. This suggests a modest effect of sleep on oral somatosensory perception which may only affect sweet non-likers in a solid food model.

In order to sweeten complex foods across a wide range of sweetness levels, high-intensity non-nutritive sweeteners, such as sucralose [61], can be used to minimize collinear changes in texture [62], aroma [63], and appearance [64] that could occur if iso-sweet quantities of sucrose were used. In a previous study, the effect of sleep curtailment on hedonic response to sucrose and sucralose solutions was determined [21]. Sleep curtailment resulted in an increase in preferred sweetener concentration for both sucrose and sucralose. However, while the effect of sleep curtailment on the slope of sweet liking across a range of sweetness concentrations increased for both sweeteners, the increase was only significant for sucrose. This difference indicates that sucralose perception may be affected by sleep curtailment to a lesser extent than sucrose. This discrepancy may be due to reduced reactivity of brain reward centers in response to non-nutritive sweeteners (NNS) [65]. However, the advantage of controlling the non-taste sensory properties in order to isolate taste changes, which is the main purpose of this study, outweighs potential differences in reward processing between nutritive and NNS. In addition, sucralose is used widely and increasingly in the developed world food supply [66], which means that a large portion of the population is exposed to it on a daily basis.

The observed increase in flavor and overall liking of the sweeter versions of each food products may play a role in determining food choice and intake after a night of insufficient sleep and suggest that sweeter products were preferable to less sweet products after sleep curtailment. Given that flavor is often the primary determinant of food choice [67] and that the increase in steepness of the flavor liking slope occurred in tandem with a similar shift in overall liking slope, insufficient sleep shows potential to shape both food choice and food intake through changes in hedonic perception. While preferred sweetener concentration was not measured in the oat products, previous work in model systems found that increased slope steepness occurred in tandem with increases in preferred sucralose and sucrose concentration after sleep curtailment [21] and, therefore, it is likely that participants would have selected sweeter versions of the product to consume given the opportunity, although this should be confirmed by future studies. The current study did not test the effects of sucrose in the food systems, but our previous work in model systems suggests that the effects of insufficient sleep are more pronounced for sucrose compared to sucralose, as the sweet liking slope increased significantly after curtailment for sucrose but not sucralose [21]. This discrepancy between the two sweeteners could be due to differential neural processing of nutritive and non-nutritive sweeteners [21,65], which makes hedonic evaluation of sucralose less susceptible to the effect of sleep curtailment. While both sucrose and sucralose activate higher-order brain reward centers in the brain [68], the magnitude of this activation is greater with sucrose exposure [65]. Thus, the results presented likely underestimate the effects of insufficient sleep on sweet taste hedonic responses where nutritively sweetened foods are concerned. In the case of sucrose-sweetened foods, as sleep-curtailed individuals select sweeter foods, these foods tend to be more energy-dense [13] and more likely to promote weight gain. Thus, the observed change in hedonic perception of complex food in this study may contribute to explaining the well-supported relationship between short sleep and obesity [69].

Due to the fact that the two oat products were not perceived as iso-sweet, directly comparing the two products to each other, especially in the context of hedonic responses over a range of sweetness levels, is not recommended. The oat beverage was perceived as more sweet compared to the oat crisp regardless of sweetness level, although the differences are much lower than what has been previously reported in similar comparisons between model and complex food systems [25,70], where sweetness intensity perception differed by nearly double. The difference in sweetness intensity perception between the products is likely a result of differences in oral processing of liquid and solid food. Liquids are able to fully and rapidly coat the tongue and, therefore, contact greater numbers of taste receptors, whereas, solids must be masticated and may be swallowed before being fully tasted [70]. Sweetness intensity has also been shown to differ significantly in food products of starkly contrasting temperatures (7 °C and 37 °C) [71]; however, others have concluded that temperature has little effect on perceived sweetness intensity [72], and when it does, the differences are very small. Therefore, the relatively modest difference in the serving temperatures of products in this study (7 °C and 23 °C) make it unlikely that temperature played a significant role in the effects observed.

Sleep curtailment negatively affected texture liking but only for oat crisps and only among non-likers. Sleep curtailment may have decreased texture liking of the oat crisps due to increased oro-facial somatosensory sensitivity after sleep curtailment [33], which may make beverages, semi-solid, or "soft" foods more appealing after sleep curtailment compared to "hard" solid foods. Previous studies reported that sleep restriction increased nociceptor reactivity [73] and oral somatosensory sensitivity [30]. While mechanoreceptors in the mouth are likely the primary contributors to the sense of texture, nociceptors also play an important role, particularly in the instance of "intense pressure," which may be experienced when consuming foods which shatter or fracture during mastication [74], as with the oat crisps. Beverages and softer foods require less oral processing time and, therefore, decrease satiety compared to foods that necessitate more oral processing, which may lead to excess energy intake and weight gain [75,76]. Therefore, food form could be one factor that mediates the relationship between insufficient sleep and weight gain.

Why the change in texture liking was restricted to sweet non-likers is not known. However, it could be the case that these individuals have increased attention towards the texture of food. Sweet liking patterns might be a single component within a multifaceted collection of attribute liking patterns which determine an individual's overall liking of a complex food. Overall liking has been described as a function comprised of interactions between hedonic response to individual sensory attributes which are each weighted differently across individuals [77,78]. For example, in one study, while most individuals weighted taste most heavily when considering overall liking, others placed the most importance on texture [77]. It is possible that sweet likers weigh sweetness as a more important factor when considering overall liking and therefore, focus less on other attributes such as texture. This finding suggests that hedonic response to sweet taste may predict hedonic response to other sensory attributes. For example, one study illustrated that a portion of consumers who preferred sweeter chocolate also preferred less cocoa flavor [78]. Furthermore, individual differences in importance placed on specific sensory attributes may moderate the effect of sleep curtailment on food perception, as sleep curtailment affected texture liking for the oat crisps but not the oat beverage. In summary, while SLP does not directly moderate the effect of sleep curtailment on sweet taste, these findings suggest that SLP may be an indicator of other sensory preferences that could be related to changes in food choice after sleep curtailment.

*Strengths and Limitations*

The strengths of this study include the use of novel oat products and sucralose to deliver varying levels of sweetness while minimizing non-sweet sensory differences. The randomized crossover design with a one-week washout period and testing sessions held within 30 min of the previous session on the same day under fasted conditions were also strengths. Additionally, the use of the Zmachine EEG to non-intrusively collect at-home sleep data from participants provided an objective measurement of each sleep condition and confirmation of participant adherence to the prescribed sleep treatment. Limitations of this study include possible fatigue effects from the large sample tasting load per lab visit, the use of sucralose as the sweetener, as opposed to the commonly employed nutritive sweetener, sucrose. Two-minute breaks were instituted between every five samples to minimize fatigue effects. Our findings can only be generalized to foods sweetened with sucralose, which might not represent the primary contributors to weight gain after sleep curtailment. No screening of taste acuity was performed and panelists were not trained to recognize changes in sweetness. Furthermore, the sleep curtailment strategy poses several limitations. Sleep duration in the nights prior to testing was not measured and therefore, the effect of cumulative nights of suboptimal sleep cannot be discounted [79]. Given that participants were not instructed as to how to use time gained by sleep curtailment, differences in the use of light-producing devices during that time may have resulted in differences in circadian rhythm disruption [80]. Finally, the majority of participants in this study were sweet likers, whereas sweet non-likers (comprised of sweet U-shape responders and sweet dislikers) were not well represented. Therefore, it was not possible to compare sweet non-liking phenotypes. A larger sample of sweet U-shape and disliking phenotypes is needed to determine whether these two groups are differentially affected by sleep curtailment. Whether these results apply to individuals with obesity remains to be tested.

## 5. Conclusions

Changes in hedonic responses to both sucralose solutions and sucralose-sweetened oat products were observed after sleep curtailment. In solutions, the sweet liking slope was not significantly increased, but the preferred sucralose concentration was increased after sleep curtailment. In oat products, in agreement with the solution data, the sweetness liking slope did not change, but the slopes of flavor and overall liking functions were steeper after sleep curtailment, corresponding with an increasing sucralose concentration. Given that sucralose concentration and therefore, sweetness, was the only difference between the products, the difference in flavor and overall liking slope suggests that participants felt that the oat products with greater sweetness were holistically preferable. The two SLPs used in this study, likers and non-likers, showed similar changes in hedonic response after sleep curtailment, suggesting that sleep does not differentially affect hedonic responses by phenotype; however, there was one exception. After sleep curtailment, texture liking for sweet non-likers was decreased in oat crisps only, which may point to altered oral somatosensory sensitivity and particular texture salience in sweet non-likers and suggest alternative means by which sleep may influence food choice. These findings represent possible mechanisms by which insufficient sleep leads to weight gain and obesity and signify a possible need to control for the previous night's sleep quality in affective food sensory studies.

**Author Contributions:** Conceptualization, E.J.S. and R.M.T.; methodology, E.J.S., R.M.T., S.C.; formal analysis: E.J.S.; writing—original draft preparation, E.J.S. and R.M.T.; writing—review and editing, E.J.S., R.M.T., S.C.; funding acquisition, R.M.T.

**Acknowledgments:** The authors are grateful to Ellie Haun for her assistance during data collection and CANR Biometry Statistical Consulting Center and Filipe Couto Alves for his guidance in the analysis of the data.

## References


1. Szczygiel, E.J.; Cho, S.; Tucker, R.M. Characterization of the relationships between sleep duration, quality, architecture and chemosensory function in non-obese females. *Chem. Senses* **2018**, *43*, 223–228. [CrossRef] [PubMed]
2. Szczygiel, E.J.; Cho, S.; Snyder, M.K.; Tucker, R.M. Associations between chemosensory function, sweet taste preference, and the previous night's sleep in non-obese males. *Food Qual. Prefer.* **2019**, *75*, 105–112. [CrossRef]
3. Smith, S.L.; Ludy, M.-J.; Tucker, R.M. Changes in taste preference and steps taken after sleep curtailment. *Physiol. Behav.* **2016**, *163*, 228–233. [CrossRef] [PubMed]
4. Lv, W.; Finlayson, G.; Dando, R. Sleep, food cravings and taste. *Appetite* **2018**, *125*, 210–216. [CrossRef] [PubMed]
5. Simon, S.L.; Field, J.; Miller, L.E.; DiFrancesco, M.; Beebe, D.W. Sweet/dessert foods are more appealing to adolescents after sleep restriction. *PLoS ONE* **2015**, *10*, e0115434. [CrossRef]
6. Nedeltcheva, A.V.; Kilkus, J.M.; Imperial, J.; Kasza, K.; Schoeller, D.A.; Penev, P.D. Sleep curtailment is accompanied by increased intake of calories from snacks. *Am. J. Clin. Nutr.* **2009**, *89*, 126–133. [CrossRef]
7. Markwald, R.R.; Melanson, E.L.; Smith, M.R.; Higgins, J.; Perreault, L.; Eckel, R.H.; Wright, K.P. Impact of insufficient sleep on total daily energy expenditure, food intake, and weight gain. *Proc. Natl. Acad. Sci. USA* **2013**, *110*, 5695–5700. [CrossRef]
8. Hanlon, E.C.; Andrzejewski, M.E.; Harder, B.K.; Kelley, A.E.; Benca, R.M. The effect of REM sleep deprivation on motivation for food reward. *Behav. Brain Res.* **2005**, *163*, 58–69. [CrossRef]
9. Greer, S.M.; Goldstein, A.N.; Walker, M.P. The impact of sleep deprivation on food desire in the human brain. *Nat. Commun.* **2013**, *4*, 2259. [CrossRef]
10. Benedict, C.; Brooks, S.J.; O'Daly, O.G.; Almèn, M.S.; Morell, A.; Åberg, K.; Gingnell, M.; Schultes, B.; Hallschmid, M.; Broman, J.-E.; et al. Acute sleep deprivation enhances the brain's response to hedonic food stimuli: An fMRI study. *J. Clin. Endocrinol. Metab.* **2012**, *97*, E443–E447. [CrossRef]
11. Demos, K.E.; Sweet, L.H.; Hart, C.N.; McCaffery, J.M.; Williams, S.E.; Mailloux, K.A.; Trautvetter, J.; Owens, M.M.; Wing, R.R. The effects of experimental manipulation of sleep duration on neural response to food cues. *Sleep* **2017**, *40*, zsx125. [CrossRef] [PubMed]
12. St-Onge, M.-P.; Wolfe, S.; Sy, M.; Shechter, A.; Hirsch, J. Sleep restriction increases the neuronal response to unhealthy food in normal-weight individuals. *Int. J. Obes.* **2014**, *38*, 411–416. [CrossRef] [PubMed]
13. Drewnowski, A. Intense sweeteners and energy density of foods: Implications for weight control. *Eur. J. Clin. Nutr.* **1999**, *53*, 757–763. [CrossRef] [PubMed]
14. Drewnowski, A.; Mennella, J.A.; Johnson, S.L.; Bellisle, F. Sweetness and food preference. *J. Nutr.* **2012**, *142*, 1142S–1148S. [CrossRef] [PubMed]
15. Yamamoto, T. Brain Mechanisms of sweetness and palatability of sugars. *Nutr. Rev.* **2003**, *61*, S5–S9. [CrossRef] [PubMed]
16. Chen, X.; Gelaye, B.; Williams, M.A. Sleep characteristics and health-related quality of life among a national sample of American young adults: Assessment of possible health disparities. *Qual. Life Res.* **2014**, *23*, 613–625. [CrossRef] [PubMed]
17. Hales, C.M.; Carroll, M.D.; Fryar, C.D.; Ogden, C.L. *Prevalence of Obesity among Adults and Youth: United States, 2015–2016*; U.S Department of Health & Human Services: Atlanta, GA, USA, 2017.
18. Furihata, R.; Uchiyama, M.; Suzuki, M.; Konno, C.; Konno, M.; Takahashi, S.; Kaneita, Y.; Ohida, T.; Akahoshi, T.; Hashimoto, S.; et al. Association of short sleep duration and short time in bed with depression: A Japanese general population survey. *Sleep Biol. Rhythm.* **2015**, *13*, 136–145. [CrossRef]
19. Lin, C.-L.; Lin, C.-P.; Chen, S.-W.; Wu, H.-C.; Tsai, Y.-H. The association between sleep duration and overweight or obesity in Taiwanese adults: A cross-sectional study. *Obes. Res. Clin. Pract.* **2018**, *12*, 384–388. [CrossRef]
20. Hogenkamp, P.S.; Nilsson, E.; Chapman, C.D.; Cedernaes, J.; Vogel, H.; Dickson, S.L.; Broman, J.-E.; Schiöth, H.B.; Benedict, C. Sweet taste perception not altered after acute sleep deprivation in healthy young men. *Somnologie* **2013**, *17*, 111–114. [CrossRef]
21. Szczygiel, E.J.; Cho, S.; Tucker, R.M. Multiple dimensions of sweet taste perception altered after sleep curtailment. *Nutrients* **2019**, *11*, 2015. [CrossRef]

22. Tanaka, T.; Hong, G.; Tominami, K.; Kudo, T. Oral fat sensitivity is associated with social support for stress coping in young adult men. *Tohoku J. Exp. Med.* **2018**, *244*, 249–261. [CrossRef] [PubMed]
23. Huber, J. *The Psychophysics of Taste: Perceptions of Bitterness and Sweetness in Iced Tea*; Advances in Consumer Research: Ann Arbor, MI, USA, 1974; Volume NA-01, pp. 166–181.
24. Mazur, J.; Drabek, R.; Goldman, A. Hedonic contrast effects in multi-product food evaluations differing in complexity. *Food Qual. Prefer.* **2018**, *63*, 159–167. [CrossRef]
25. Drewnowski, A.; Shrager, E.E.; Lipsky, C.; Stellar, E.; Greenwood, M.R.C. Sugar and fat: Sensory and hedonic evaluation of liquid and solid foods. *Physiol. Behav.* **1989**, *45*, 177–183. [CrossRef]
26. Tan, S.-Y.; Tucker, R.M. Sweet taste as a predictor of dietary intake: A systematic review. *Nutrients* **2019**, *11*, 94. [CrossRef] [PubMed]
27. Dhillon, J.; Running, C.A.; Tucker, R.M.; Mattes, R.D. Effects of food form on appetite and energy balance. *Food Qual. Prefer.* **2016**, *48*, 368–375. [CrossRef]
28. Gujar, N.; Yoo, S.-S.; Hu, P.; Walker, M.P. Sleep deprivation amplifies reactivity of brain reward networks, biasing the appraisal of positive emotional experiences. *J. Neurosci.* **2011**, *31*, 4466–4474. [CrossRef] [PubMed]
29. Krause, A.J.; Simon, E.B.; Mander, B.A.; Greer, S.M.; Saletin, J.M.; Goldstein-Piekarski, A.N.; Walker, M.P. The sleep-deprived human brain. *Nat. Rev. Neurosci.* **2017**, *18*, 404. [CrossRef]
30. Kamiyama, H.; Iida, T.; Nishimori, H.; Kubo, H.; Uchiyama, M.; Laat, A.D.; Lavigne, G.; Komiyama, O. Effect of sleep restriction on somatosensory sensitivity in the oro-facial area: An experimental controlled study. *J. Oral Rehabil.* **2019**, *46*, 303–309. [CrossRef]
31. Schloegl, H.; Percik, R.; Horstmann, A.; Villringer, A.; Stumvoll, M. Peptide hormones regulating appetite—Focus on neuroimaging studies in humans. *Diabetes/Metab. Res. Rev.* **2011**, *27*, 104–112. [CrossRef]
32. Auvray, M.; Spence, C. The multisensory perception of flavor. *Conscious. Cogn.* **2008**, *17*, 1016–1031. [CrossRef]
33. Roehrs, T.; Hyde, M.; Blaisdell, B.; Greenwald, M.; Roth, T. Sleep loss and REM sleep loss are Hyperalgesic. *Sleep* **2006**, *29*, 145–151. [CrossRef] [PubMed]
34. Barr, R.G.; Pantel, M.S.; Young, S.N.; Wright, J.H.; Hendricks, L.A.; Gravel, R. The response of crying newborns to sucrose: Is it a "sweetness" effect? *Physiol. Behav.* **1999**, *66*, 409–417. [CrossRef]
35. Iatridi, V.; Hayes, J.E.; Yeomans, M.R. Reconsidering the classification of sweet taste liker phenotypes: A methodological review. *Food Qual. Prefer.* **2019**, *72*, 56–76. [CrossRef]
36. Iatridi, V.; Hayes, J.E.; Yeomans, M.R. Quantifying sweet taste liker phenotypes: Time for some consistency in the classification criteria. *Nutrients* **2019**, *11*, 129. [CrossRef] [PubMed]
37. Yeomans, M.R.; Tepper, B.J.; Rietzschel, J.; Prescott, J. Human hedonic responses to sweetness: Role of taste genetics and anatomy. *Physiol. Behav.* **2007**, *91*, 264–273. [CrossRef]
38. Asao, K.; Miller, J.; Arcori, L.; Lumeng, J.C.; Han-Markey, T.; Herman, W.H. Patterns of sweet taste liking: A pilot study. *Nutrients* **2015**, *7*, 7298–7311. [CrossRef]
39. Kim, J.-Y.; Prescott, J.; Kim, K.-O. Patterns of sweet liking in sucrose solutions and beverages. *Food Qual. Prefer.* **2014**, *36*, 96–103. [CrossRef]
40. Mennella, J.A.; Pepino, M.Y.; Reed, D.R. Genetic and environmental determinants of bitter perception and sweet preferences. *Pediatrics* **2005**, *115*, e216–e222. [CrossRef]
41. Bachmanov, A.A.; Bosak, N.P.; Floriano, W.B.; Inoue, M.; Li, X.; Lin, C.; Murovets, V.O.; Reed, D.R.; Zolotarev, V.A.; Beauchamp, G.K. Genetics of sweet taste preferences. *Flavour Fragr. J.* **2011**, *26*, 286–294. [CrossRef]
42. Garneau, N.L.; Nuessle, T.M.; Mendelsberg, B.J.; Shepard, S.; Tucker, R.M. Sweet liker status in children and adults: Consequences for beverage intake in adults. *Food Qual. Prefer.* **2018**, *65*, 175–180. [CrossRef]
43. Holt, S.H.A.; Cobiac, L.; Beaumont-Smith, N.E.; Easton, K.; Best, D.J. Dietary habits and the perception and liking of sweetness among Australian and Malaysian students: A cross-cultural study. *Food Qual. Prefer.* **2000**, *11*, 299–312. [CrossRef]
44. Wiet, S.G.; Beyts, P.K. Sensory characteristics of sucralose and other high intensity sweeteners. *J. Food Sci.* **1992**, *57*, 1014–1019. [CrossRef]
45. Kim, M.-Y.; Cho, H.-Y.; Park, J.-Y.; Lee, S.-M.; Suh, D.-S.; Chung, S.-J.; Kim, H.-S.; Kim, K.-O. Relative sweetness of sucralose in beverage systems and sensory properties of low calorie beverages containing sucralose. *Korean J. Food Sci. Technol.* **2005**, *37*, 425–430.

46. Reis, F.; De Andrade, J.; Deliza, R.; Ares, G. Comparison of two methodologies for estimating equivalent sweet concentration of high-intensity sweeteners with untrained assessors: Case study with orange/pomegranate juice. *J. Sens. Stud.* **2016**, *31*, 341–347. [CrossRef]
47. Dinges, D.F.; Pack, F.; Williams, K.; Gillen, K.A.; Powell, J.W.; Ott, G.E.; Aptowicz, C.; Pack, A.I. Cumulative sleepiness, mood disturbance, and psychomotor vigilance performance decrements during a week of sleep restricted to 4–5 hours per night. *Sleep* **1997**, *20*, 267–277.
48. Buysse, D.J.; Reynolds, C.F.; Monk, T.H.; Berman, S.R.; Kupfer, D.J. The Pittsburgh sleep quality index: A new instrument for psychiatric practice and research. *Psychiatry Res.* **1989**, *28*, 193–213. [CrossRef]
49. Cohen, S.; Kamarck, T.; Mermelstein, R. A Global measure of perceived stress. *J. Health Soc. Behav.* **1983**, *24*, 385–396. [CrossRef]
50. Cepeda-Benito, A.; Gleaves, D.H.; Williams, T.L.; Erath, S.A. The development and validation of the state and trait food-cravings questionnaires. *Behav. Ther.* **2000**, *31*, 151–173. [CrossRef]
51. Meule, A.; Kübler, A. Double trouble. Trait food craving and impulsivity interactively predict food-cue affected behavioral inhibition. *Appetite* **2014**, *79*, 174–182. [CrossRef]
52. Kaplan, R.F.; Wang, Y.; Loparo, K.A.; Kelly, M.R.; Bootzin, R.R. Performance evaluation of an automated single-channel sleep–wake detection algorithm. *Nat. Sci. Sleep* **2014**, *6*, 113–122. [CrossRef]
53. Kaida, K.; Takahashi, M.; Åkerstedt, T.; Nakata, A.; Otsuka, Y.; Haratani, T.; Fukasawa, K. Validation of the Karolinska sleepiness scale against performance and EEG variables. *Clin. Neurophysiol.* **2006**, *117*, 1574–1581. [CrossRef]
54. Watson, D.; Clark, L.A.; Tellegen, A. Development and validation of brief measures of positive and negative affect: The PANAS scales. *J. Personal. Soc. Psychol.* **1988**, *54*, 1063–1070. [CrossRef]
55. Merrill, E.P.; Kramer, F.M.; Cardello, A.; Schutz, H. A comparison of satiety measures. *Appetite* **2002**, *39*, 181–183. [CrossRef]
56. Franzen, P.L.; Siegle, G.J.; Buysse, D.J. Relationships between affect, vigilance, and sleepiness following sleep deprivation. *J. Sleep Res.* **2008**, *17*, 34–41. [CrossRef]
57. Noel, C.; Dando, R. The effect of emotional state on taste perception. *Appetite* **2015**, *95*, 89–95. [CrossRef]
58. Popper, R.; Rosenstock, W.; Schraidt, M.; Kroll, B.J. The effect of attribute questions on overall liking ratings. *Food Qual. Prefer.* **2004**, *15*, 853–858. [CrossRef]
59. Mennella, J.A.; Lukasewycz, L.D.; Griffith, J.W.; Beauchamp, G.K. Evaluation of the Monell forced-choice, paired-comparison tracking procedure for determining sweet taste preferences across the lifespan. *Chem. Senses* **2011**, *36*, 345–355. [CrossRef]
60. Glickman, M.E.; Rao, S.R.; Schultz, M.R. False discovery rate control is a recommended alternative to Bonferroni-type adjustments in health studies. *J. Clin. Epidemiol.* **2014**, *67*, 850–857. [CrossRef]
61. Binns, N.M. Sucralose—All sweetness and light. *Nutr. Bull.* **2003**, *28*, 53–58. [CrossRef]
62. Cheer, R.L.; Lelievre, J. Effects of sucrose on the rheological behavior of wheat starch pastes. *J. Appl. Polym. Sci.* **1983**, *28*, 1829–1836. [CrossRef]
63. Van Boekel, M.A.J.S. Formation of flavour compounds in the Maillard reaction. *Biotechnol. Adv.* **2006**, *24*, 230–233. [CrossRef] [PubMed]
64. Ashoor, S.H.; Zent, J.B. Maillard browning of common amino acids and sugars. *J. Food Sci.* **1984**, *49*, 1206–1207. [CrossRef]
65. Frank, G.K.W.; Oberndorfer, T.A.; Simmons, A.N.; Paulus, M.P.; Fudge, J.L.; Yang, T.T.; Kaye, W.H. Sucrose activates human taste pathways differently from artificial sweetener. *NeuroImage* **2008**, *39*, 1559–1569. [CrossRef] [PubMed]
66. Sylvetsky, A.C.; Rother, K.I. Trends in the consumption of low-calorie sweeteners. *Physiol. Behav.* **2016**, *164*, 446–450. [CrossRef] [PubMed]
67. MacFie, H.J.H.; Meiselman, H.L. *Food Choice, Acceptance and Consumption*; Springer Science & Business Media: New York, NY, USA, 2012; ISBN 978-1-46-131221-5.
68. Green, E.; Murphy, C. Altered processing of sweet taste in the brain of diet soda drinkers. *Physiol. Behav.* **2012**, *107*, 560–567. [CrossRef] [PubMed]
69. Cappuccio, F.P.; Taggart, F.M.; Kandala, N.-B.; Currie, A.; Peile, E.; Stranges, S.; Miller, M.A. Meta-analysis of short sleep duration and obesity in children and adults. *Sleep* **2008**, *31*, 619–626. [CrossRef] [PubMed]
70. Alley, R.L.; Alley, T.R. The influence of physical state and color on perceived sweetness. *J. Psychol.* **1998**, *132*, 561–568. [CrossRef] [PubMed]

71. Calvino, A.M. Perception of sweetness: The effects of concentration and temperature. *Physiol. Behav.* **1986**, *36*, 1021–1028. [CrossRef]
72. Schiffman, S.S.; Sattely-Miller, E.A.; Graham, B.G.; Bennett, J.L.; Booth, B.J.; Desai, N.; Bishay, I. Effect of temperature, pH, and ions on sweet taste. *Physiol. Behav.* **2000**, *68*, 469–481. [CrossRef]
73. Kundermann, B.; Spernal, J.; Huber, M.T.; Krieg, J.-C.; Lautenbacher, S. Sleep deprivation affects thermal pain thresholds but not somatosensory thresholds in healthy volunteers. *Psychosom. Med.* **2004**, *66*, 932–937. [CrossRef]
74. Engelen, L.; Bilt, A.V.D. Oral physiology and texture perception of semisolids. *J. Texture Stud.* **2008**, *39*, 83–113. [CrossRef]
75. De Graaf, C. Why liquid energy results in overconsumption. *Proc. Nutr. Soc. USA* **2011**, *70*, 162–170. [CrossRef]
76. James, B. Oral processing and texture perception influences satiation. *Physiol. Behav.* **2018**, *193*, 238–241. [CrossRef]
77. Moskowitz, H.R.; Krieger, B. The contribution of sensory liking to overall liking: An analysis of six food categories. *Food Qual. Prefer.* **1995**, *6*, 83–90. [CrossRef]
78. De Kermadec, F.H.; Durand, J.F.; Sabatier, R. Comparison between linear and nonlinear PLS methods to explain overall liking from sensory characteristics. *Food Qual. Prefer.* **1997**, *8*, 395–402. [CrossRef]
79. Van Dongen, H.P.A.; Maislin, G.; Mullington, J.M.; Dinges, D.F. The cumulative cost of additional wakefulness: Dose-response effects on neurobehavioral functions and sleep physiology from chronic sleep restriction and total sleep deprivation. *Sleep* **2003**, *26*, 117–126. [CrossRef]
80. Wever, R.A.; Polášek, J.; Wildgruber, C.M. Bright light affects human circadian rhythms. *Pflug. Arch.* **1983**, *396*, 85–87. [CrossRef]

# 12

# Factors Influencing Purchase Intention for Low-Sodium and Low-Sugar Products

**Yeowoon Park [1], Dongmin Lee [2], Seoyoung Park [1] and Junghoon Moon [1,*]**

1 Department of Agricultural Economics & Rural Development, Seoul National University, Seoul 08826, Korea; qkrdudns1101@snu.ac.kr (Y.P.); syp1130@gmail.com (S.P.)

2 Department of Food Processing and Distribution, Gangneung-Wonju National University, Gangneung 25457, Korea; dongminlee@gwnu.ac.kr

* Correspondence: moonj@snu.ac.kr.

**Abstract:** As sodium and sugar intake in South Korea exceed recommended levels, the government and food industry have been attempting to reduce the amount of sodium and sugar in the food products. In line with these efforts, this study sought to examine how the purchase intention for low-sodium/low-sugar products vary based on consumers' previous choices of low-sodium/low-sugar products and other consumer-related factors. For this study, two online survey-based experiments were conducted: one using soy sauce to represent a sodium-based product and the other using yogurt to represent a sugar-based product. The significant variables that influenced the purchase intention for both were the consumers' previous low-sodium/low-sugar product choices and their propensity for food neophobia. Consumers who had previously selected regular products showed a lower intention to purchase low-sodium soy sauce or low-sugar yogurt. In addition, those who had a strong tendency toward food neophobia also had a significantly lower purchase intention for these products. Moreover, the lower the consumer's unhealthy = tasty intuition (UTI), the higher the purchase intention for the low-sodium soy sauce, but UTI did not act as a significant variable for the low-sugar yogurt. These results demonstrate that government interventions for low-sodium products and low-sugar products should be differentiated.

**Keywords:** cognitive dissonance theory; unhealthy = tasty intuition; food neophobia; low-sodium; low-sugar

---

## 1. Introduction

Sodium is an essential nutrient for the maintenance of human health. The recommended minimum daily intake of sodium is 500 mg [1]. However, excessive sodium intake has been identified as a health risk as it can lead to health problems such as stroke and cardiovascular disease [2]. The issue of excessive sodium intake is particularly problematic in South Korea since it has one of the highest sodium consumption rates in the world [1].

Similarly, while sugar has now become a major component of our daily nutrition [3], the excessive consumption of sugar is associated with the risk of weight gain and cardiometabolic problems [4]. And excessive sugar intake, as well as sodium intake is also a health problem in South Korea as the sugar-sweetened beverage (SSB) consumption rate among adults increased from 66% to 69% between 2001 and 2009 [5].

In response to these situations, the food industry has been reducing the sodium and sugar contents of products to meet government health guidelines [6]. However, food reformulation efforts alone are insufficient for inducing healthy consumption patterns among consumers. Depending on external and consumer-related factors, the results of these efforts can vary. Previous studies have shown that variables such as age, gender, income, health (whether they smoke, whether they have certain diseases, etc.), special diet, and price sensitivity have a significant impact on consumer attitudes toward low-sodium or-low-sugar products [7–12].

Therefore, this study investigates how consumers' purchase intention for low-sodium/low-sugar products differs depending on their previous choices, level of unhealthy = tasty intuition (UTI), and food neophobia. As stimuli, we use soy sauce products to represent a sodium-based food category and yogurt products to represent a sugar-based food category. In comparison to regular products, we assume that the low-sodium and low-sugar versions of both the soy sauce and yogurt are relatively new to consumers, while the regular versions are familiar to consumers in South Korea.

## 2. Theoretical Framework

### 2.1. Cognitive Dissonance Theory

According to Festinger's dissonance theory, two cognitions that are related to each other can be either consonant or dissonant. If the two cognitions are consonant, then one follows consistently with the other. However, if the two cognitions are dissonant, the opposite of one cognition is consistent with the other. When dissonance occurs, an individual experiences psychological discomfort, and in order to reduce this discomfort, the individual is motivated to avoid any information that is likely to aggravate this dissonance [13]. Some of the important research generated by cognitive dissonance theory can be classified as based on the free-choice paradigm, belief-disconfirmation paradigm, effort-justification paradigm, and induced-compliance paradigm [14].

Among these paradigms, we will focus on the free-choice paradigm and belief-disconfirmation paradigm in this study. According to the free-choice paradigm, the choice between similarly valued alternatives triggers a psychological tension mediated by the preferred aspects of the rejected alternative and the unpreferred aspects of the selected alternative [13]. This tension is relieved by reassessing the options after the choice has been made [15,16]. The concept that choice shapes preference has been widely accepted by many researchers for many years [17].

This idea was demonstrated in 1956 by an experiment conducted by psychologist Jack Brehm, and the results have been replicated numerous times since [18,19]. In Brehm's experiment, participants were required to select one out of two similarly valued items and then evaluate the selected item. as a result, they rated the selected item as better than they did initially, while they rated the rejected item as worse [20].

According to the belief-disconfirmation paradigm, dissonance occurs when people are exposed to information that does not match their beliefs. If the dissonance is not diminished by altering one's own beliefs, then the dissonance can lead to the rejection or refutation of the new information [14].

Following dissonance theory, the prior choice of a particular soy sauce or yogurt product will lead to a preference for this previously selected product. Therefore, even if new information is provided later regarding a low-sodium or low-sugar alternative, those who have already chosen a regular product will disregard this new information as it is dissonant to their existing preference. By referring to previous research that has stated that preference is almost identical to purchase intention and is a good predictor of it [21], it can be predicted that:

- H1a: Consumers who once chose a regular soy sauce will have a lower purchase intention for a low-sodium soy sauce regardless of additional information about the low-sodium soy sauce product.
- H1b: Consumers who once chose a regular yogurt will have a lower purchase intention for a low-sugar yogurt regardless of additional information about the low-sugar yogurt product.

### 2.2. Unhealthy = Tasty Intuition (UTI)

Many researchers have posited and identified the positive effects of providing consumers with easily identified healthy foods [22–24]. However, not all consumers will necessarily be more likely to buy certain products because they seem healthier [25]. For example, some consumers will not choose healthier products over less healthy ones within the same food categories. For those consumers, a food with relatively healthy names, such as salad are perceived as less tasty than the same food with relatively unhealthy names, such as pasta [26].

Connected to this rationale is the UTI that some people hold [27]. According to Raghunathan et al., UTI is a belief held by consumers that the healthiness and tastiness of a food product have a negative correlation. In other words, when information is provided to consumers indicating that a product is healthy, the worse they infer its taste. There are two main sources for this belief. First, this intuition is generated internally from a more general belief that there is an inverse relationship between utilitarian and hedonic values. Along with this internal source of UTI, mass media and personal communications instill consumers with views that are compatible with this intuition.

However, the relationship between UTI and its consequences may be more complicated than we think [26]. Furthermore, it is also important to highlight that the degree of UTI varies among consumers. Previous studies have indicated that there are consumers who are less likely to believe that unhealthy foods are tastier [26,28]. For instance, health-conscious consumers are less likely to base food decisions on UTI because they may place greater importance on health over taste [29,30]. To illustrate this, the taste expectations of dieters for healthy foods are more positive than those of non-dieters [26]. In fact, one previous study found an intuition opposite to UTI that exists in France in which healthy foods are believed to be tastier than unhealthy foods [31].

in the present study, UTI is indicated by the consumer tendency to evaluate healthy food as less tasty when given product information (nutrient contents, labels on the package, etc.). In this study, we consider healthy foods as those that are low in sodium or sugar. Therefore, it can be assumed that consumers with a lower UTI will have a higher purchase intention for low-sodium/low-sugar products. Thus, it is predicted that:

- H2a: Consumers with a lower UTI will have a higher purchase intention for a low-sodium soy sauce product.
- H2b: Consumers with a lower UTI will have a higher purchase intention for a low-sugar yogurt product

*2.3. Food Neophobia*

Underlying the relationship between a change in the taste of a product and the consumer acceptance of it is the feeling that consumers have when they are unfamiliar with the changed taste. In this context, the concept of food neophobia can be described as a consumer reluctance or unwillingness to try an unfamiliar product [32]. People expect novel foods to be unfavorable or even dangerous as compared to familiar ones, and this tendency lowers their willingness to try novel foods [33].

One of the ways to reduce food neophobia is to expose individuals to and increase familiarity with unfamiliar foods [34–36]. However, since the degree to which people are exposed to novel foods varies, attempts to reduce food neophobia by increasing familiarity or exposure should be made with caution [37]. Previous studies have revealed that there is a positive relationship between age and food neophobia [32,38,39]. Other variables, such as culture, education, and income, can also influence the degree of individual food neophobia [38,40–43]

In addition, the tendency for food neophobia can vary from person to person [32,44]. While food neophobics tend to reject and avoid unfamiliar foods without even trying them [45], food neophilics enjoy trying unknown foods [46,47].

in the food industry, it is sometimes inevitable that alterations in flavor will result from changes in the sodium/sugar content of a food product, and consumers must make a hedonic adjustment to the change in flavor in order to become accustomed to the reformulated (low-sodium/low-sugar) product [48]. as a result, this could influence the long-term consumer acceptance of a product [49].

Marketing strategies related to increasing consumer familiarity with novel foods have shown the potential to introduce these products into the market successfully [50]. Therefore, in order to ensure the market success of reformulated food products, such as with sodium or sugar-reduced foods, it is imperative to verify the purchase intentions of consumers for these products in consideration of their level of food neophobia. Therefore, we suggest the following hypotheses:

- H3a: Consumers with a lower tendency toward food neophobia will have a higher purchase intention for a low-sodium soy sauce product.
- H3b: Consumers with a lower tendency toward food neophobia will have a higher purchase intention for a low-sugar yogurt product.

## 3. Materials and Methods

### 3.1. Stimuli Materials

The two products chosen for the study were (1) soy sauce as a sodium-based product and (2) yogurt as a sugar-based product (Appendix A). The experiments were designed to reflect all the brands of products sold in the actual markets to increase external validity. To do so, pictures of the real products, as well as their information (brand, price, and volume) were given to the participants at the early stage of the experiment. Soy sauce was selected as a stimulus in the first experiment and used as a representative product for the sodium-based product category because it has traditionally been used in South Korea and is assumed to be one of the main reasons why Korea's sodium consumption is one of the highest in the world [1]. Although soy sauce is recognized worldwide for its many health benefits as a fermented food, it also has the potential to cause high blood pressure and diabetes since it accounts for 20.8% of the daily sodium consumption in Korea [51]. In the case of sugar intake, yogurt is selected as a stimulus for this study because it is one of Korea's major sugar sources, according to the Korea Ministry of Food and Drug Safety [52]. Therefore, yogurt will be presented to experiment participants as a sugar-based product in the second experiment.

In order to select the soy sauce products for this study, brands with the highest sales were selected based on a report from the Korea Agro-Fisheries and Food Trade Corporation [53]. For the low-sugar yogurt products, on the other hand, the products with the highest sales volume on the E-Mart Mall, one of the largest Korean retail chains, were selected.

### 3.2. Participants

The effects of consumers' previous choices, UTI, and food neophobia on the purchase intention for a low-sodium soy sauce product and a low-sugar yogurt product were studied among Korean housewives. Housewives were selected as the subject of our study because they are the main decision-makers for soy sauce and yogurt purchases in a market shopping context.

The data was collected from an online panel established by a leading online research company in the winter of 2018. Ages of the members of the panel ranged from their 20s to 50s and were all legally capable of consenting to participation. Participants of this study were at first gathered by answering the e-mail whether to participate in the study.

in the soy sauce experiment, respondents were asked whether they had bought soy sauce in the prior 6 months in order to screen out respondents who do not have recent experience of purchasing soy sauce products. Of the 2989 respondents, 1913 answered that they have experiences of buying soy sauce products in the prior 6 months. Then, 1913 respondents, who have bought soy sauce in the prior 6 months, were asked to choose one of the choice-set of regular soy sauce and low-sodium soy sauce products. Of the 1913 respondents, 1628 (85.10%) had chosen a regular soy sauce product and 285 (14.90%) had chosen a low-sodium soy sauce product. For the convenience and balance of the demographic composition of the two sample groups, however, 233 respondents out of 285 who had selected the low-sodium product and 233 out of 1628 respondents who had selected the regular product (total 466) were sampled and studied. The number of samples for each group was assigned to distribute the demographic data, such as age, evenly. The average age of the 466 respondents was 39.4 years old (SD = 10.8).

Likewise, for the yogurt experiment, respondents were also asked whether they had bought yogurt in the prior 6 months. 1070 out of 1401 answered that they have experiences of buying yogurt products in the prior 6 months. Of the 1070 respondents, 662 (61.87%) had chosen a regular yogurt product and 408 (38.13%) had chosen a low-sugar yogurt product. Among these, 239 out of 662 respondents who had selected a low-sugar yogurt product and 246 out of 1070 respondents who had selected a regular yogurt product (total 485) were sampled and studied. The number of samples for each group was assigned for the same reason as in the first experiment. The average age of the participants was 39.4 years old (SD = 10.6).

### 3.3. Procedure

Both experiments were performed by an online survey. Prior to the survey, all of the respondents were told that the data and opinions gathered through the survey were confidential and to be used for research purposes only, following Section 33 of the Korean statistical law. Only participants who gave consent to participate in the survey pressed the "Next" button. Gathered data of the participants were non-identifiable and stored in the company's database for 1 year.

The survey of the soy sauce experiment was performed based on the scenario that the respondent had to buy soy sauce at the market for preparing meals. The 466 respondents were provided with images of eight soy sauce products currently sold at the market, with each including brief information on its volume, price, and unit price. Among the eight products, two were low-sodium and six were regular products, and respondents were asked to choose one.

A similar design and procedure were applied in the yogurt experiment. The survey was also carried out based on the scenario that the respondent had to buy yogurt at the market. The 485 respondents were provided with images of seven yogurt products currently sold at the market, with each including brief information on its volume, price, and unit price. Among the seven products, three were low-sugar and four were regular products, and respondents were asked to choose one.

All of the respondents from both experiments were asked about their intention to purchase a low-sodium/low-sugar product. The study aimed to see how the sodium/sugar content of the product affects the consumer's purchase intention of the product by providing them with additional information about low-sodium/sugar products immediately after the first choice of the participants. Those who had selected a low-sodium/low-sugar product in each experiment were given the information that the product they chose was low in sodium/sugar and were asked to evaluate their intention to purchase the product. on the other hand, those who had selected a regular product in each experiment were shown pictures and information for low-sodium/low-sugar products and were asked to evaluate their intention to purchase these products. The quality of the information given to both respondents in each experiment was controlled to be the same.

Next, the respondents from both experiments completed questionnaires regarding their UTI and food neophobia. Finally, basic demographic and personal information about the participants were taken in order to control the main independent variables. Information was collected about their interest in diet, family size, whether they have a family member suffering from atopy, high blood pressure, diabetes, or cancer, income, how often they cook at home, education level, and age.

### 3.4. Measures

Purchase intention was the main dependent variable in the study. To measure the purchase intention of a low-sodium/low-sugar product for each experiment, three items were rated on a 5-point scale (1 = "very low" and 5 = "very high"; Table 1) [54].

**Table 1.** Items used in the purchase intention (PI) scale.

| PI | Description |
|---|---|
| PI 1 | I am considering purchasing the product. |
| PI 2 | I will purchase the product as soon as possible. |
| PI 3 | if there is a chance, I have the intention to buy the product. |

There were three main independent variables in this study: previous choice, UTI, and food neophobia. First, in order to measure the respondents' previous choice, we divided the respondents into two groups based on whether they had chosen a low-sodium/low-sugar product or not. According to their choice, we assigned a dummy variable (PC) (Table 8).

Next, to measure the participants' intuitive belief that the less healthy the food, the better its inferred taste, that is the UTI, they were asked to rate three items [30] on a 5-point scale (1 = "strongly disagree" and 5 = "strongly agree"; Table 2).

**Table 2.** Items used in the unhealthy = tasty intuition (UTI) scale.

| UTI | Description |
|---|---|
| UTI 1 | Things that are good for me rarely taste good. |
| UTI 2 | There is no way to make food healthier without sacrificing taste. |
| UTI 3 | Healthy food is usually less tasty. |

Finally, the participants answered 10 items for a food neophobia scale (FNS) [32] to measure their reluctance to eat and/or avoidance of novel foods based on a 5-point scale (1 = "strongly disagree" and 5 = "strongly agree"; Table 3). The higher the FNS score, the weaker the consumer's food neophobia, and the lower the FNS score, the stronger the food neophobia. All items were written in Korean.

**Table 3.** Items used in the food neophobia scale (FNS).

| FNS | Description |
|---|---|
| FNS 1 | I am constantly sampling new and different foods. |
| FNS 2 | I don't trust new foods. (R) |
| FNS 3 | if I don't know what is in a food, I won't try it. (R) |
| FNS 4 | I like foods from different countries. |
| FNS 5 | Ethnic foods look too weird to eat. (R) |
| FNS 6 | At different parties, I will try new foods. |
| FNS 7 | I am afraid to eat things I have never had before. (R) |
| FNS 8 | I am very particular about the foods I will eat. (R) |
| FNS 9 | I will eat almost anything. |
| FNS 10 | I like to try new ethnic restaurants. |

Items for which scoring is reversed are marked (R).

### 3.5. Data Analysis

Before conducting a multiple regression analysis to analyze the effects of the independent variables on the consumers' purchase intention for the products of the study, the validity of each individual item from the questionnaire was assessed. To test the validity of our constructs, we used structural equation modeling to perform a confirmatory factor analysis with SmartPLS 2.0 and a bootstrap resample procedure [55].

Reliability was confirmed with a Cronbach's alpha. Generally, values that are greater than 0.70 represent a good measure [56]. Then, individual item loadings were used to check convergent validity. A loading of 0.70 or higher from a partial least square method is considered adequate.

The validity of the measurement model was evaluated by the composite reliability (CR) index, partial least squares (PLS) factor loadings, and average variance extracted (AVE). A value greater than 0.70 CR means that a construct maintains both its internal consistency and convergent validity [57]. The recommended levels for PLS factor loadings and AVE are 0.70 and 0.50, respectively.

As can be seen in Tables 4 and 5, the constructs from both experiments passed the reliability test since all of the Cronbach's alpha values were higher than 0.70, indicating that the constructs have adequate internal consistency [56]. In addition, values greater than 0.70 for CR show that the constructs have both internal consistency and convergent validity [57].

**Table 4.** Reliability and validity values from the first experiment: soy sauce products.

| Constructs | Item | Factor Loading | Average Variance Extracted | Composite Reliability | Cronbach's Alpha |
|---|---|---|---|---|---|
| Unhealthy = tasty intuition (UTI) | UTI 1 | 0.9021 | 0.8339 | 0.9377 | 0.9013 |
| | UTI 2 | 0.9298 | | | |
| | UTI 3 | 0.9074 | | | |
| Food neophobia (FNS) | FNS 2 | 0.8295 | 0.7184 | 0.8841 | 0.7184 |
| | FNS 4 | 0.9131 | | | |
| | FNS 5 | 0.7959 | | | |
| Purchase intention for a low-sodium soy sauce product (PI) | PI 1 | 0.8542 | 0.7504 | 0.9000 | 0.7504 |
| | PI 2 | 0.8216 | | | |
| | PI 3 | 0.9201 | | | |

**Table 5.** Reliability and validity values from the second experiment: yogurt products.

| Constructs | Item | Factor Loading | Average Variance Extracted | Composite Reliability | Cronbach's Alpha |
|---|---|---|---|---|---|
| Unhealthy = tasty intuition (UTI) | UTI 1 | 0.9216 | 0.7784 | 0.9129 | 0.8773 |
| | UTI 2 | 0.7956 | | | |
| | UTI 3 | 0.9234 | | | |
| Food neophobia (FNS) | FNS 2 | 0.7278 | 0.6126 | 0.8629 | 0.7882 |
| | FNS 4 | 0.8751 | | | |
| | FNS 5 | 0.7579 | | | |
| | FNS 8 | 0.7621 | | | |
| Purchase intention for a low-sodium soy sauce product (PI) | PI 1 | 0.9055 | 0.7869 | 0.9171 | 0.8656 |
| | PI 2 | 0.9018 | | | |
| | PI 3 | 0.8528 | | | |

This study used the square root of AVE (Tables 6 and 7) and a cross-loading matrix to verify the discriminant validity of the measurement model. The square root should be larger than the correlations between the constructs [58]. The amount of variance shared between the latent variables indicates the presence of a strong discriminant validity [59]. Tables 6 and 7 show that for both experiments, each construct has a higher correlation with the measures than the other.

**Table 6.** Correlations of the latent variables and the square root of the average variance extracted from the first experiment: soy sauce products.

| | (1) | (2) | (3) |
|---|---|---|---|
| (1) Unhealthy = tasty intuition | (0.9132) | | |
| (2) Food neophobia | −0.1857 | (0.8476) | |
| (3) Purchase intention for a low-sodium soy sauce product | −0.1482 | 0.0983 | (0.8663) |

**Table 7.** Correlations of the latent variables and the square root of the average variance extracted from the second experiment: yogurt products.

| | (1) | (2) | (3) |
|---|---|---|---|
| (1) Unhealthy = tasty intuition | (0.8823) | | |
| (2) Food neophobia | −0.1126 | (0.7827) | |
| (3) Purchase intention for a low-sodium soy sauce product | −0.0503 | 0.1723 | (0.8871) |

Subsequently, a multiple regression analysis was conducted with the three main independent variables: whether the respondents had previously selected a low-sodium soy sauce or not (previous choice), UTI, and food neophobia. Calculations were performed using RStudio version 3.4.4.

## 4. Results

The results from the soy sauce experiment and the yogurt experiment are as follows (Table 8). First, the results from the soy sauce experiment revealed that among 466 respondents, those consumers who had previously chosen a regular soy sauce over low-sodium versions tended to have a lower purchase intention for a low-sodium product. Meanwhile, consumers who had a weaker intuition that delicious foods are not necessarily unhealthy (lower UTI) tended to have a higher purchase intention for a low-sodium soy sauce product. In addition, consumers who were less fearful of trying new foods (low food neophobia) tended to have a higher purchase intention for a low-sodium soy sauce product. These results support H1a, H2a, and H3a.

**Table 8.** Multiple regression analysis results.

| | | Purchase Intention for a Low-Sodium Soy Sauce Product | | Purchase Intention for a Low-Sugar Yogurt Product | |
|---|---|---|---|---|---|
| **Type** | **Independent Variable** | **Stan.B** | ***p*-Value** | **Stan.B** | ***p*-Value** |
| Previous Choice | PC | 0.2905 | 0.0000 *** | 0.3554 | 0.0000 *** |
| Unhealthy = tasty intuition | UTI | −0.1067 | 0.0163 * | 0.0007 | 0.9875 |
| Food neophobia | FNS | 0.0955 | 0.0334 * | 0.1214 | 0.0048 ** |
| Control Variable | Age_20 [a] | −0.2106 | 0.0003 *** | −0.1730 | 0.0011 ** |
| | Age_30 [a] | −0.1409 | 0.0119 * | −0.0580 | 0.2756 |
| | Age_40 [a] | −0.0665 | 0.2195 | −0.0895 | 0.0826 |
| | Income | 0.0863 | 0.0780 | −0.0416 | 0.3644 |
| | Diet Interest | 0.1440 | 0.0011 ** | 0.1019 | 0.0166 * |
| | Home Cooking interest | 0.0365 | 0.4376 | 0.0593 | 0.1757 |
| | Education | −0.0453 | 0.3293 | 0.0225 | 0.5960 |
| | Grocery shopping frequency | 0.0492 | 0.2580 | 0.1372 | 0.0011 ** |
| | Family size | 0.0098 | 0.8434 | 0.1364 | 0.0039 ** |
| | **$R^2$** | 0.1530 | | 0.1851 | |
| | **F-statistic** | 8.0000 | | 10.1600 | |
| | ***p*-value** | 0.0000 | | 0.0000 | |

*** $p < 0.001$, ** $p < 0.01$, * $p < 0.05$; [a] Based on a dummy variable that was coded as 1 if the respondent belongs to this age range. All control variables except for the age variables are continuous variables.

There were three control variables found in the results from the first experiment significant to the purchase intention for a low-sodium soy sauce: age, income, and diet interest. Consumers in their 20s and 30s had a lower purchase intention for a low-sodium soy sauce product compared to those who were in their 50s. on the other hand, consumers with a higher household income and consumers who were more interested in their diet had a higher purchase intention for a low-sodium soy sauce product.

The results from the yogurt experiment show that among 485 respondents, those consumers who had previously chosen a regular yogurt product over low-sugar versions tended to have a lower purchase intention for a low-sugar yogurt product. In addition, consumers who were less afraid of trying novel foods (low food neophobia) had a higher purchase intention for a low-sugar yogurt product. However, UTI was not a significant independent variable for the purchase intention for a low-sugar yogurt product. Therefore, the results support H1b and H3b but do not support H2b.

in the yogurt experiment, there were four control variables significant to the purchase intention for a low-sugar yogurt: age, diet interest, grocery shopping frequency, and family size. Consumers in their 20s had a lower purchase intention for a low-sugar yogurt product compared to those who were in their 50s. Older consumers, consumers who were more interested in their diet, and consumers who did their grocery shopping more frequently had a higher purchase intention for a low-sugar yogurt product. In addition, the larger the size of a consumer's family, the higher the purchase intention for a low-sugar product.

## 5. Discussion

### *5.1. Consumers' Previous Choice and the Purchase Intention for a Low-Sodium/Low-Sugar Product*

The results from our study demonstrate that the previous choice shapes the preference of consumers. In other words, whether consumers had chosen a low-sodium/low-sugar product in the past or not make a significant impact on their purchase intention for these products. From the findings, we can draw some important implications.

In this study, respondents were asked to choose a product first from a choice set and were then informed about the low-sodium/low-sugar products. They were then asked to evaluate their purchase intention for these products. It should be noted that the presented information did not play a significant role in increasing the purchase intention for the low-sodium/low-sugar products. This is in line with our conjecture that those who had chosen regular products would disregard the information as it was dissonant with their preferences [14].

The findings suggest that governmental agencies should divide consumers into segments and implement different strategies. We suggest dividing consumers into segments of (1) consumers who have never purchased sodium/sugar-based products, (2) consumers who buy low-sodium/low-sugar products, and (3) consumers who buy regular products.

First, for the (1) consumers who have never purchased sodium/sugar-based products but are prospective consumers, governmental agencies should consider strategies to raise awareness of these products. It is imperative to add low-sodium/low-sugar products to the consumer choice set. Next, there is a need to encourage (2) consumers who are already a customer of low-sodium/low-sugar products to repurchase them. In order to do that, it is suggested that governmental agencies introduce recipes and put labels on the products to make consumers more informed and increase their involvement.

Finally, strategies that instill dissonant cognitions to the behavior of consuming high-sodium/low-sugar products can be considered for (3) consumers who are already consuming regular products. Following an example used by Festinger (1957), some consumers may experience dissonance and divert their purchasing behaviors to reduce dissonance when they are faced with campaigns and public advertisements promoting that excessive sodium and sugar intake is bad for their health.

### *5.2. Consumers' Unhealthy = Tasty Intuition and the Purchase Intention of a Low-Sodium/Low-Sugar Product*

Consumers' UTI made a significant impact on their purchase intention for a low-sodium soy sauce product. However, contrary to our expectations, it did not significantly influence the purchase intention for a low-sugar yogurt product. There may be some explanations for this. Soy sauce has a relatively high switching cost as it has a longer consumption period per purchase as compared to yogurt. In addition, soy sauce is not a product that is consumed directly, but rather, it is seasoning that adds saltiness during the cooking process. Thus, this saltiness itself is a highlight of soy sauce. Therefore, consumers with a high UTI will tend to have a significantly lower intention to purchase a low-sodium soy sauce product because purchasing it entails high uncertainty. on the other hand, yogurt is not a product eaten for sweetness alone, and therefore, UTI may not make a significant impact on consumer purchase intention in this case.

In addition, consumer perceptions of tastiness and healthiness can differ depending on the food product category [60]. Food products can be divided into a tastiness-associated food category and a healthiness-associated food category [61], yet some products can be perceived as having both tasty and healthy dimensions. In line with this idea, it is assumed that yogurt, unlike soy sauce, can be perceived as neutral in terms of healthiness and tastiness for consumers in Korea.

We can also speculate that Korea's market environment affected the experiment results. Our data shows that 14.90% of the respondents selected a low-sodium soy sauce product and 85.10% selected a regular product. Meanwhile, the yogurt choices were more evenly distributed with 38.13% of respondents having selected a low-sugar yogurt product and 61.87% having selected a regular product. This suggests that consumer preconceptions about the taste of low-sugar yogurt have already disappeared or that there are already good tasting low-sugar yogurt products on the market, while there still exist preconceptions regarding the taste of low-sodium soy sauce.

Therefore, in order to increase consumers' purchase intention for low-sodium products, especially for the soy sauce category, it is necessary to change the behavior of consumers in the long term by utilizing media-based promotion strategies. For example, media promotion strategies using contents such as specific and interesting sensory experiments and blind tests can be efficacious for overcoming consumer preconceptions that using low-sodium soy sauce will negatively affect the taste of food.

### 5.3. *Consumers' Food Neophobia and the Purchase Intention for a Low-Sodium/Low-Sugar Product*

This study assumes that when a food product with lower sodium or sugar content is introduced by a widely accepted and long-established food brand, the consumer perceives the new product as novel relative to the original. Based on this assumption, we hypothesized that the weaker the food neophobia tendency of consumers, the stronger their purchase intention for low-sodium and low-sugar products. The results confirmed our hypothesis (H1a and H1b confirmed). One of the implications of this study is the interpretation of the consumer the consumption of soy sauce and yogurt products from a different perspective than previous studies. The study assumes that regular soy sauce and yogurt products are familiar to consumers, but are recognized as unfamiliar to consumers when the sodium and sugar contents are lower, respectively.

Although yogurt is a widely-consumed product category worldwide, soy-based foods, including soy sauce, are only recently being introduced to the Western world [62]. For this reason, many prior studies have been conducted in consideration of the factors that affect Western consumers (i.e., food neophobia) in their consumption of soy products, which would be novel foods from their perspective [63–69]. However, soybeans are a traditional food source in East and Southeast Asia [70]. In particular, soy sauce is a product that has traditionally been used in many dishes in Korea. Therefore, it is a more reasonable explanation that, while soy sauce itself is not a new food for Korean consumers, low-sodium soy sauce reformulated from the original is considered a new product for Koreans.

There have been other studies that have considered consumers' purchase intention of yogurt products and the effects of food neophobia [71]. The yogurt items considered in those studies, however, were mainly functional products, such as probiotic yogurts [72,73]. However, the novelty covered in this study was not a novelty from adding a new nutritional ingredient to an existing product but rather from lowering the sugar content of a product already familiar to consumers.

The relationship between food neophobia and purchase intention for the products of our study indicates that alleviating consumer food neophobia for low-sodium/low-sugar content products may result in more positive consumer responses. Since a number of prior studies have demonstrated that repeated exposure to novels foods and flavors tends to enhance preferences [34–36], increasing the amount of exposure to these products can be one way to reduce food neophobia and encourage consumers to acquire preferences for low-sodium/low-sugar products. Therefore, giving out food samples or, as in the case of low-sodium soy sauce, making and distributing recipe books using the product can be suggested.

### 5.4. *Control Variables*

Control Variables that had significant impacts on the purchase intention of low-sodium soy sauce products were age, income, and diet interest. The result is in line with previous studies. According to the previous research [7], sodium content is considered significantly important in food purchases to older and middle-high income consumers. In addition, when it comes to the perception of low-sodium products, age, income, and family concern about diet-related problems are significant factors. Aged consumers with higher income levels and diet-related concerns within the family are more likely to recognize low-sodium products [8].

on the other hand, control variables that had significant impacts on the purchase intention of low-sugar yogurt products were age, diet interest, grocery shopping frequency, and family size. While consumers in their 20s and 30s had significantly lower purchase intention for low-sodium soy sauce products compared to those in their 40s and 50s, only consumers in their 20s had significantly lower purchase intention for low-sugar yogurt product. on the other hand, consumers with bigger family size had significantly higher purchase intention for low-sugar yogurt product while they were not for a low-sodium soy sauce product. According to the report from the Korea Agro-Fisheries & Food Trade Corporation report, the ages of the yogurt product consumers are evenly distributed compared to other dairy products [74]. As the number of family members increases, the age range of family members will also vary. In the case of a yogurt product, since the products are consumed by various ages compared to a soy sauce product, the increase in the number of family members would mean the increased demand for yogurt per household. This may explain why the Family size was a significant variable in the purchase intention of a low-sugar yogurt product compared to a low-sodium soy sauce product. In addition, according to a previous study, individuals who were more interested in losing their weights consumed low-sugar carbonated drinks such as Coca-Cola Light and Fanta Light significantly more than those who were not [75]. It is consistent with our finding that as consumers are more interested in their diet, their purchase intention for a low-sugar yogurt product is significantly higher. Furthermore, previous researches show that consumers who do grocery shopping more often tend to buy healthier foods [76], and college students who do not do grocery shopping tend to have poor eating habits, such as drinking more carbonated drinks or eating high-calorie food products compared to those who do grocery shopping at a convenience store or supermarkets [77]. These findings explain why grocery shopping frequency had a significant impact on the purchase intention of low-sugar yogurt products in this study.

### *5.5. Limitations and Further Research*

Many different factors influence consumers' purchase intention for low-sodium/low-sugar products, and it is impossible to consider all of these factors in a single study. This study focused on the influence of external factors, specifically previous choice experience, and consumer-related factors (UTI, food neophobia) on the purchase intention for these products. Other factors, such as product-related factors, need further research.

In addition, many different low-sodium food products exist in the sodium-based food product category, and there are also many low-sugar food products on the market. As stimuli, the present study used soy sauce to represent the sodium-based food product category and yogurt to represent the sugar-based food product category. More food product categories should be considered in order to generalize our findings in future research.

Furthermore, sodium-based and sugar-based food products differ not only in their sensory characteristics but also in the situations where they are consumed and the people who consume them. Further research is needed to determine the factors that influence the purchase intention of low-sodium and low-sugar food products based on the qualitative differences between these two product categories. For instance, comparing sodium-based products and sugar-based products that are perceived in terms of the same dimension (healthiness vs. healthiness, tastiness vs. tastiness) can be one way to consider the qualitative differences of these two food product categories.

Moreover, in conducting online purchasing experiments and surveys with low-sodium soy sauce and low-sugar yogurt products and analyzing the results, it was difficult to control for variables such as brand, price, and volume due to the structure of the survey. There was a trade-off between internal validity on the one hand and external validity on the other hand in the experimental design process. In future experiments, there is a need to control the impact of these variables on consumer choice and purchase intention while conducting the surveys and analyzing the results.

Finally, this study was conducted mainly on the group that can be surveyed online easily due to the nature of the online survey. Therefore, there is a possibility that the research results could be age-biased because the panel members who participated in the survey were limited to their 20s to 50s. Future research needs to focus on recruiting samples to represent all ages.

## 6. Conclusions

This study has its originality and value in that the study has interpreted the consumer consumption of soy sauce and yogurt products from a different perspective than previous studies. The study begins with the assumption that regular soy sauce and yogurt products are familiar to consumers, but are recognized as unfamiliar to consumers when the sodium and sugar content is reformulated, respectively. This way, the present work helps expand the view of defining a new food product.

This study aimed to examine the consumer's previous choice of low-sodium/sugar products, unhealthy = tasty intuition and the propensity of food neophobia influence their intention to purchase low-sodium/sugar products. The result turned out that the previous choice of regular soy sauce and yogurt products had a negative effect on the purchase intention for low-sodium soy sauce and low-sugar yogurt products. In addition, consumer's strong UTI had a negative influence on the purchase intention for low-sodium soy sauce products, but no significant influence on that of low-sugar yogurt products. Furthermore, consumer's stronger propensity for food neophobia had a negative effect on the purchase intention for both low-sodium soy sauce products and low-sugar products. The result also suggests that not only consumer's previous experience and their disposition, but also demographic factors such as age, income, diet interest, and family size have an impact on purchase intention for these low-sodium/sugar products. Furthermore, it can be said that the factors have different impacts on the purchase intention depending on the product category and whether sodium or sugar is the main content of the product.

**Author Contributions:** Conceptualization, Y.P., D.L., S.P., and J.M.; methodology, Y.P. and D.L.; analysis, Y.P.; writing—original draft preparation, Y.P.; writing—review and editing, D.L. and J.M.; supervision, J.M. All authors have read and agreed to the published version of the manuscript.

## Appendix A

**Figure A1.** Soy sauce products used as stimuli in the survey.

**Figure A2.** Yogurt products used as stimuli in the survey.

## References

1. Kim, S.; Kim, M.; Min, J.; Yoo, J.; Kim, M.; Kang, J.; Won, C.W. How Much Intake of Sodium Is Good for Frailty? the Korean Frailty and Aging Cohort Study (KFACS). *J. Nutr. Health Aging* **2019**, *23*, 503–508. [CrossRef] [PubMed]
2. Strazzullo, P.; D'Elia, L.; Kandala, N.B.; Cappuccio, F.P. Salt intake, stroke, and cardiovascular disease: Meta-analysis of prospective studies. *Br. Med. J.* **2009**, *339*, 1–9. [CrossRef] [PubMed]
3. Hagmann, D.; Siegrist, M.; Hartmann, C. Taxes, labels, or nudges? Public acceptance of various interventions designed to reduce sugar intake. *Food Policy* **2018**, *79*, 156–165. [CrossRef]
4. Popkin, B.M.; Hawkes, C. Sweetening of the global diet, particularly beverages: Patterns, trends, and policy responses. *Lancet Diabetes Endo* **2016**, *4*, 174–186. [CrossRef]
5. Han, E.; Kim, T.H.; Powell, L.M. Beverage consumption and individual-level associations in South Korea. *BMC Public Health* **2013**, *13*, 195. [CrossRef]
6. Kremer, S.; Shimojo, R.; Holthuysen, N.; Koster, E.P.; Mojet, J. Consumer acceptance of salt-reduced "soy sauce" bread over repeated in home consumption. *Food Qual. Prefer.* **2013**, *28*, 484–491. [CrossRef]
7. Bryła, P. Selected Predictors of the Importance Attached to Salt Content Information on the Food Packaging (a Study among Polish Consumers). *Nutrients* **2020**, *12*, 293. [CrossRef]
8. Büyükkaragöz, A.; Baş, M.; Sağlam, D.; Cengiz, E. Consumers' awareness, acceptance and attitudes towards functional foods in Turkey. *Int. J. Consum. Stud.* **2014**, *38*, 628–635. [CrossRef]
9. Guardia, M.D.; Guerrero, L.; Gelabert, J.; Gou, P.; Arnau, J. Consumer attitude towards sodium reduction in meat products and acceptability of fermented sausages with reduced sodium content. *Meat Sci.* **2006**, *73*, 484–490. [CrossRef]
10. Kim, M.K.; Lee, K.G. Consumer awareness and interest toward sodium reduction trends in Korea. *J. Food Sci.* **2014**, *79*, S1416–S1423. [CrossRef]
11. Saleh, M.; Alhaidari, A.; Al-Mansour, A.; Al-Khudair, A. Health Awareness and Price Sensitivity as Predictors of Consumers' Purchase Attitude towards Soft Drinks. *Expert J. Mark.* **2018**, *6*, 22–32.
12. Nakmongkol, A. The Study of Consumer's Attitudes and Behaviors towards Carbonate Soft Drinks. Master's Thesis, Bangkok University, Bangkok, Thailand, 2009.
13. Festinger, L. *A Theory of Cognitive Dissonance*; Stanford University Press: Stanford, CA, USA, 1957.
14. Harmon-Jones, E.E.; Mills, J.E. *Cognitive Dissonance: Progress on a Pivotal Theory in Social Psychology*; American Psychological Association: Arlington, TX, USA, 1999.
15. Bem, D.J. Self-Perception—An Alternative Interpretation of Cognitive Dissonance Phenomena. *Psychol. Rev.* **1967**, *74*, 183. [CrossRef] [PubMed]
16. Bem, D.J. Self-perception theory. In *Advances in Experimental Social Psychol*; Elsevier: Amsterdam, The Netherlands, 1972; Volume 6, pp. 1–62.
17. Ariely, D.; Norton, M.I. How actions create—Not just reveal—Preferences. *Trends Cogn. Sci.* **2008**, *12*, 13–16. [CrossRef] [PubMed]

18. Lieberman, M.D.; Ochsner, K.N.; Gilbert, D.T.; Schacter, D.L. Do amnesics exhibit cognitive dissonance reduction? the role of explicit memory and attention in attitude change. *Psychol. Sci.* **2001**, *12*, 135–140. [CrossRef] [PubMed]
19. Egan, L.C.; Santos, L.R.; Bloom, P. The origins of cognitive dissonance: Evidence from children and monkeys. *Psychol. Sci.* **2007**, *18*, 978–983. [CrossRef] [PubMed]
20. Brehm, J.W. Postdecision changes in the desirability of alternatives. *J. Abnorm. Soc. Psychol.* **1956**, *52*, 384. [CrossRef]
21. Banks, S. The Relationships between Preference and Purchase of Brands. *J. Mark.* **1950**, *15*, 145–157. [CrossRef]
22. Rogers, P.J.; Hogenkamp, P.S.; de Graaf, C.; Higgs, S.; Lluch, A.; Ness, A.R.; Penfold, C.; Perry, R.; Putz, P.; Yeomans, M.R.; et al. Does low-energy sweetener consumption affect energy intake and body weight? a systematic review, including meta-analyses, of the evidence from human and animal studies. *Int. J. Obes.* **2016**, *40*, 381–394. [CrossRef]
23. Liu, P.J.; Roberto, C.A.; Liu, L.J.; Brownell, K.D. A test of different menu labeling presentations. *Appetite* **2012**, *59*, 770–777. [CrossRef]
24. Lee, W.C.J.; Shimizu, M.; Kniffin, K.M.; Wansink, B. You taste what you see: Do organic labels bias taste perceptions? *Food Qual. Prefer.* **2013**, *29*, 33–39. [CrossRef]
25. Huang, Y.H.; Wu, J. Food pleasure orientation diminishes the "healthy = less tasty" intuition. *Food Qual. Prefer.* **2016**, *54*, 75–78. [CrossRef]
26. Irmak, C.; Vallen, B.; Robinson, S.R. The Impact of Product Name on Dieters' and Nondieters' Food Evaluations and Consumption. *J. Consum. Res.* **2011**, *38*, 390–405. [CrossRef]
27. Raghunathan, R.; Naylor, R.W.; Hoyer, W.D. The unhealthy equal tasty intuition and its effects on taste inferences, enjoyment, and choice of food products. *J. Mark.* **2006**, *70*, 170–184. [CrossRef]
28. Howlett, E.A.; Burton, S.; Bates, K.; Huggins, K. Coming to a Restaurant Near You? Potential Consumer Responses to Nutrition Information Disclosure on Menus. *J. Consum. Res.* **2009**, *36*, 494–503. [CrossRef]
29. Naylor, R.W.; Droms, C.M.; Haws, K.L. Eating with a Purpose: Consumer Response to Functional Food Health Claims in Conflicting Versus Complementary Information Environments. *J. Public Policy Mark.* **2009**, *28*, 221–233. [CrossRef]
30. Mai, R.; Hoffmann, S. How to Combat the Unhealthy = Tasty Intuition: The Influencing Role of Health Consciousness. *J. Public Policy Mark.* **2015**, *34*, 63–83. [CrossRef]
31. Werle, C.O.C.; Trendel, O.; Ardito, G. Unhealthy food is not tastier for everybody: The "healthy = tasty" French intuition. *Food Qual. Prefer.* **2013**, *28*, 116–121. [CrossRef]
32. Pliner, P.; Hobden, K. Development of a Scale to Measure the Trait of Food Neophobia in Humans. *Appetite* **1992**, *19*, 105–120. [CrossRef]
33. Pliner, P.; Pelchat, M.; Grabski, M. Reduction of Neophobia in Humans by Exposure to Novel Foods. *Appetite* **1993**, *20*, 111–123. [CrossRef]
34. Pliner, P. The Effects of Mere Exposure on Liking for Edible Substances. *Appetite* **1982**, *3*, 283–290. [CrossRef]
35. Birch, L.L.; Marlin, D.W. I don't like it; I never tried it: Effects of exposure on two-year-old children's food preferences. *Appetite* **1982**, *3*, 353–360. [CrossRef]
36. Birch, L.L.; McPhee, L.; Shoba, B.C.; Pirok, E.; Steinberg, L. What kind of exposure reduces children's food neophobia? Looking vs. tasting. *Appetite* **1987**, *9*, 171–178. [CrossRef]
37. Siegrist, M.; Hartmann, C.; Keller, C. Antecedents of food neophobia and its association with eating behavior and food choices. *Food Qual. Prefer.* **2013**, *30*, 293–298. [CrossRef]
38. Meiselman, H.; King, S.; Gillette, M. The demographics of neophobia in a large commercial US sample. *Food Qual. Prefer.* **2010**, *21*, 893–897. [CrossRef]
39. Fernández-Ruiz, V.; Claret, A.; Chaya, C. Testing a Spanish-version of the food neophobia scale. *Food Qual. Prefer.* **2013**, *28*, 222–225. [CrossRef]
40. Hursti, U.K.K.; Sjoden, P.O. Food and general neophobia and their relationship with self-reported food choice: Familial resemblance in Swedish families with children of ages 7–17 years. *Appetite* **1997**, *29*, 89–103. [CrossRef]
41. Olabi, A.; Najm, N.E.; Kebbe Baghdadi, O.; Morton, J.M. Food neophobia levels of Lebanese and American college students. *Food Qual. Prefer.* **2009**, *20*, 353–362. [CrossRef]
42. Ritchey, P.N.; Frank, R.A.; Hursti, U.K.; Tuorila, H. Validation and cross-national comparison of the food neophobia scale (FNS) using confirmatory factor analysis. *Appetite* **2003**, *40*, 163–173. [CrossRef]

43. Tuorila, H.; Lahteenmaki, L.; Pohjalainen, L.; Lotti, L. Food neophobia among the Finns and related responses to familiar and unfamiliar foods. *Food Qual. Prefer.* **2001**, *12*, 29–37. [CrossRef]
44. Knaapila, A.; Tuorila, H.; Silventoninen, K.; Keskitalo, K.; Kallela, M.; Wessman, M.; Peltonen, L.; Cherkas, L.F.; Spector, T.D.; Perola, M. Food neophobia shows heritable variation in humans. *Physiol. Behav.* **2007**, *91*, 573–578. [CrossRef]
45. Heath, P.; Houston-Price, C.; Kennedy, O.B. Increasing food familiarity without the tears. A role for visual exposure? *Appetite* **2011**, *57*, 832–838. [CrossRef] [PubMed]
46. Henriques, A.S.; King, S.C.; Meiselman, H.L. Consumer segmentation based on food neophobia and its application to product development. *Food Qual. Prefer.* **2009**, *20*, 83–91. [CrossRef]
47. Schickenberg, B.A. Towards Strategies to Stimulate First Time Trial of Unfamiliar Healthful Food Products. Ph.D. Thesis, Maastricht University, Maastricht, The Netherlands, 2010.
48. Kinnear, M.; de Kock, H.L. Would repeated consumption of sports drinks with different acidulants lead to hedonic adjustment? *Food Qual. Prefer.* **2011**, *22*, 340–345. [CrossRef]
49. Chung, S.-J.; Vickers, Z. Influence of sweetness on the sensory-specific satiety and long-term acceptability of tea. *Food Qual. Prefer.* **2007**, *18*, 256–264. [CrossRef]
50. Barcellos, M.D.d.; Aguiar, L.K.; Ferreira, G.C.; Vieira, L.M. Willingness to try innovative food products: a comparison between British and Brazilian consumers. *BAR Braz. Adm. Rev.* **2009**, *6*, 50–61. [CrossRef]
51. Kim, S.-H.; Jeong, Y.-J. Domestic and international trends in sodium reduction and practices. *Food Sci. Ind.* **2016**, *49*, 25–33.
52. Jung, J. Action plan for sugars reduction. *Food Sci. Ind.* **2016**, *49*, 12–16.
53. Korea Agro-Fisheries&Food Trade Corporation. *2018 Processed Food Segment Market Status: Sauce and Dressing Market*; 11-1543000-002434-01; Korea Agro-Fisheries&Food Trade Corporatio: Naju, Korea, 2018.
54. Gefen, D. E-commerce: the role of familiarity and trust. *Omega* **2000**, *28*, 725–737. [CrossRef]
55. Ringle, C.M.; Wende, S.; Will, A. *SmartPLS 2.0 (M3) Beta*; University of Hamburg: Hamburg, Germany, 2005.
56. Knoke, D.; Bohrnstedt, G.W.; Mee, A.P. *Statistics for Social Data Analysis*; FE Peacock Publishers, Inc: Itasca, IL, USA, 1994.
57. Werts, C.E.; Linn, R.L.; Joreskog, K.G. Intraclass Reliability Estimates—Testing Structural Assumptions. *Educ. Psychol. Meas.* **1974**, *34*, 25–33. [CrossRef]
58. Fornell, C.; Larcker, D.F. Evaluating Structural Equation Models with Unobservable Variables and Measurement Error. *J. Mark. Res.* **1981**, *18*, 39–50. [CrossRef]
59. Gefen, D.; Straub, D. A practical guide to factorial validity using PLS-Graph: Tutorial and annotated example. *Commun. Assoc. Inf. Syst.* **2005**, *16*, 5. [CrossRef]
60. Cheong, Y.; Kim, K. The Interplay Between Advertising Claims and Product Categories in Food Advertising: a Schema Congruity Perspective. *J. Appl. Commun. Res.* **2011**, *39*, 55–74. [CrossRef]
61. Wertenbroch, K. Consumption self-control by rationing purchase quantities of virtue and vice. *Mark. Sci.* **1998**, *17*, 317–337. [CrossRef]
62. Fenko, A.; Backhaus, B.W.; van Hoof, J.J. The influence of product- and person-related factors on consumer hedonic responses to soy products. *Food Qual. Prefer.* **2015**, *41*, 30–40. [CrossRef]
63. Roininen, K. Evaluation of Food Choice Behavior: Development and Validation of Health and Taste Attitude Scales. Ph.D. Thesis, University of Helsinki, Helsinki, Finland, 2001.
64. Choe, J.Y.; Cho, M.S. Food neophobia and willingness to try non-traditional foods for Koreans. *Food Qual. Prefer.* **2011**, *22*, 671–677. [CrossRef]
65. Schyver, T.; Smith, C. Reported attitudes and beliefs toward soy food consumption of soy consumers versus nonconsumers in natural foods or mainstream grocery stores. *J. Nutr. Educ. Behav.* **2005**, *37*, 292–299. [CrossRef]
66. Moon, W.; Balasubramanian, S.K.; Rimal, A. Health claims and consumers' behavioral intentions: the case of soy-based food. *Food Policy* **2011**, *36*, 480–489. [CrossRef]
67. Martins, Y.; Pliner, P. Human food choices: an examination of the factors underlying acceptance/rejection of novel and familiar animal and nonanimal foods. *Appetite* **2005**, *45*, 214–224. [CrossRef]
68. Hoek, A.C.; Luning, P.A.; Weijzen, P.; Engels, W.; Kok, F.J.; de Graaf, C. Replacement of meat by meat substitutes. A survey on person- and product-related factors in consumer acceptance. *Appetite* **2011**, *56*, 662–673. [CrossRef]
69. Chang, J.B.; Moon, W.; Balasubramanian, S.K. Consumer valuation of health attributes for soy-based food: a choice modeling approach. *Food Policy* **2012**, *37*, 335–342. [CrossRef]

70. Tu, V.P.; Husson, F.; Sutan, A.; Ha, D.T.; Valentin, D. For me the taste of soy is not a barrier to its consumption. And how about you? *Appetite* **2012**, *58*, 914–921. [CrossRef]
71. Bimbo, F.; Bonanno, A.; Nocella, G.; Viscecchia, R.; Nardone, G.; De Devitiis, B.; Carlucci, D. Consumers' acceptance and preferences for nutrition-modified and functional dairy products: A systematic review. *Appetite* **2017**, *113*, 141–154. [CrossRef] [PubMed]
72. Urala, N.; Lahteenmaki, L. Consumers' changing attitudes towards functional foods. *Food Qual. Prefer.* **2007**, *18*, 1–12. [CrossRef]
73. Siegrist, M.; Stampfli, N.; Kastenholz, H. Consumers' willingness to buy functional foods. The influence of carrier, benefit and trust. *Appetite* **2008**, *51*, 526–529. [CrossRef] [PubMed]
74. Korea Agro-Fisheries&Food Trade Corporation. *2018 Processed Food Segment Market Report: Butter, Cheese, Fermented Milk*; 11-1543000-002320-01; Korea Agro-Fisheries&Food Trade Corporation: Naju, Korea, 2018.
75. Mullie, P.; Aerenhouts, D.; Clarys, P. Demographic, socioeconomic and nutritional determinants of daily versus non-daily sugar-sweetened and artificially sweetened beverage consumption. *Eur. J. Clin. Nutr.* **2012**, *66*, 150–155. [CrossRef]
76. Moore, L.V.; Pinard, C.A.; Yaroch, A.L. Features in grocery stores that motivate shoppers to buy healthier foods, ConsumerStyles 2014. *J. Community Health* **2016**, *41*, 812–817. [CrossRef]
77. VanKim, N.A.; Erickson, D.J.; Laska, M.N. Food shopping profiles and their association with dietary patterns: a latent class analysis. *J. Acad. Nutr. Diet.* **2015**, *115*, 1109–1116. [CrossRef]

# Permissions

All chapters in this book were first published in MDPI; hereby published with permission under the Creative Commons Attribution License or equivalent. Every chapter published in this book has been scrutinized by our experts. Their significance has been extensively debated. The topics covered herein carry significant findings which will fuel the growth of the discipline. They may even be implemented as practical applications or may be referred to as a beginning point for another development.

The contributors of this book come from diverse backgrounds, making this book a truly international effort. This book will bring forth new frontiers with its revolutionizing research information and detailed analysis of the nascent developments around the world.

We would like to thank all the contributing authors for lending their expertise to make the book truly unique. They have played a crucial role in the development of this book. Without their invaluable contributions this book wouldn't have been possible. They have made vital efforts to compile up to date information on the varied aspects of this subject to make this book a valuable addition to the collection of many professionals and students.

This book was conceptualized with the vision of imparting up-to-date information and advanced data in this field. To ensure the same, a matchless editorial board was set up. Every individual on the board went through rigorous rounds of assessment to prove their worth. After which they invested a large part of their time researching and compiling the most relevant data for our readers.

The editorial board has been involved in producing this book since its inception. They have spent rigorous hours researching and exploring the diverse topics which have resulted in the successful publishing of this book. They have passed on their knowledge of decades through this book. To expedite this challenging task, the publisher supported the team at every step. A small team of assistant editors was also appointed to further simplify the editing procedure and attain best results for the readers.

Apart from the editorial board, the designing team has also invested a significant amount of their time in understanding the subject and creating the most relevant covers. They scrutinized every image to scout for the most suitable representation of the subject and create an appropriate cover for the book.

The publishing team has been an ardent support to the editorial, designing and production team. Their endless efforts to recruit the best for this project, has resulted in the accomplishment of this book. They are a veteran in the field of academics and their pool of knowledge is as vast as their experience in printing. Their expertise and guidance has proved useful at every step. Their uncompromising quality standards have made this book an exceptional effort. Their encouragement from time to time has been an inspiration for everyone.

The publisher and the editorial board hope that this book will prove to be a valuable piece of knowledge for researchers, students, practitioners and scholars across the globe.

# List of Contributors

**Niculina Iudita Sampalean and Daniele Rama**
Department of Agri-Food Economics, Faculty of Agriculture, Food and Environmental Sciences, Campus Piacenza-Cremona, Catholic University of Sacred Heart, via Emilia Parmense, 84, 29122 Piacenza (PC), Italy

**Tiziana de-Magistris**
Unidad de Economía Agroalimentaria y de los Recursos Naturales, Centro de Investigación y Tecnología, Agroalimentaria de Aragón, 50059 Zaragoza, Spain
Instituto Agroalimentario de Aragón (IA2), CITA-Universidad de Zaragoza, 50013 Zaragoza, Spain

**Ching-Hua Yeh**
Institute for Food and Resource Economics, University of Bonn, 53115 Bonn, Germany

**Davide Menozzi**
Department of Food and Drug, University of Parma, 43124 Parma, Italy

**Áron Török**
Department of Agricultural Economics and Rural Development, Corvinus University of Budapest, 1093 Budapest, Hungary

**Mette Duerlund, Barbara Vad Andersen and Derek Victor Byrne**
Department of Food Science, Faculty of Science and Technology, Aarhus University, Kirstinebjergvej 10, DK-5792 Aarslev, Denmark

**Stefano Massaglia, Danielle Borra, Cristiana Peano and Valentina Maria Merlino**
Department of Agricultural, Forest and Food Sciences, University of Turin, Largo Paolo Braccini 2, 10095 Grugliasco, Italy

**Francesco Sottile**
Department of Architecture, University of Palermo, Viale delle Scienze 19, 90128 Palermo, Italy

**Pascal Ohlhausen and Nina Langen**
Department Education for Sustainable Nutrition and Food Science, Institute of Vocational Education and Work Studies, Technische Universität Berlin, 10587 Berlin, Germany

**Anne Sjoerup Bertelsen, Line Ahm Mielby, Niki Alexi and Ulla Kidmose**
Department of Food Science, Faculty of Technical Sciences, Aarhus University, 8200 Aarhus N, Denmark

**Boban Melovic, Dragana Cirovic and Tamara Backovic Vulic**
Faculty of Economics, University of Montenegro, 81000 Podgorica, Montenegro

**Branislav Dudic**
Faculty of Management, Comenius University in Bratislava, 82005 Bratislava, Slovakia
Faculty of Economics and Engineering Management, University Business Academy, 21000 Novi Sad, Serbia

**Michal Gregus**
Faculty of Management, Comenius University in Bratislava, 82005 Bratislava, Slovakia

**Katarzyna Włodarska, Katarzyna Pawlak-Lemańska and Ewa Sikorska**
Institute of Quality Science, Poznań University of Economics and Business, al. Niepodległości 10, 61-875 Poznań, Poland

**Tomasz Górecki**
Faculty of Mathematics and Computer Science, Adam Mickiewicz University, Uniwersytetu Poznańskiego 4, 61-614 Poznań, Poland

**Monika Hartmann**
Department of Agricultural and Food Market Research, Institute for Food and Resource Economics, University of Bonn, 53115 Bonn, Germany

**Nina Langen**
Department of Education for Sustainable Nutrition and Food Science, Institute of Vocational Education and Work Studies, Technical University of Berlin, 10587 Berlin, Germany

**Qian Janice Wang**
Department of Food Science, Faculty of Science and Technology, Aarhus University, 5792 Aarslev, Denmark
Crossmodal Research Laboratory, Department of Experimental Psychology, University of Oxford, Oxford OX2 6GG, UK

**Line Ahm Mielby, Jonas Yde Junge and Anne Sjoerup Bertelsen**
Department of Food Science, Faculty of Science and Technology, Aarhus University, 5792 Aarslev, Denmark

**Charles Spence**
Crossmodal Research Laboratory, Department of Experimental Psychology, University of Oxford, Oxford OX2 6GG, UK

**Edward J. Szczygiel, Sungeun Cho and Robin M. Tucker**
Department of Food Science and Human Nutrition, Michigan State University, East Lansing, MI 48824, USA

**Yeowoon Park, Seoyoung Park and Junghoon Moon**
Department of Agricultural Economics & Rural Development, Seoul National University, Seoul 08826, Korea

**Dongmin Lee**
Department of Food Processing and Distribution, Gangneung-Wonju National University, Gangneung 25457, Korea

# Index

Printed in the USA
CPSIA information can be obtained
at www.ICGtesting.com
LVHW081633241023
761885LV00006B/117

9 781641 168397